Arms and Armour

History, Conservation and Analysis

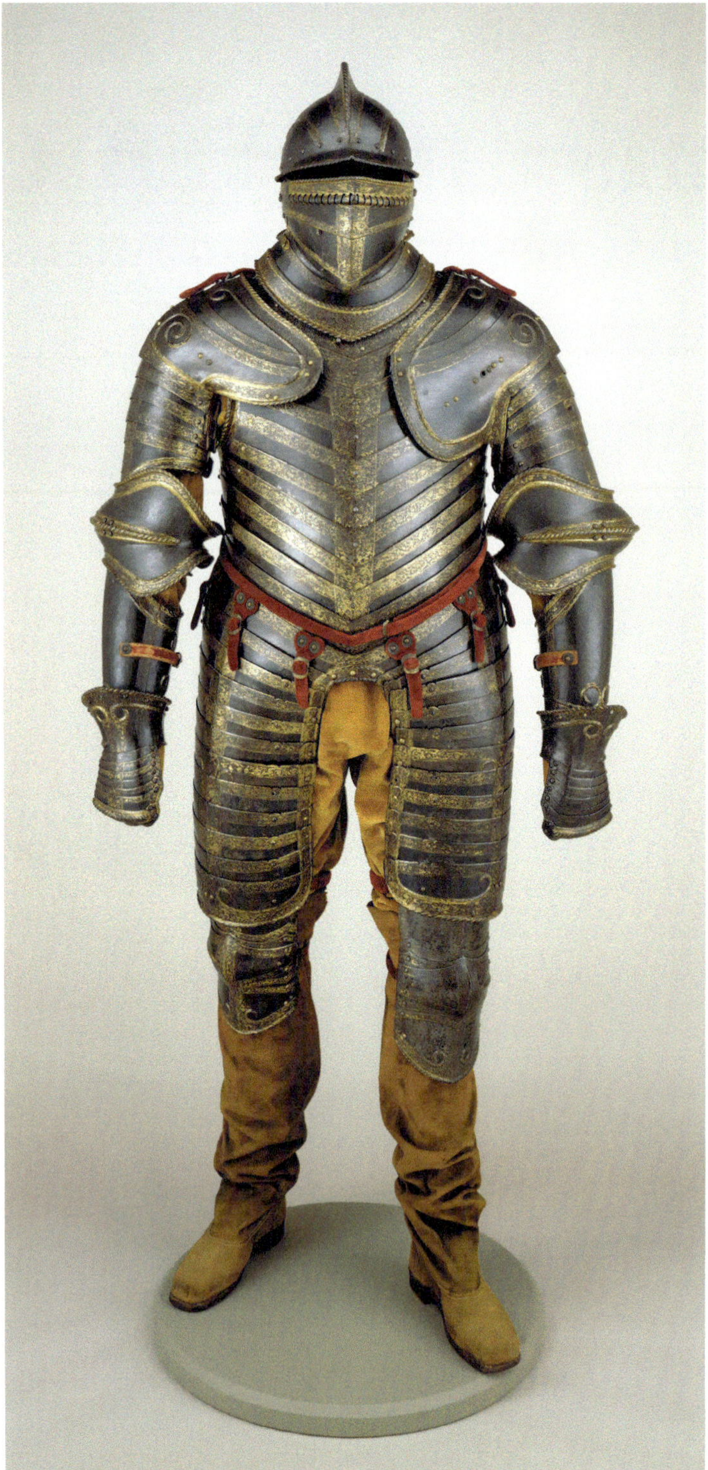

The Italian armour of Henry VIII c.1544. The Metropolitan Museum of Art, New York, 32.130.7. Harris Brisbane Dick Fund, 1932.

Arms and Armour

History, Conservation and Analysis

Essays in Honour of David Edge

Edited by Alan Williams
and Keith Dowen

Archetype Publications
www.archetype.co.uk

in association with the
Arms and Armour Society
www.thearmsandarmoursociety.org.uk

First published 2021 by Archetype Publications Ltd in association with the Arms and Armour Society

Archetype Publications Ltd
c/o International Academic Projects
1 Birdcage Walk
London SW1H 9JJ
www.archetype.co.uk

ISBN: 978-1-909492-82-0

British Library Cataloguing in Publication Data
A catalogue record for this book is available from the British Library.

Front cover: Indo-Persian axe 0A1594. © Trustees of the Wallace Collection.
Back cover: Burgonet, Franz and Wolfgang Grosschedel, c.1558, Real Armería, cat. A-274, Patrimonio Nacional, inv. no. 10054452-10054453.

Designed and typeset by Marcus Nichols at PDQ Digital Media Solutions Ltd, Bungay
Printed in the UK by Hobbs the Printers Ltd, Totton, Hampshire

FSC www.fsc.org MIX Paper from responsible sources FSC® C020438

Contents

Foreword

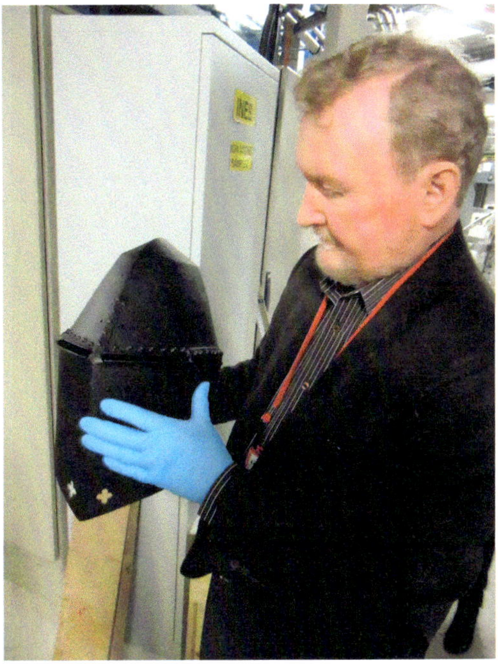

Over the course of his 45 years at the Wallace Collection, first as assistant to Nick Norman, then Curator and finally Head of Conservation and Armourer, David Edge did not merely act as the keeper of a collection – he also aimed to disseminate knowledge of it worldwide. His desire to share his experience and enthusiasm for the subject led to him create a network of fellow students, scholars and enthusiasts. Many of the contributors to this volume owe their later academic and professional achievements to his earlier mentoring. Many of these contributions owe their origins to a question first asked by David. For example, Chapter 4 features a lance rest (A228) in the Wallace Collection, which had been recognised by Vesey Norman as part of a garniture in the Real Armería, Madrid, possibly belonging to King Philip II. David made a fibreglass replica and took it to Madrid where it was found to fit perfectly on the breastplate of another royal garniture belonging not to Philip, but to his son, Prince Don Carlos. Chapter 12 describes the group of Indian or Persian saddle-axes in the Wallace Collection that were studied at the Rutherford Appleton Laboratory, Harwell, because although they appeared to be made of *wootz* ('Damascus steel'), David considered that their size and weight seemed to make them unlikely objects to be formed from this difficult metal. His suspicions were confirmed when many turned out to be manufactured of simulated rather than genuine *wootz*.

Following in the footsteps of the late Russell Robinson, David was not simply satisfied by studying the outer forms of museum exhibits: he also inquired into the manufacture and function of arms and armour. Although not a scientist himself, he was the steersman of a team of scientists, historians and conservators who pioneered the application of scientific analytical techniques to the study of arms and armour in order to attempt to answer the questions that he posed. This field of study is not one that can be pursued by any single individual, regardless of their background: it requires a team. It should come as no surprise therefore that several of the chapters in this volume are multi-authored and even include the subject of the book himself.

Preface and acknowledgements

The Arms and Armour Society is pleased to publish this book in association with Archetype Publications, a leading publisher in the fields of the conservation of art and technical art history. The Society's contribution has been underwritten by the North Fund, the result of a generous bequest to the Society by Tony North, a well-known collector and museum professional. With the oversight of its Council, the Society uses this bequest in part to help fund publications, to support new entrants to the field and to promote a better understanding of the conservation of arms and armour. This book, its authors and the articles it contains are well aligned with these objectives: they demonstrate the importance of the scientific study of arms and armour and the benefits of collaboration between scientists and historians.

This volume additionally demonstrates and recognises the significant and particular influence that David Edge has had on both the field of arms and armour and key individuals within it. It is authored by many of those whom he has mentored, encouraged and influenced. David has inspired many people, both professional and amateur, to become interested in the fascinating subject of arms and armour by his publications, his open and engaging personality, and his energy and unquenchable curiosity. He has actively encouraged the younger generation through education activities at the Wallace Collection involving both schools and less formal 'family events', and he has been a strong and practical supporter of the work of the Arms and Armour Society.

The Society would like to thank Alan Williams and Keith Dowen for their editorial efforts, the authors for their excellent contributions, and the team at Archetype Publications. I would also like to thank David Penn, former Society President, for piloting the project through our Council – and its members for keeping it a 'surprise' – and the Oakeshott Institute for their support.

Special thanks are due to the Wallace Collection for permission to use their exquisite images taken by Cassandra Parsons and to Xavier Bray for his support of the project.

David Williams
President of the Arms and Armour Society

List of contributors

Arthur Bijl, Hutton McRoberts Assistant Curator of Ottoman, Middle Eastern and Asian Arms and Armour, the Wallace Collection, London, UK

Tobias Capwell, Curator of Arms and Armour, The Wallace Collection, London, UK

Yannick Chastang, Conservator and Cabinetmaker, Yannick Chastang Conservation, Faversham, UK

Keith A. Dowen, Assistant Curator, Royal Armouries, Leeds, UK

David Edge, Former Head of Conservation and Armourer, The Wallace Collection, London, UK

Petra Fuchs, Head of Conservation, Hofjagd und Rüstkammer, Kunsthistorisches Museum, Vienna, Austria

Richard Furrer, Door County Forgeworks, Wisconsin, USA

Brian Gilmour, Research Associate, Department of Archaeology, University of Oxford, UK

Francesco Grazzi, Professor of Physics, Consiglio Nazionale delle Ricerche, Florence, Italy

Robert Hill, Independent Researcher, London, UK

Donald J. La Rocca, Curator Emeritus, Department of the Arms and Armor, The Metropolitan Museum, New York, USA

Francesca Levy, Arms and Armour Conservator, The Royal Collection, Windsor, UK

Lasse Matilla, Conservator, National Museum of Finland, Helsinki, Finland

Simon Metcalf, The Queen's Armourer, The Royal Collection, Windsor, UK

Antonella Scherillo, Instrument Scientist, Rutherford Appleton Laboratory, Didcot, UK

Álvaro Soler del Campo, Director, Real Armería, Madrid, Spain

Alan Williams, Archaeometallurgist, The Wallace Collection, London, UK

Introduction

The study of European and Oriental arms and armour has developed greatly over recent decades. More archival sources have come to light and pieces re-examined using both traditional and cutting-edge approaches. Increasingly the subject is becoming more interdisciplinary as curators, conservators and scientists each apply their own knowledge and skills to this fascinating and rewarding subject.

Few armours draw more attention than those of Henry VIII and are addressed here in two separate studies. In Chapter 1, Keith Dowen provides a thorough in-depth examination of the two remarkable foot combat armours made for the tournament at the Field of Cloth of Gold in 1520. His comprehensive investigation places the armours within their wider European context and delves into details of their design and technology as well as emphasising their role in shaping Henry's identity. The essay by Simon Metcalf and Alan Williams (Chapter 3) examines a pair of Italian vambraces belonging to an armour of Henry VIII in the Metropolitan Museum of Art, New York, which are now in the Royal Collection, Windsor. Significantly, through metallographic analysis, they show that the vambraces were modified at Greenwich probably for Henry's military campaign to France in 1544.

It has long been recognised that armour and fashion go hand in hand and in Chapter 2, Tobias Capwell presents an absorbing and comprehensive examination of a remarkable yet previously overlooked 'costume armour' at the Wallace Collection, London. Modelled on the clothing of the colourful German *Landsknechte* of the early 16th century, he clearly demonstrates that armour of this type was not only meant for a parade but, by providing detailed comparisons with other similar armours, also fulfilled a variety of different functions.

From the overlooked to the misidentified, Álvaro Soler del Campo (Chapter 4) provides an engaging study of a beautifully decorated lance rest at the Wallace Collection previously thought to have belonged to a garniture of Philip II of Spain. Through careful observation of the decoration and diligent consultation of the archival records in Madrid, his examination reveals a different but no less interesting royal provenance.

Donald La Rocca leads the reader from Europe to Asia as in Chapter 5 he discusses the remains of a rare Tibetan shaffron in the Metropolitan Museum of Art, New York. His expert and erudite study outlines the lasting significance of the Younghusband Expedition of 1903 and analyses in detail the various constituent parts of this fascinating piece of horse armour. Notably, he also highlights the use of modern scientific techniques to date pieces that otherwise lack contextual sources.

The conservation of arms and armour presents many unique challenges, but it also provides opportunities for new insights and the development of novel and innovative techniques that reveal long-hidden details. The conservator's intimate understanding of an object's constituent materials and construction contributes greatly to our understanding of arms and armour and informs new ways in which pieces can be displayed and interpreted. An awareness of this is presented by Petra Fuchs (Chapter 6) who provides a very interesting look into the process of making display mannequins more realistic by modelling them on the bodies of real-life individuals. The result is a superior more natural appearance, giving a much better impression of how armour is carefully crafted to fit the human body. The project's success will doubtless be followed by other museums.

Lasse Mattila's in-depth study of six important Northern European armours in the National Museum of Finland (Chapter 7) provides a fascinating insight into the lives of these objects. By highlighting their original purpose, use as funeral achievements and latterly as museum objects, this engaging paper clearly shows how each of these functions have left their mark, and the various treatments they require.

From Scandinavia to the Middle East, Francesca Levy returns to the theme of Asian arms and armour as she discusses in detail the various processes and challenges involved in conserving complex pierced metalwork found on Iranian vambraces at the Wallace Collection (Chapter 8). Her considered and invaluable contribution to the field of conservation outlines newly developed techniques which are likely to influence the work of other arms and armour conservators.

In recent years the development of new technology, such as CT scanning and neutron diffraction, has allowed researchers to gain new knowledge of the construction and technology of arms and armour that previously would not have been possible. In many instances the Wallace Collection, alongside other institutions, has been at the forefront of these pioneering methods. The use of CT scanning and X-radiography is the focus of Robert Hill's comprehensive and illuminating investigation into Viking-era swords (Chapter 9). His research clearly outlines the marked advantages these investigative methods bring over other approaches and, importantly, proposes a new pommel typology based on the thorough and careful analysis of 51 swords from various collections. The study also provides some very interesting observations into 'Vlfberht' blades.

Another promising technique, which is becoming more widely employed in the study of arms and armour, is neutron diffraction. Neutrons, unlike X-rays, are able to pass through metal and thus allow its atomic structure to be identified. Unlike metallography, this non-destructive technique can be used to analyse a range of objects without causing damage. This pioneering approach has been used by David Edge, Francesco Grazzi, Antonella Scherillo and Alan Williams in their highly important study of the Black Prince's helmet at Canterbury Cathedral (Chapter 10). The intriguing results suggest that the helmet has an English rather than Continental provenance as has sometimes been thought. Their findings will doubtless lead to further questions being asked of similar helmets in other collections.

Among the many treasures of the Wallace Collection, the remarkable jewel-studded dagger of Shah Jahan is one of the most impressive. However, it is the seemingly unprepossessing blade, not the hilt, which is the subject of the next paper authored by the same team with the addition of Arthur Bijl (Chapter 11). Their innovative investigation into the composition of the metal by neutron diffraction reveals it to be made of high-carbon crucible steel which, before over-cleaning took its toll, would have had the appearance of 'watered-silk'. Interestingly, the same results were also seen in other Mughal blades at the Wallace Collection, suggesting that many apparently plain blades might in fact be hiding their true identity.

The team's final paper (Chapter 12), co-authored with Richard Furrer, examines a group of Indo-Persian axes at the Wallace Collection and the process of faking *wootz*. Using neutron diffraction the results of their detailed study showed that contrary to expectations, many of the axes were made of simulated *wootz*. How this was achieved is a fascinating question and is addressed in the second part of their paper, which suggests possible methods based on the written account of the early medieval Iranian scholar Abu Rayhan al-Biruni.

The complex subject of fakes, which both plague and intrigue every private collector and museum, is the focus of Brian Gilmour's study (Chapter 13). Using a number of case studies, he clearly demonstrates how traditional metallography, alongside other analytical methods, can be used to determine the production techniques involved in the creation of these objects and how this can provide an insight into their date of manufacture.

Lastly, in Chapter 14 Yannick Chastang presents a thorough and comprehensive study of metal woodscrews from c.1650 to 1830, the result of many years of practical experience working with furniture. His impressive knowledge of the subject is evident throughout this paper which looks into the design, technology and manufacture of traditionally made screws as well as their use by gunsmiths. By sharing certain characteristics he concludes by highlighting their value as a help in dating the construction of flintlock firearms.

As noted elsewhere, the essays in this book reflect David Edge's many and varied interests covering the history, conservation and analysis of European and Oriental arms and armour from the early Middle Ages right through to the 19th century. The authors number among some of David's closest friends and colleagues, all of whom share his passion for the subject and a drive to 'push back the frontiers of knowledge'. This book offers new insights into a range of unique objects, many of which can be found in the Wallace Collection.

1

The 1520 foot combat armours of Henry VIII: design, technology and meaning

Keith A. Dowen

Introduction

The Field of Cloth of Gold was one of the greatest diplomatic and sporting events of the Renaissance. Held between the towns of Ardres and Guines on the border of the English-held enclave of Calais in June 1520, the 'summit' was designed to strengthen Anglo-French relations following the peace treaty of 1514. For roughly two weeks, between 7 and 24 June, the flower of English and French chivalry demonstrated their martial prowess through a series of jousts, tourneys and foot combats.[1] Its 500th anniversary in 2020 and the redisplay of the tournament gallery at the Royal Armouries, Leeds, prompted further research into the two foot combat armours of Henry VIII (RA II.6 and II.7) made for this spectacular event. Although both armours have been studied previously, most notably by Claude Blair and Thom Richardson, this article aims to build upon these foundations and provide closer observation and new insights into the design, technology and meaning of these intriguing pieces.[2] It offers a more detailed comparative analysis of these two armours by placing both within the wider context of foot combat armours and armour technology of the period.

Background

During the often tense negotiations in the build-up to the Field of Cloth of Gold, it had been agreed by both sides that as the event was to be held on English soil, the French would be responsible for setting the rules of the tournament, which consequently followed French fashion. Fortunately, these detailed ordinances survive and provide a wealth of information for all three sporting events including the foot combat over a barrier. Published by the Parisian Jean Lescaille, the 'Ordinances and Instructions of the Tourney, Jousts and Combats on Foot and on Horse' (*'L'ordonnance et ordre du tourney, ioustes, & combat a pied, & a cheval'*) not only specify the weapons used but also the type of armour the participants were required to wear.[3] As was customary among the French, individuals were instructed to be equipped in all events with armour for war with additional reinforcing pieces: 'in field armour with pieces of advantage' rather than specialised tournament armour. In case there was any doubt as to what this meant, the French were careful to specify what was required:

> so that everybody can understand and know what the aforesaid organisers mean when so often they talk about field armour with pieces of advantage, they mean that everyone should come with an armet [a term also used to refer to a close helmet] without wearing any other kind of head gear. That is to say no helm, half-helm nor bascinet. And concerning the other pieces of advantage everybody, *tenans* and *venans* (defenders

and challengers respectively), is free to wear whatever they want.[4]

While the decision not to use tournament armour might appear surprising, the use of field armour at the tournament was widespread in Western Europe, including at the French court where it had been worn at the tournaments held at Vincelles (1512) and Nozeroy (1519).[5] This decision may well have had the additional benefit of attracting more participants, as only the wealthiest of individuals would have owned costly tournament armour whereas field armour was more widely available.[6]

Having recently established Henry's own court armoury (in 1511/15), Henry was able to call upon the services of a team of highly skilled craftsmen to produce a suitably magnificent armour for the foot combat at the Field of Cloth of Gold. Not only did it have to be functional, it had to enhance and, in part, fashion Henry's own image as King of England. This task was entrusted to Master Armourer Martin van Royne, who oversaw the design and construction of the king's new armour at Southwark where the armoury had been based since 1516 while building work was undertaken at Greenwich.[7] Unfortunately, nothing is known of Van Royne prior to 1515 other than the fact that he must have possessed outstanding practical and organisational skills to be appointed to such a prestigious post.[8] He headed a group of German-speaking armourers, known as 'Almains', who had first arrived at Greenwich in 1515. According to the lists of armourers for 1515 and 1517/18, their number included men from the great armour-producing centres of Europe such as Nuremberg (Conrad van Nuremberg), Augsburg (Dionys Holzmann) and Landshut (Wolfgang Groschedell and Anthony Suytser).

The next list, dated April 1520, shows that Conrad van Nuremberg, Dionys Holzmann and Wolfgang Groschedell had left, but that Martin van Royne remained head of the workshop.[9] The Milanese armourers Filippo de Grampis, Giovanni Angela de Littis and three other unnamed masters, who had been brought over in 1511, do not appear to have stayed at Greenwich after their two-year contract expired, although the presence of Italian armourers there in 1559 may suggest that others remained on the staff in 1520.[10] In 1511, Henry had also contracted two armourers, Peter Fevers and Jacob 'Copyn' de Watte, from Brussels. Peter Fevers died in 1518, however, Jacob 'Copyn' de Watte continued to work at Greenwich until 1524.[11] Unfortunately there are no accounts specifying either when work began on Henry's foot combat armours or who worked on them but it is likely that it had commenced by February 1520, if not before, and lasted until mid-March when work came to a halt (see below).[12]

Foot combat armour

Although very rarely resulting in death, the foot combat was still a dangerous activity and competitors faced the ever-present risk of serious injury.[13] In an attempt to mitigate some of the risk, the combat at the Field of Cloth of Gold was fought over a horizontal wooden barrier which separated the opposing teams.[14] Nevertheless, it was imperative that armour provided the maximum amount of protection even if that meant sometimes sacrificing a certain degree of mobility. Throughout the 15th century, the foot combat had been fought in a range of armour styles, the choice of which was sometimes left to the individual participants. In 1442, for example, the third article of the challenge between Sir John Astley and the Aragonese knight Felip Boïl noted that 'the harness and the weapons that is left unto us (the competitors)'.[15]

The most common type of armour for the foot combat incorporated a deep skirt, known as a 'tonlet' (from the Old French *tonnel* – a barrel), and a large globose great bascinet (Figure 1).[16] Providing a very good level of protection, the design of this style of armour had changed relatively little from the mid-15th century when armour for the foot combat had started to become formalised.[17] It was based on contemporary field armour which also incorporated a plate skirt or 'paunce' to protect the lower body, and a great bascinet that was often firmly secured to the cuirass (Figure 2). Numerous contemporary references to this style of armour exist including, most famously, the English treatise of c.1450

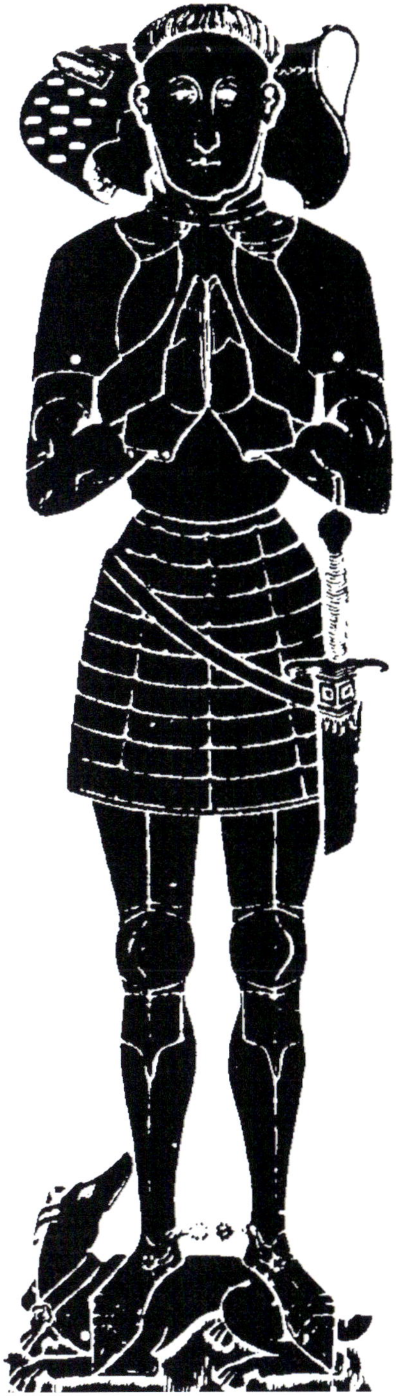

Figure 1 Funerary brass of Thomas Reynes (d.1451) depicting a knee-length plate skirt and great bascinet with bulbous visor behind his head. © *The Monumental Brasses of Bedfordshire*, W. Lack, H.M. Stuchfield and P. Whittemore (Monumental Brass Society 1992).

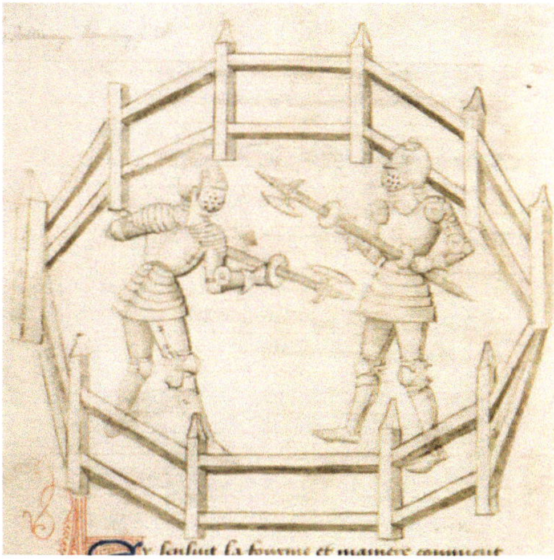

Figure 2 Foot combat with pollaxes c.1430. Both combatants are wearing typical field armours of the period. © Bodleian Library University of Oxford (MS. Douce 271 f.027v)

entitled 'How a man schall be armyd at his ese when he schal fighte on foote':

> First Þe must sette on Sabatones and tye hem (sic) up on the shoo with smale poyntis and wol [not?] breke And then griffus & then quisses & Þe breche of mayle And the tonletis And the brest And Þe vambras And Þe rerebras And then glouys.....And then his bascinet pynid up on two gret staplis before the breste with a dowbell bukill behynde up on the bak for to make the basinet sitte iuste.[18]

Only one 'tonlet armour' from the late 15th century survives, namely that made for Claude de Vaudrey (c.1485–1495) (Figure 3), preserved today in the Kunsthistorisches Museum Vienna (B 33). Whereas earlier skirts fitted relatively close to the body – due to the practical necessities of fighting both on horse and foot – those designed purely for the tournament, such as de Vaudrey's, were often deep and bell-shaped, becoming somewhat shorter and more flared as the 16th century progressed.[19]

The long-skirted, fully developed tonlet of the late 15th century may have emerged first in the Habsburg territories, where its use certainly

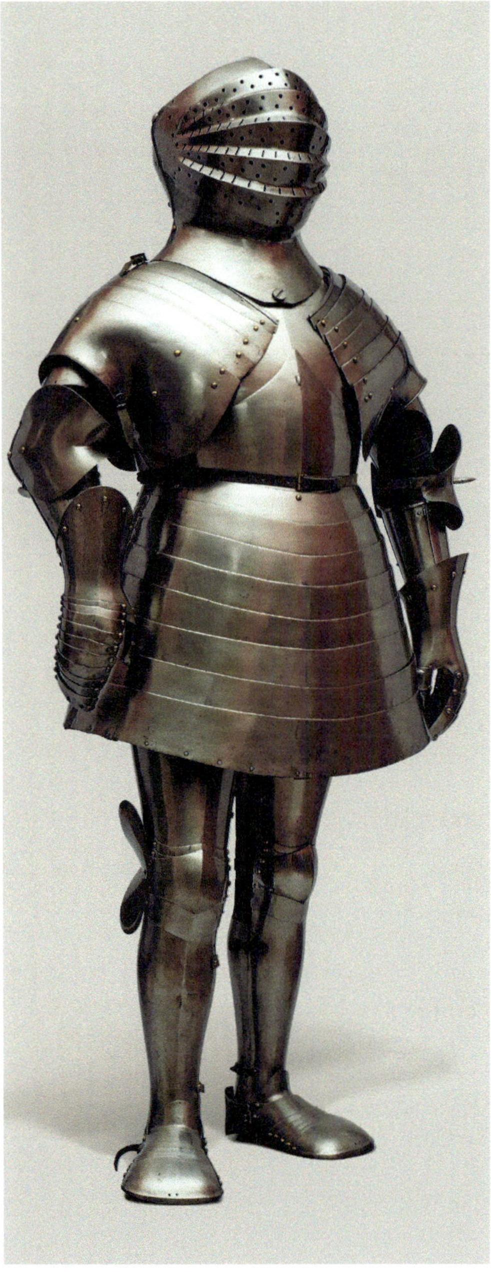

Figure 3 Tonlet armour of Claude de Vaudrey, c.1485–95. © KHM Museumsverband, Vienna (B 33).

appears to have been the most widespread.[20] However, its popularity elsewhere in Europe is hard to estimate, partly due to a lack of surviving records but also to the use of general descriptions such as 'armour for the foot combat'.[21] In 1497, Charles VIII of France ordered two large lames for his tonlet, while in 1514, Francis I of France placed an order for a tonlet armour from his armourer, Jacques Merveilles of Tours, for the tournament in Paris. The following year, a list of armour kept by Alan Cochrane on behalf of James IV of Scotland listed 'thre pair of tumbletis for fute' (three pairs of tonlets for the foot combat). Henry VIII's 1547 inventory also records 'twoo Tunletts wt two Basenetts' in addition to that made for the Field of Cloth of Gold (see below).[22] Thereafter, tonlets continued to be worn until the third quarter of the century, particularly in the Holy Roman Empire, despite having become outdated.[23] The latest surviving example was made for Emperor Maximilian II as part of the 'Rosenblatt Garniture' ('Roseleaf Garniture') of 1571.[24]

By the turn of the 16th century, another type of foot combat armour had been developed that was also based on the standard field armour. Offering a greater degree of protection, but sacrificing a certain amount of mobility, these highly sophisticated armours incorporated numerous narrow articulating lames covering those parts of the body that were particularly vulnerable to injury, such as the insides of the elbow joints and around the groin.[25] Laminated defences were not an innovation of the 16th century but had begun to be used to a limited degree by the 1410s if not before.[26] No early 15th-century examples survive, although they do occasionally appear in contemporary art and sculpture. One of the best depictions can be seen on the painting of *St George and the Dragon* of c.1432–35 by Rogier van der Weyden (Figure 4), which shows a series of narrow articulated lames protecting the inside bend of the saint's right elbow, the upper part of which is partially covered by a mail sleeve. Another slightly earlier example of c.1430 can be found on the figure of St Maurice on the 'Merchants Altar' in St George's Church, Wismar.[27] Somewhat later, laminated vambraces are also depicted in Rene of Anjou's 'Livre des Tournois' of c.1460.[28]

Laminated defences do, however, feature on a number of surviving 15th-century tournament pieces made at Augsburg for the Holy Roman Emperor Maximilian I. Indeed, it is likely that

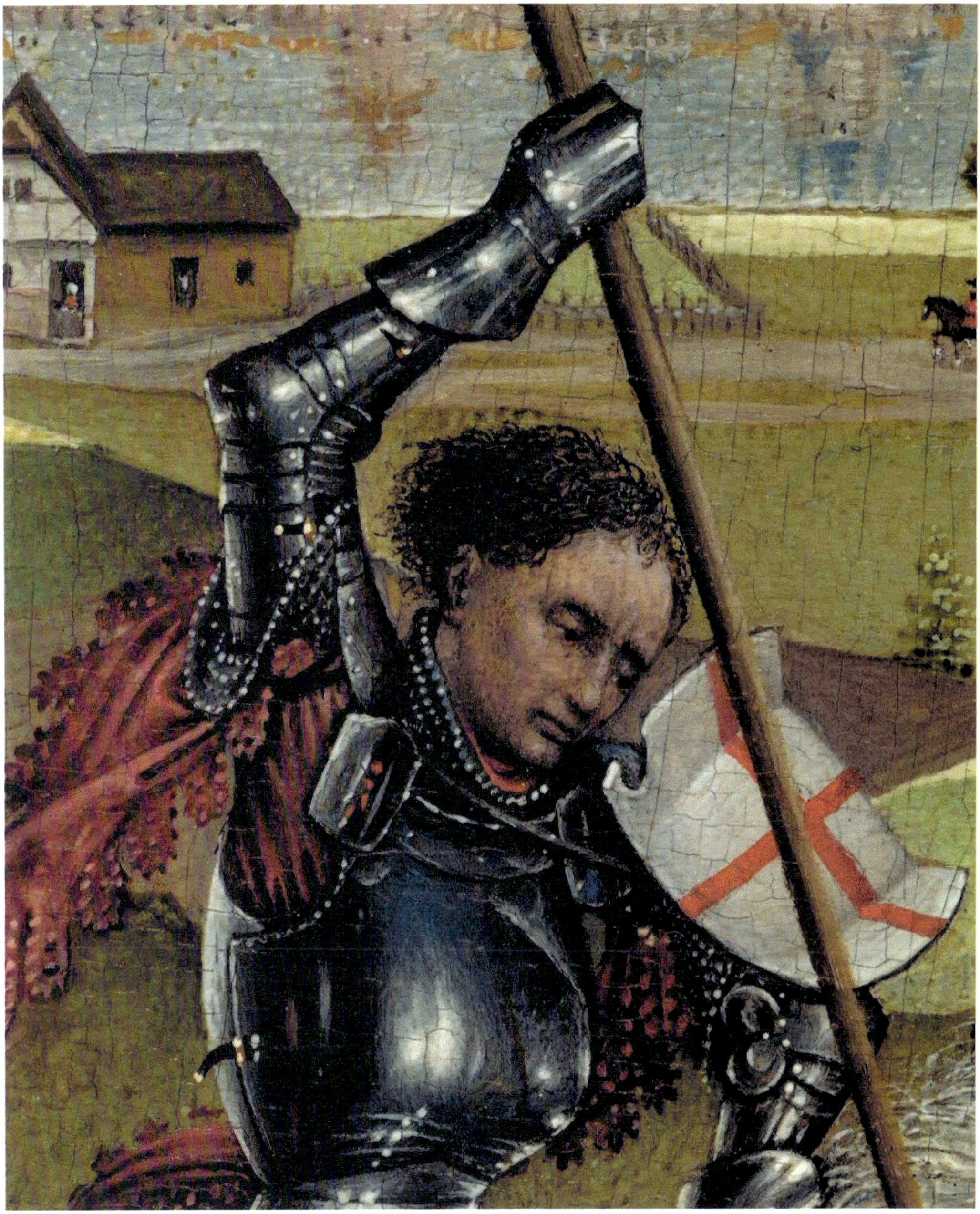

Figure 4 Detail from *St George and the Dragon* by Rogier van der Weyden showing a laminated defence protecting the inner bend of the elbow, c.1432–35. © National Gallery of Art Washington (1966.11).

given the level of experimentation in armour design under Lorenz Helmschmied, the first steps towards producing enclosed armours were taken there.[29] As early as 1477, Helmschmied had made a horse armour for Emperor Frederick III which is believed to have originally comprised laminated defences for the animal's girth, abdomen and legs. More famously in 1480, he produced another horse armour for Maximilian that also incorporated numerous articulating

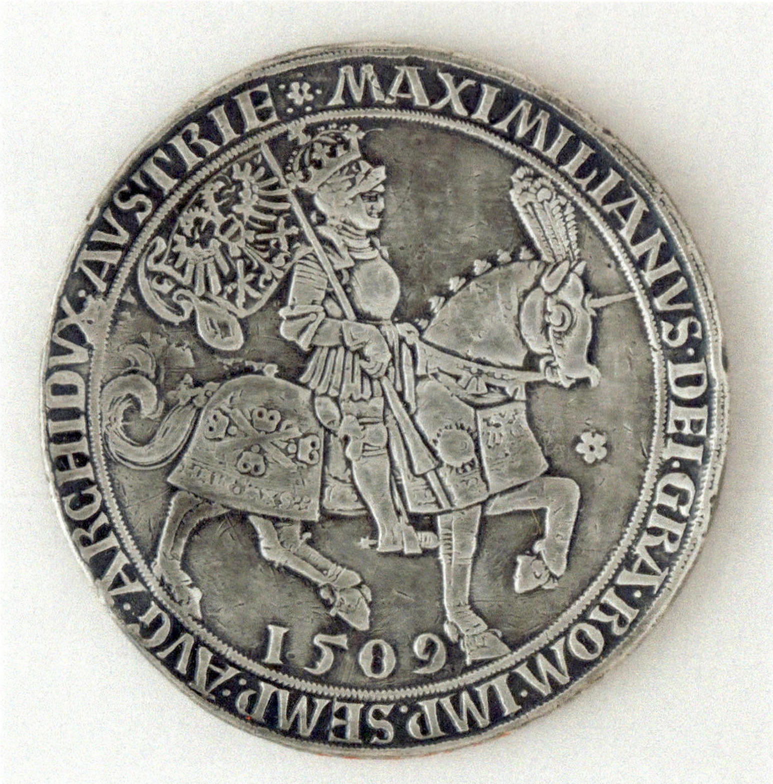

Figure 5 Presentation coin of Maximilian I depicting the emperor riding a fully armoured horse. The laminated defences are visible on the animal's legs. Die cut 1509, minted 1517. © The Metropolitan Museum of Art, New York (26.261.14).

lames covering much of the animal's body (Figure 5).[30] Approximately five years later, Helmschmied made an armour for the Kolbenturnier of which only the fully enclosed vambraces survive (Kunsthistorisches Museum Vienna A 79) (Figure 6). They incorporate articulating narrow lames or hoops to protect the shoulders and armpits while further lames guard the inner bend of the elbow. The gauntlets (missing), however, were worn separately. The 'Thun Sketchbooks' also include a drawing of a foot combat armour of c.1500 featuring laminated defences protecting the armpits, inner bend of the elbows as well as a compact enclosed helmet designed to rotate on the gorget, although the design of the rest of the armour remains conventional.[31]

Given that four of the five surviving fully enclosed armours have been attributed to Conrad Seusehofer, it is possible that armours of this type were produced by him at Innsbruck. Brought from Augsburg to work at Maximilian's court armoury in 1504 (and appointed its head in 1509), Seusenhofer would have possessed the requisite skills and experience to develop this style of armour under the patronage of a ruler whose knowledge and passion for the subject is well known.[32] However, the knowledge and ability to produce pieces of enclosed armour was not confined to the German-speaking lands of Central Europe. We know this firstly because of one early written reference to two sets of what may be enclosed plate breeches (formerly known as a 'hoguine') in the Lord High Treasurer's Account for Scotland for 1506/7. It records that James IV ordered 'one breik of plait to the King, vij Franch (sic) crounis' and one to the French knight Sir Antoine d'Arces from his armourers at Dundee.[33] Secondly, the existence of the enclosed

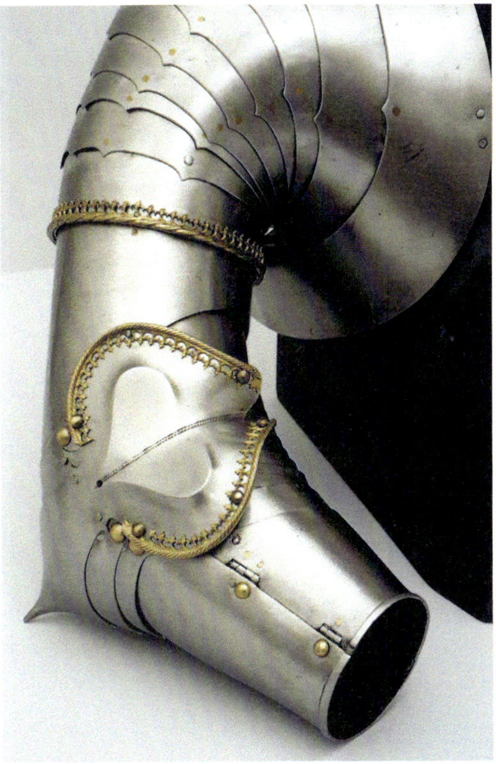

Figure 6 Right vambrace for the Kolbenturnier by Lorenz Helmschmeid. A complex system of articulating hooped lames protect the shoulder. © KHM Museumsverband, Vienna (A 79).

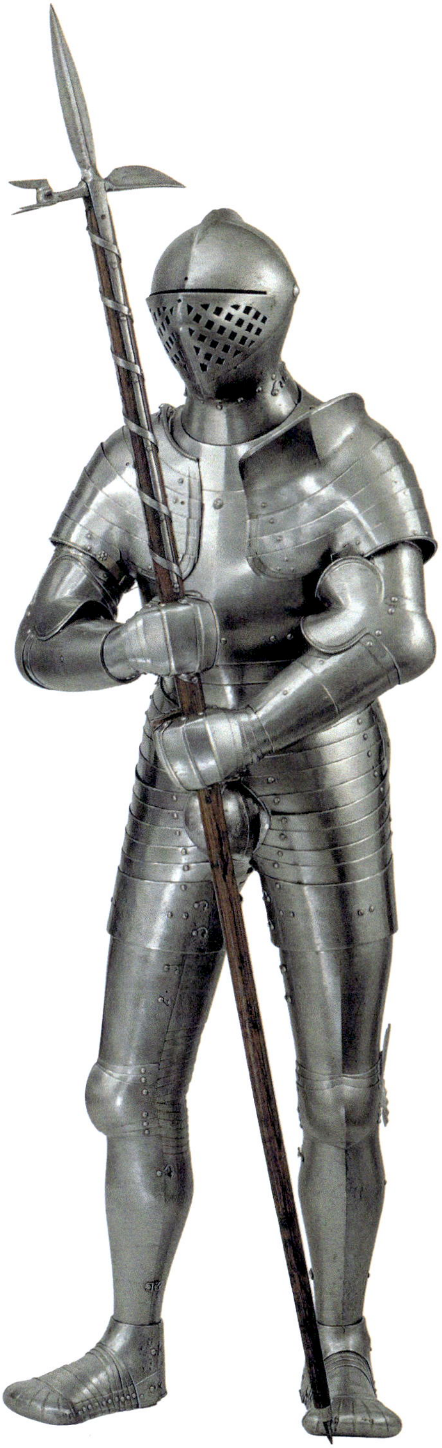

Figure 7 Henry's fully enclosed armour for the Field of Cloth of Gold is the most technically advanced armour of its type. © Royal Armouries (II.6).

Italian armour of c.1510, G 178 at the Musée de l'Armée in Paris (see below), is tangible evidence of this. Consequently, the precise chronology of the development of fully enclosed armour remains unclear.[34]

The creation of these fully enclosed armours seems to reflect a desire to push armour technology to the limits not simply for the sake of protection, but more as a reflection of princely patronage and court spectacle.[35] As Pierre Terjanian has observed, armour had become 'the currency of power'. These remarkable and sophisticated armours were therefore ideally suited as diplomatic gifts, such as G 179 (Musée de l'Armée, Paris) which was presented by Emperor Maximilian to Guiliano de Medici in c.1514.[36] No doubt by commissioning his own fully enclosed armour for the Field of Cloth of Gold, Henry VIII wished to join what

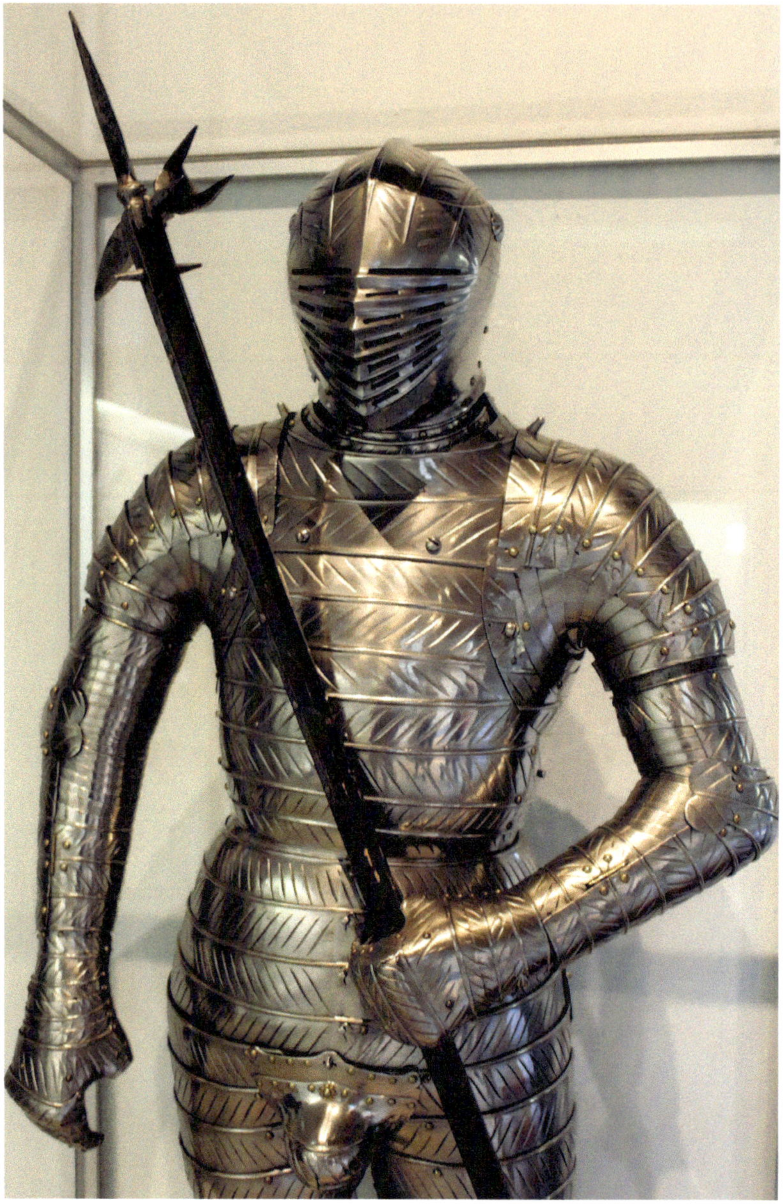

Figure 8 This fully enclosed armour is attributed to Conrad Seusenhofer. In common with a number of his other pieces, it features tubular-like upper arm defences (G 180).

appears to have been an exclusive group of European princely patrons who had the ambition, resources and skilled manpower available to invest in the creation of these awe-inspiring 'luxury' pieces (Figure 7). As a self-proclaimed warrior king, armour was part of the language of knightly power and was therefore the ideal medium through which Henry could forge his own identity. In addition, its close fit would also serve to highlight his youthful athletic frame, while the codpiece emphasised his masculinity. Importantly it also enabled him to demonstrate that his armourers were as skilled as the very best in Europe.

Unlike the relatively simple tonlet, which would continue in use until the late 16th century, these technologically demanding fully enclosed armours were a relatively short-lived phenomenon that only appear to have been in fashion for a brief time – probably no more than 20 years between c.1500 and 1520. However, various elements such as plate breeches and lames over the inner bend of the elbow would continue in use for some time. For example, the c.1525–30 'Hunt' armour of Charles V at the Real Armería, Madrid (A 93), made in Augsburg by Colman Helmschmied, incorporates laminated defences over the inner bend of the elbows. According to the 'Inventario Illuminado' (Madrid N 18), the same armour was also equipped with a pair of laminated exchange pieces for the armpits and a set of plate breeches – in this case, to be worn beneath the tonlet.[37]

Of the three enclosed armours attributed to Seusenhofer, two are in the possession of the Musée de l'Armée in Paris, namely G 179 the armour belonging to Giuliano de Medici, and G 180 of c.1510–15 (Figure 8), on loan to the Musée national de la Renaissance, Château de Écouen. The third, now on display at Schloss Ambras, Innsbruck (E 1), was made for young Louis II of Hungary. Although incomplete this armour is unique in consisting of two halves joined together at the sides. Produced in the German style (*'alla Tedesca'*), a further armour, G 178 at the Musée de l'Armée – previously incorrectly believed to have belonged to Francis I – is signed by the Milanese armourer, Niccolo Silva.[38] Due to the lack of written records, it is not known whether these armours were designed exclusively for the foot combat over a barrier (the use of the barrier had become popular in the early 16th century) or if they were also worn in the *champ clos* (from the French meaning 'an enclosed field').

Henry's enclosed armour (II.6)

Henry's fully enclosed armour comprises 235 separate pieces. Fitting close to the body, articulation is provided by a system of internal leather straps and sliding rivets.[39] In 1761, one visitor to the Tower of London marvelled at the technical sophistication of the numerous overlapping lames which 'play like the joints of a rattle snake and are moved with all the facility imaginable'.[40] Yet despite the complexity of its design, it is identifiably a field armour, albeit one with extensive reinforcing. It could therefore be reasonably argued that the armour still conformed to the specifications outlined by the French. All these extra pieces, however, came at a cost: at 94 lb (42.6 kg), Henry's armour is roughly twice the weight of a field armour.

The helmet features a one-piece visor with a narrow horizontal vision slit and is pierced with numerous square or diamond-shaped holes for breathing (Figure 9). Two holes above and below the vision slit were probably designed for a reinforcing piece. As with the surviving enclosed armours in Paris, the extended chin and skull plates lock onto a flange on the gorget, allowing it to rotate freely. The gorget itself is made of two solid plates that sit flush with the upper edges of the breast and backplates and is secured in place by a pair of bolts. In contrast, the gorgets of the three Paris armours are of a more conventional design and comprise multiple separate lames.

Pins at each shoulder secure the pauldrons in place, that on the left also having an upright guard or 'haute-piece'. An identical guard would also have been attached to the right pauldron although work ceased before this could be completed. As would become standard at Greenwich, the pauldrons are constructed of upward overlapping lames joined together by internal leather straps. Due to the range of movement required, the underarms were one of the most difficult parts of the body to protect adequately. Traditionally, this problem was solved by the wearing of mail voiders (gussets) but in the case of Henry's armour and those at Paris (apart from G 178), these have been replaced by skilfully shaped removable sets of articulating lames (Figure 10). For all the armour's technical sophistication, the pauldrons are of a conventional design and overlap the cuirass (as on G 178). In contrast, Seusenhofer, like Helmschmied before him, constructed integrated tubular upper arm defences (see Figure 8) on his surviving foot combat armours in the Musée de l'Armée (G 179 and G 180) as well as on the armour for the future Charles V at Kunsthistorisches Museum Vienna (A 109).[41]

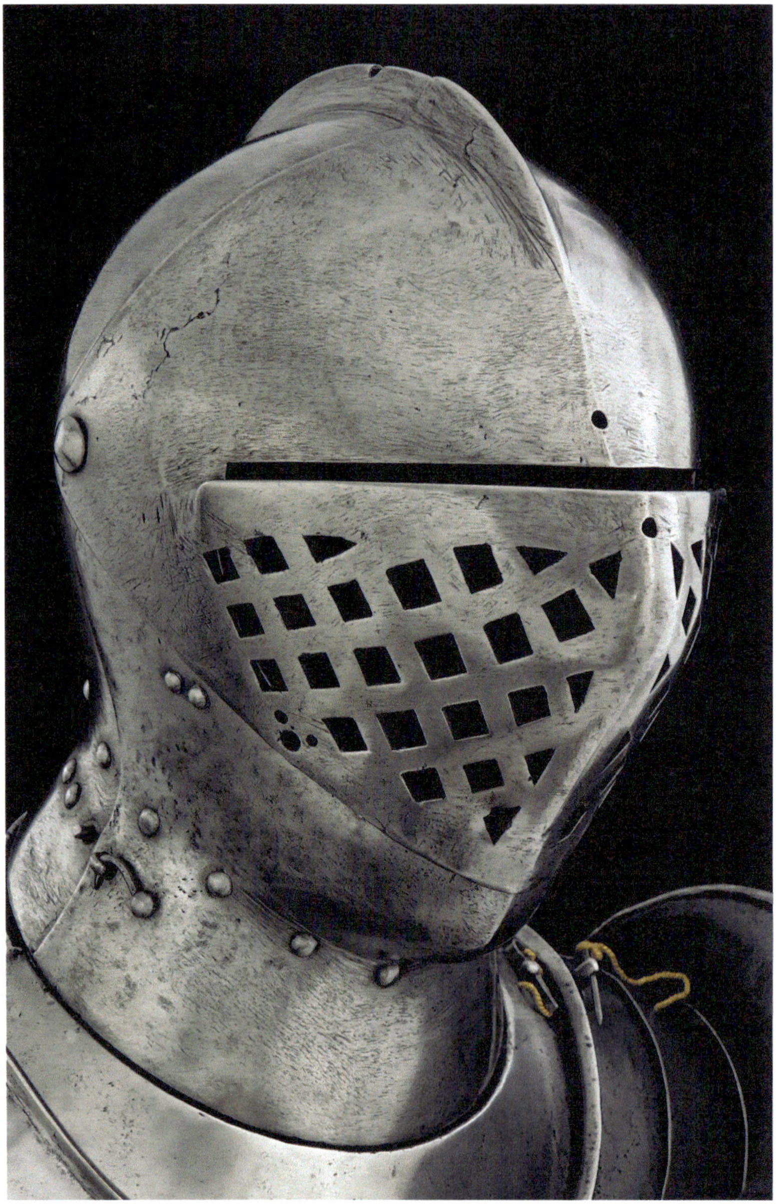

Figure 9 The helmet incorporates numerous diamond-shaped holes providing a good level of ventilation. © Royal Armouries (II.6).

The vambraces on Henry's armour feature flush turning joints and large winged bracelet couters while the inner bends of the elbows are protected by numerous articulating narrow lames. As with G 179 and G 180, the lower cannons overlap the gauntlets and incorporate flush turning joints allowing the hands to rotate (conversely, G 178 has gauntlets of a conventional design). Interestingly, the right-hand gauntlet of Henry's armour is a very early example of a 'locking gauntlet' (Figure 11) and features an extension over the fingers which can be fastened closed to the inside of the cuff.[42] The breast and backplates are of typical early 16th-century

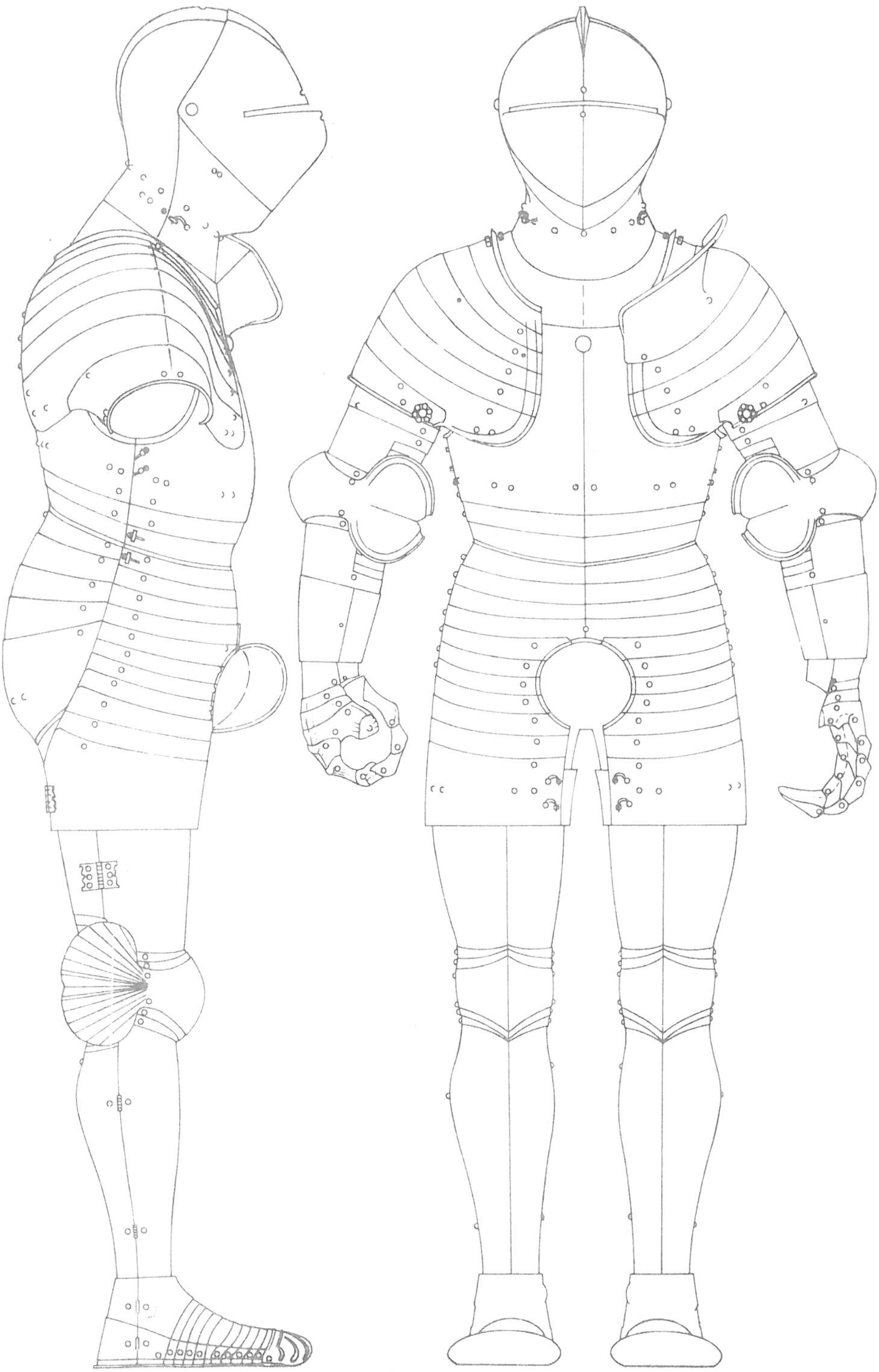

Figure 10 This technical drawing of Henry's armour shows the numerous carefully shaped lames covering every part of the body. © Royal Armouries.

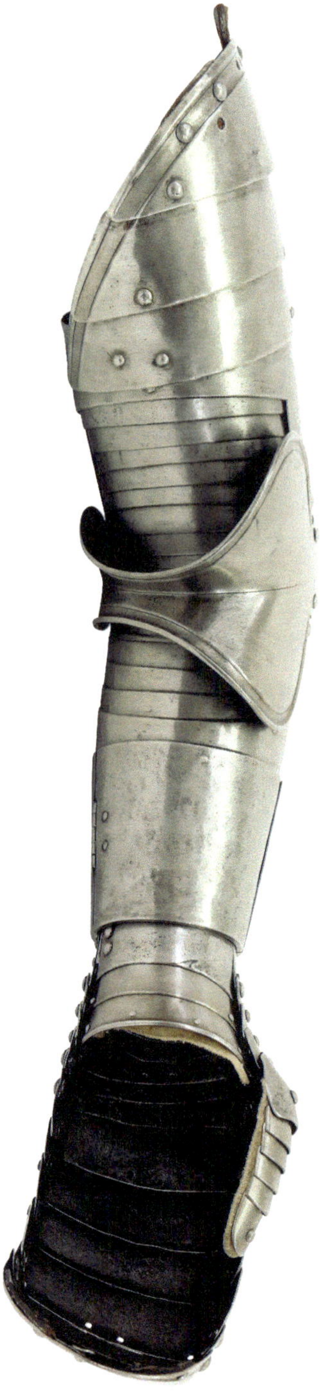

Figure 11 The right vambrace incorporates an early locking gauntlet, the cuff of which fits inside the lower cannon. © Royal Armouries (II.6).

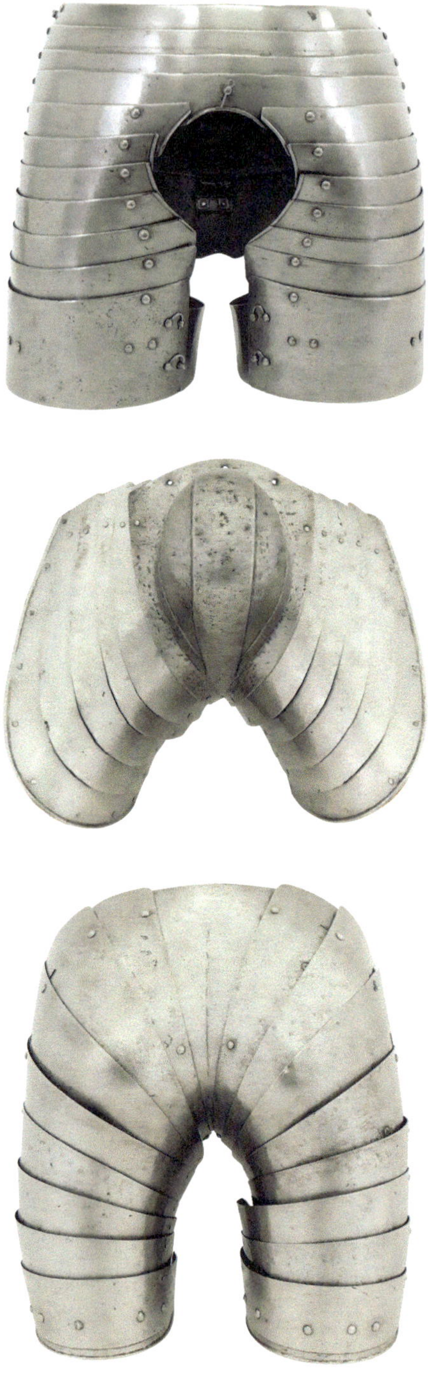

Figure 12 The multilayered design of the steel breeches prevented gaps appearing between the plates. Each lame has been carefully forged to precisely follow the contours of the body. © Royal Armouries (II.6).

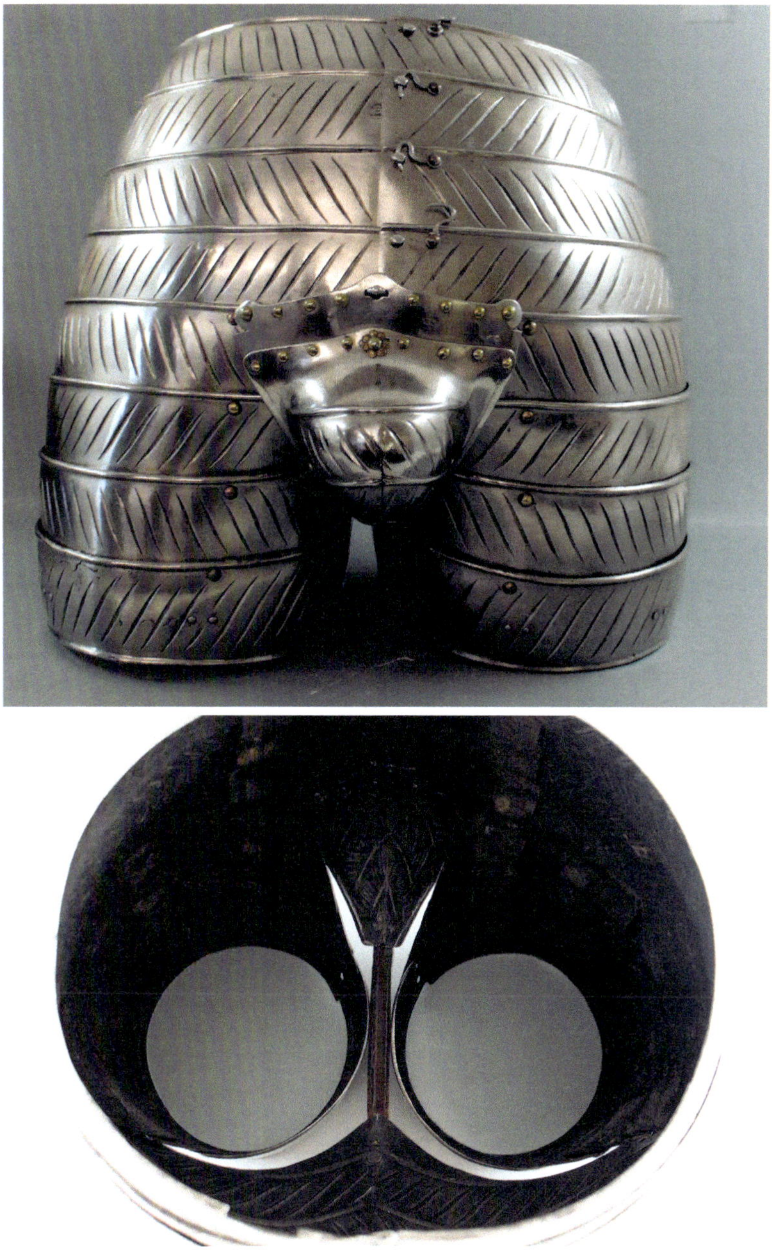

Figure 13 The breeches of G 180 are of a simpler design, leaving some areas of the lower body exposed.

form and closely mirror those on the armour of Maréchal de Fleuranges of c.1520–25 (Musée de l'Armée G 48), which was also made by Henry's Almain armourers.[43]

Undoubtedly the breeches of II.6 are the most remarkable part of the armour and are of a unique and technically sophisticated design. While it might be assumed that the use of a barrier precluded armour for the lower body, at this early date there were no specific rules against striking below the waist,[44] consequently it was imperative that this vulnerable area was adequately

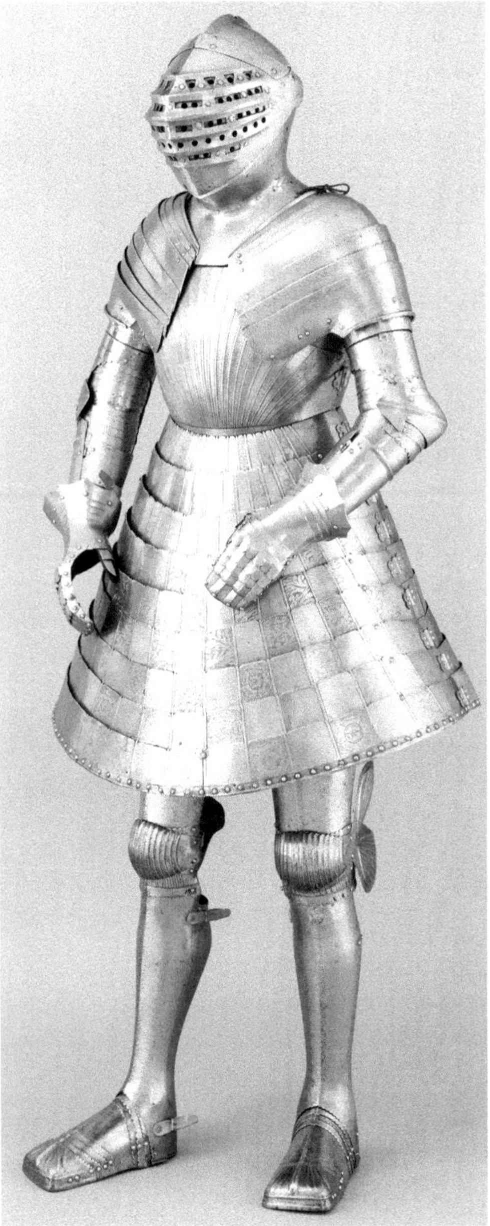

Figure 14 Henry VIII's new foot combat armour, although assembled from different parts, is nonetheless a striking piece. A panache of feathers would have made the armour more imposing. © Royal Armouries (II.7).

protected, mail alone not being sufficient. In 1547 it was recorded that Henry had six 'Curetts to fight w^{t} one fote with Basenetts and breches'.[45] The tassets and cullet are hinged together on the left and secured by pierced studs and pins on the right, forming a single assemblage. Both the codpiece and inner groin plates, together with the inner rump piece, fit inside the tassets and culet. Articulation is provided by a series of sliding rivets and internal leathers. This practical multilayer solution – which ensures that gaps do not appear as the armour moves – is not found on any of the other surviving fully enclosed armours (Figure 12).[46] For example, G 180 features a similar tasset and cullet assemblage, but rather than having separate inner groin and rump plates, the rump is protected by a shaped plate held in place by a pair of turn pins and the separate codpiece (originally) by a single-turn screw. To further secure the rump plate and codpiece together, a leather strap passes beneath the fork and is fastened by a buckle. Conversely, no provision has been made for any inner thigh plates on G 180, meaning that if the legs were to be extended, the top of the inner thighs would be exposed (Figure 13). The lowest lames of Henry's plate breeches fit over the top of the cuisses by means of flush turning joints, while the cuisses themselves feature hinged extension plates which fully enclose the lower thighs. The poleyns have fluted side wings and the backs of the knees are protected by a series of narrow lames. As on the armour of G 179, the sabatons are designed to fit over, rather than under, the greaves.

As work progressed, unwelcome news arrived on 16 March from Sir Richard Wingfield, Henry's ambassador to France.[47] In a last-minute 'clarification' to the rules, the French specified an entirely different style of armour for the foot combat from the one being made by Henry's armourers:

> As tochyng the combatt at the barrier, ther is lefft out, these wordes following, wt. pieces of avauntage, And in stede therof is sett in Tonnelett & bacinett, the consideracon was that the same might be clerly declaryd to the Anserers whiche ells shud have been in doubt, what myght have Importyd the said pieces of avauntage [which was potentially open to wide interpretation].[48]

Quite what motivated this late announcement is not altogether clear, but work on Henry's armour came to an abrupt halt. It has sometimes been explained away as no more than personal rivalry on the part of Francis I, who could not abide the thought of his rival and former enemy outshining him.[49] However, it may simply have reflected a real anxiety on the part of the French that Henry's fully enclosed armour gave him an unfair advantage over his opponents. Alternatively, albeit somewhat belatedly, the French may simply have wanted to clarify the ambiguity of the phrasing in the original ordinances (it is arguable whether this style of armour also constitutes a modified field armour). Whatever Henry's personal reaction, at the very least he appears not to have risked a major diplomatic breach by formally disputing the decision[50] – the French were, after all, responsible for setting the rules, although it now meant that English plans were in disarray.

Henry had originally been due to arrive at Guisnes by 31 May, however, after some careful negotiation – ostensibly to allow the English more time to prepare, but actually to enable a meeting with Emperor Charles V in England – a postponement to 4 June was agreed (in fact Henry arrived on 5 June).[51] This left just over two and a half months to produce a new foot combat armour for Henry and transport it across the Channel. Although overlooked by modern scholars, preparations for transporting the armoury mill at Greenwich along with all the other equipment needed to perform running repairs when in France would have seriously hampered work on Henry's new armour. Work to dismantle the mill began on 21 March but it was not re-erected at Guisnes until around 7 May (in addition, 54 wagons were needed to transport all the equipment and materials).[52] Twenty-four armourers also accompanied Henry to the Field of Cloth of Gold together with two master armourers including Erasmus Kirkner, 'armourer for the (king's) body', who eventually succeeded Martin van Royne as Master Workman;[53] the latter, it seems, remained in England, possibly to oversee the move back to Greenwich from Southwark.[54]

Henry's tonlet armour (II.7)

It was against this background that Martin van Royne and his team had to produce a new and very different foot combat armour (Figure 14).[55] With time against them and lacking resources, it proved impossible to manufacture a new armour entirely from scratch. Instead, as observed by Claude Blair, Henry's armourers were forced to assemble an armour by adapting as many pre-existing pieces as possible, including the helmet, breast and backplates, vambraces and legs.[56] Although the armour is currently displayed with a pair of gauntlets, these are not original to the armour. Indeed, if the identification is correct, by 1547 one gauntlet had already gone missing: 'Itm one Tunlett ꝑcell guilte wt a Basenett complete Lackinge one gauntlett'.[57] In fact, the only parts of the armour that were newly made in 1520 prior to the Field of Cloth of Gold were the tonlet and pauldrons, both of which are of noticeably lower quality.

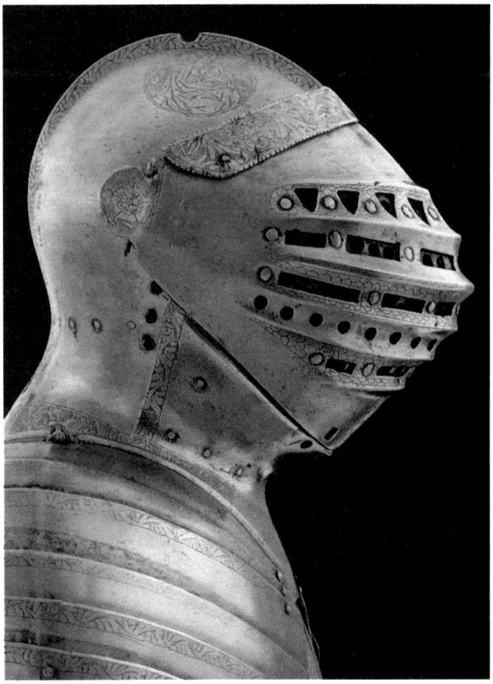

Figure 15 As an additional safety measure pierced plates were riveted to the inside of the visor to reduce the size of the apertures © Royal Armouries (II.7).

The great bascinet of Henry's tonlet armour is of a type commonly used for the foot combat.[58] Like many great bascinets, it features a bellows-type visor which, in this case, has been modified for the Field of Cloth of Gold with the addition of a series of narrow pierced plates fixed behind the

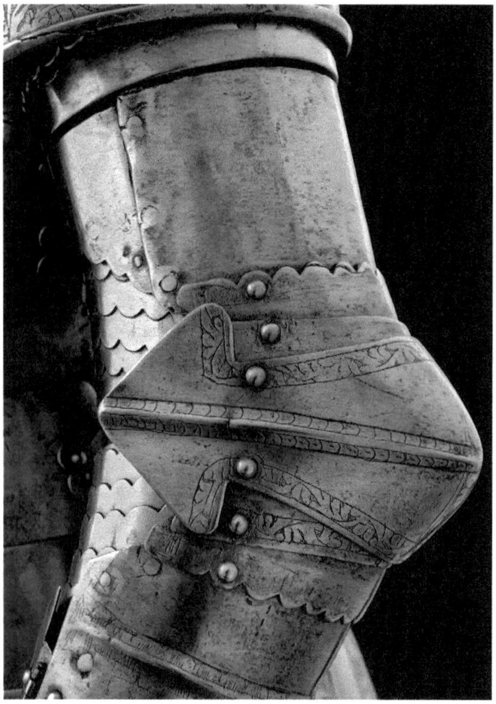

Figure 16 Both couters were cut down when the armour was assembled and laminated defences inserted to protect the inner bend of the arms © Royal Armouries (II.7).

vision and ventilation slots in order to reduce their size and provide an extra degree of safety (Figure 15). To offer further protection against overhand blows, a narrow reinforcing band was also riveted to the visor's top edge. Originally the helmet was provided with a catch to secure the visor in place and prevent it from being accidentally knocked open, but this is no longer extant. The lower edges of the gorget and neck plates (the latter having been crudely bent to fit the backplate) have been pierced with a series of screw holes to firmly secure the bascinet to the cuirass.

The cuirass itself is from a standard infantry field armour of the period and is embellished with narrow radiating flutes. The vambraces, with the exception of the pauldrons, have likewise been adapted from two existing field armours and formerly featured large fan-shaped couters that were crudely cut down as the armour was assembled (Figure 16). As with other contemporary foot combat armours, the inner bends of the arms were fitted with laminated defences (those on the right being 20th-century replacements), a practice that would continue on both tournament and field armours until the 1630s. The rather square

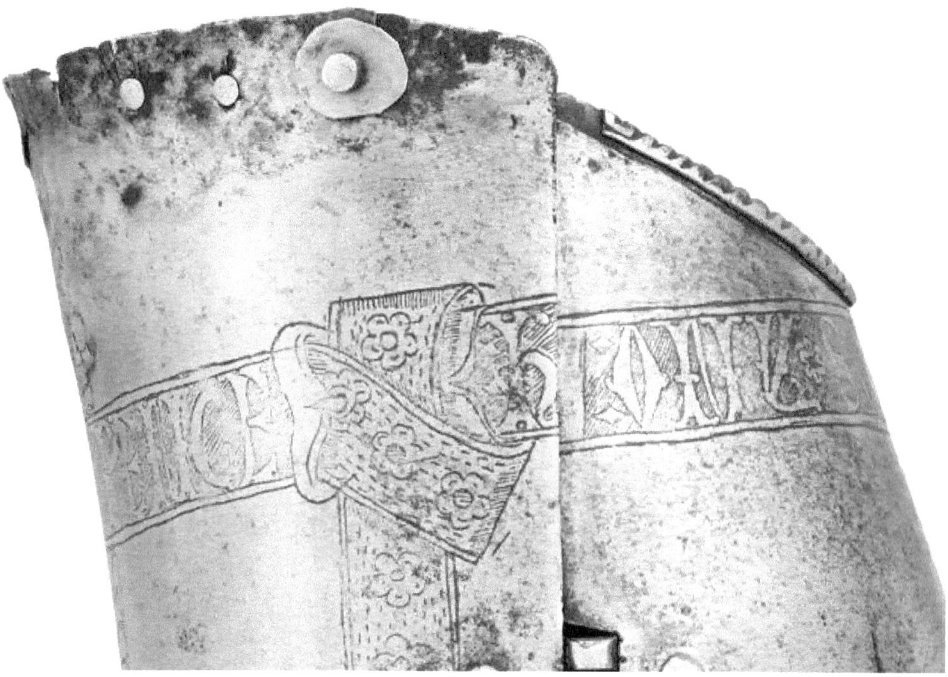

Figure 17 The greaves are the only part of the armour to feature roped edges © Royal Armouries (II.7).

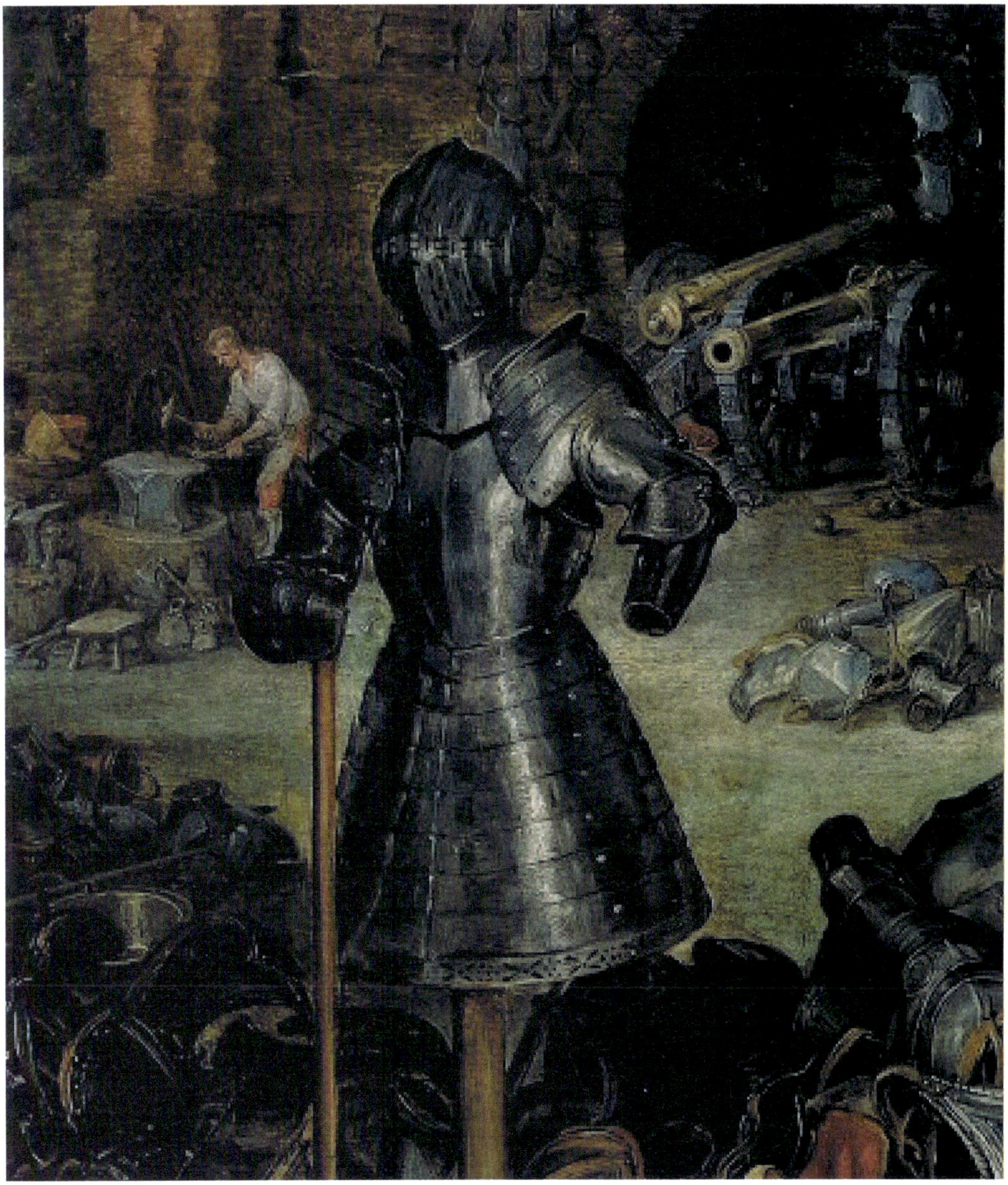

Figure 18 The fashionable deep tonlet on Philip the Handsome's armour of c.1500 is comparable to that on II.7. © Museo Nacional del Prado, Madrid (P001398).

pauldrons, newly made prior to the Field of Cloth of Gold, are constructed from a series of upward overlapping lames – seven (originally eight) on the left and eight on the right which are held together by internal leather straps and articulating rivets.

The leg defences, reunited with the rest of the armour in 1947, are designed for mounted rather than infantry use as they feature open cuisses and slots at the base of the greaves for a pair of spurs.[59] It is quite likely that the greaves and sabatons are from a different armour than the cuisses and poleyns, a suggestion first proposed by Ian Eaves based on the presence of roped turns on the upper edges of the rear greave plates (Figure 17).[60] The tonlet, also made in the weeks prior to

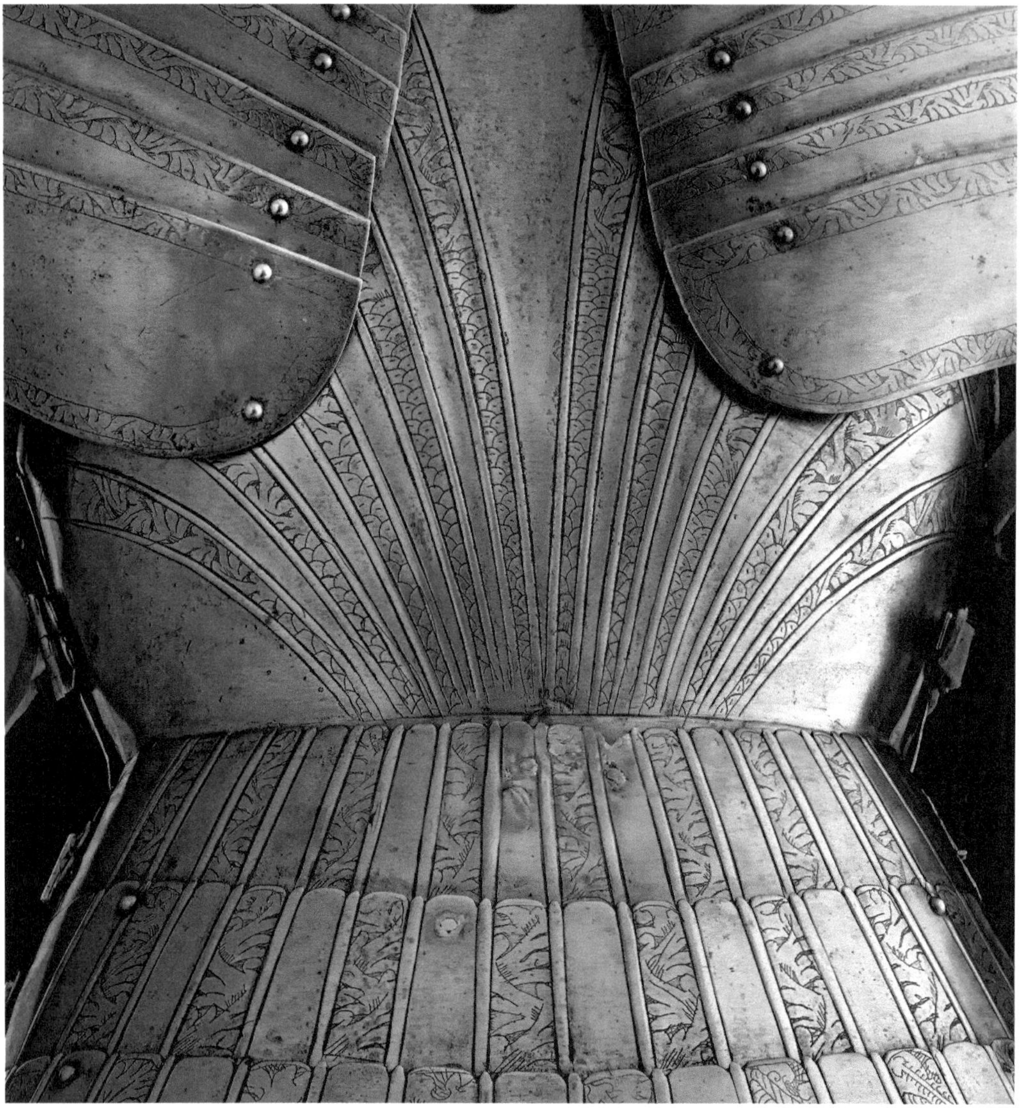

Figure 19 Henry's cuirass is decorated with Italian-style etched flutes. © Royal Armouries (II.7).

the Field of Cloth of Gold, reaches down to mid-thigh and comprises nine lames front and back, hinged on the left and simply secured with straps and buckles on the right, an arrangement also seen on Emperor Maximilian's armour (B 71) but not on Claude de Vaudrey's which, instead, uses a more unusual system of snapping pegs. Somewhat later in date, Henry's tonlet is notable for being deeper and wider than de Vaudrey's and instead is comparable to the c.1504 armour probably of Philip the Handsome (Figure 18) in the 1618 painting *The Sense of Touch* by Reubens and Brueghel 'the Elder' at the Museo Nacional del Prado in Madrid (P001398), and those erroneously depicted in the *Battle of Orsha* by Hans Krell in the National Museum, Warsaw, of c.1524–30 (MP 2475).[61]

Henry's new tonlet armour was therefore assembled from a number of different armours, perhaps as many as six.[62] However, as only the helmet bears an armourer's mark – that of the Missaglia family of Milan – establishing a

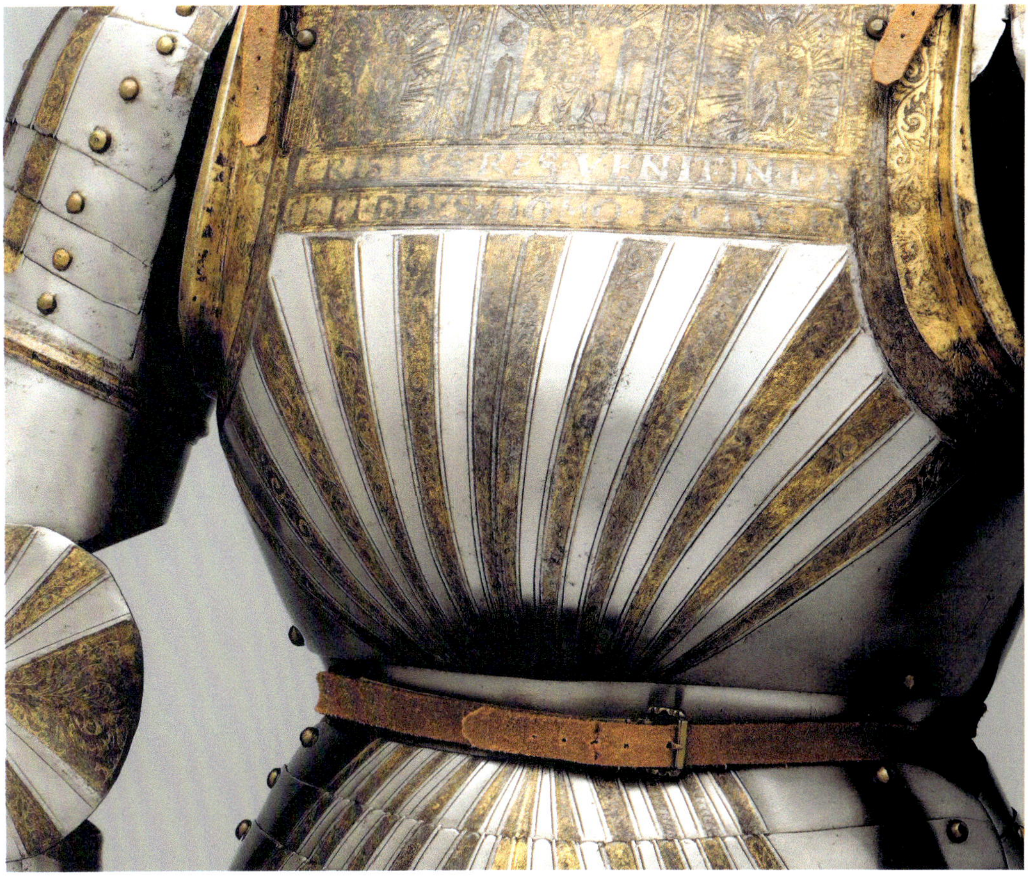

Figure 20 The flat etched flutes on this early 16th-century Italian light cavalry armour closely mirror those on II.7. © The Metropolitan Museum of Art, New York (14.25.716b-f)

provenance for the other component parts is far from simple, particularly as Western European armours of the early 16th century share many of the same Italianate characteristics.[63] Furthermore, the highly international workforce at Southwark/Greenwich (Flemish, German and Italian) also complicates the picture.

The cuirass, for example, could either be from Flanders or Italy as Henry purchased armour from both areas in the early years of his reign (Figure 19).[64] Although Claude Blair previously noted that the cuirass bears a resemblance to two Innsbruck armours, Thom Richardson has correctly observed that the flutes have been etched (Figure 20) to create flattened ribs and not embossed as was common in the German-speaking lands of Central Europe. Therefore, Blair's suggestion that the cuirass may have been repurposed from one of the two armours Henry had commissioned from Conrad Seusenhofer in 1511 is highly unlikely.[65] In fact, it should be noted that the ribs of the cuirass bear a marked similarity to those on a number of Italian examples, as can been observed on a light cavalry armour of c.1515 at the Metropolitan Museum of Art, New York (14.25.716b–f) as well as on the composite armour of B11 at the Diocesan Museum Mantova of the same date. A further cuirass at the Museo Civico, Bologna (MC 1-3), has been embellished in a similar manner; adding weight to an Italian provenance.[66]

The non-identical vambraces have also been identified as either Flemish or Italian.[67] The upper cannon flush turning joints, for example, can be

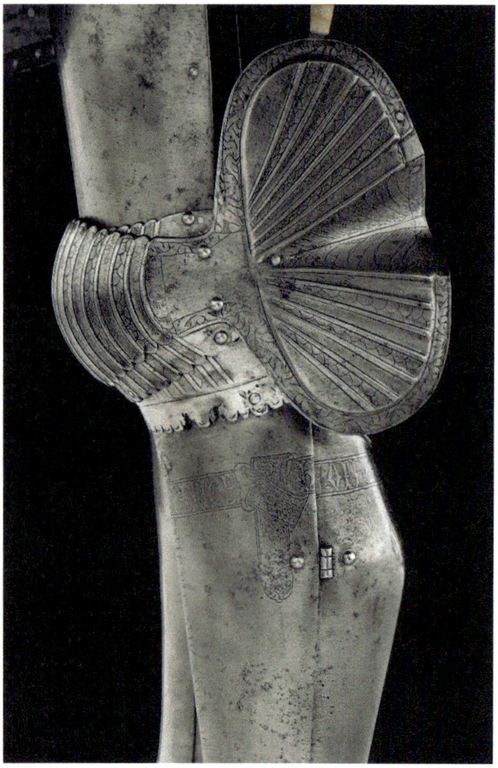

Figure 21 The elegant fluted poleyns have a number of parallels, including those found on armours made in Innsbruck. © Royal Armouries (II.7).

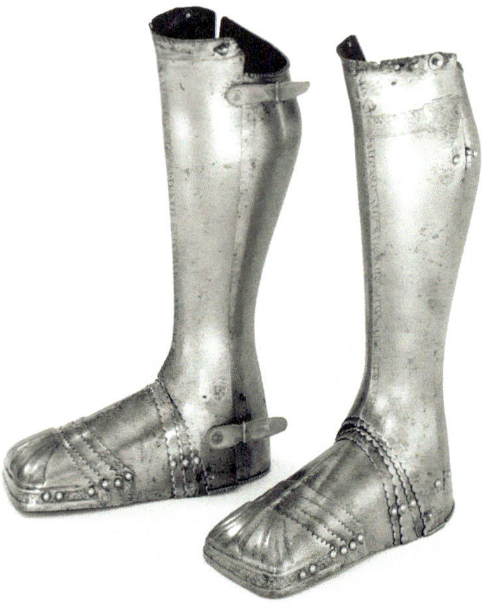

Figure 22 The sabatons of II.7 are distinctly boxed in shape and feature a central ridge. © Royal Armouries (II.7).

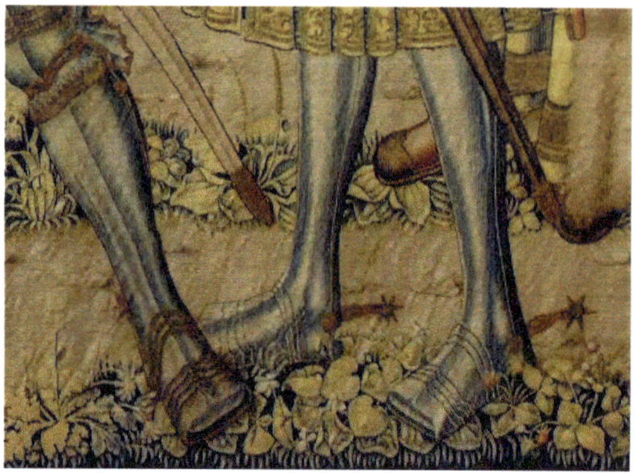

Figure 23 The sabatons on Henry VIII's 'Silver and Engraved Armour' of c.1515 are very similar to those on II.7. © Royal Armouries (II.5).

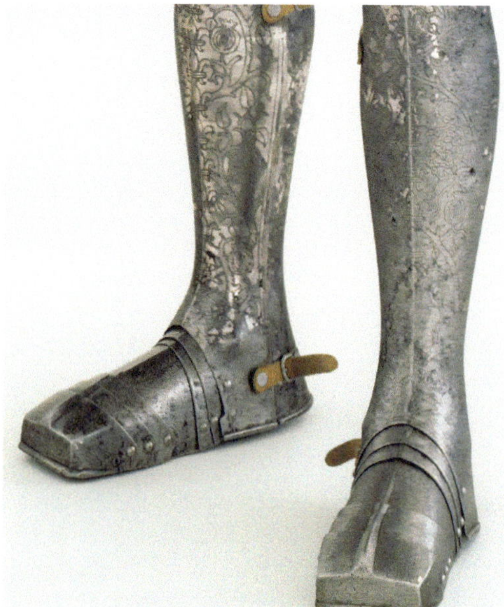

Figure 24 The sabatons depicted on the David and Bathsheba tapestries at Musée national de la Renaissance, Château de Écouen, also have a narrow boxed-shape and distinctive central ridge echoing both II.5 and II.7.

found on numerous Low Countries/Flemish armours such as that of Karel von Egmond in the Church of St Eusebius, Arnhem, and on the armour of Philip the Handsome by Anthonis van Ghindertaelen in Madrid (Real Armería, A 11). Although many Italian armours feature boxed turning joints, flush joints are certainly not unknown: they occur, for example, on G 178 at

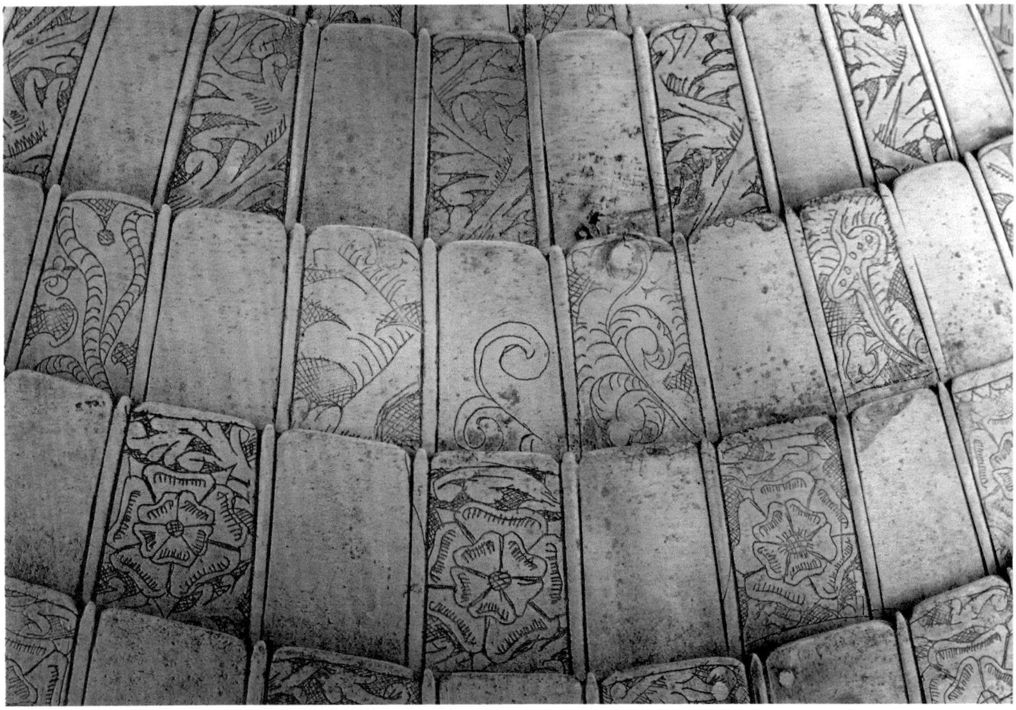

Figure 25 Henry VIII's tonlet has been etched in a chequerboard design of roses, crude leaf-like fronds and blank squares. The speed of its execution and lack of concentration is evidenced by the half-finished frond. © Royal Armouries (II.7).

the Musée de l'Armée and on the armour of Gian Campofregoso of c.1510 at Kunsthistorisches Museum Vienna (A 11). Similarly, while certainly not a uniquely Italian feature, the scalloped edges on the vambraces of Henry's armour do have a number of Italian '*alla Tedesca*' ('in the German style') parallels such as G 178, a deep poleyn from c.1500 at the Museo Nazionale del Palazzo di Venezia in Rome (inv. no. 11989) and another at the Royal Armouries, Leeds (III.1133). They can also be seen in the portraits of a *Man in Armour*, possibly by Francesco Granacci at the National Gallery, London (NG895) and the *Man in Armour with a Squire* by Giorgione at the Uffizi in Florence (inv. 1890 no.911) among others.[68] In fact, it could be reasonably argued that Henry's vambraces are more likely to be Italian, as suggested by Blair, than Flemish.[69]

The cuisses and poleyns, however, are unlikely to be Italian as fluted knee plates very rarely appear on contemporary Italian armours. Although they are probably Flemish, it is nonetheless worth noting that the poleyns are closely related to several South German examples. For instance, the scalloped edge with semicircular, incised line decoration on the lowest articulating lame matches that on the lance rest of armour B8 at the Diocesan Museum, Mantova, dated to c.1503.[70] Furthermore, the overall form of the side wings is similar to that of two German cuisses and poleyns at the Victoria and Albert Museum, London (M. 535-1927 and M.471-1927) and on a separate poleyn at the Metropolitan Museum of Art in New York (27.183.91).[71] It is therefore possible, but not certain, that the cuisses and poleyns on II.7 may have been reused from one of Henry's Innsbruck armours (Figure 20).

As has been noted previously, the distinctive boxed and ribbed sabatons of Henry's armour (Figure 22) are very similar to those on the 'Silver and Engraved Armour' (Figure 23) at the Tower of London (II.5). While this may indicate that both came from the same workshop, as suggested by Blair, comparison with those on the David and Bathsheba tapestries at Musée national de la Renaissance, Château de Écouen

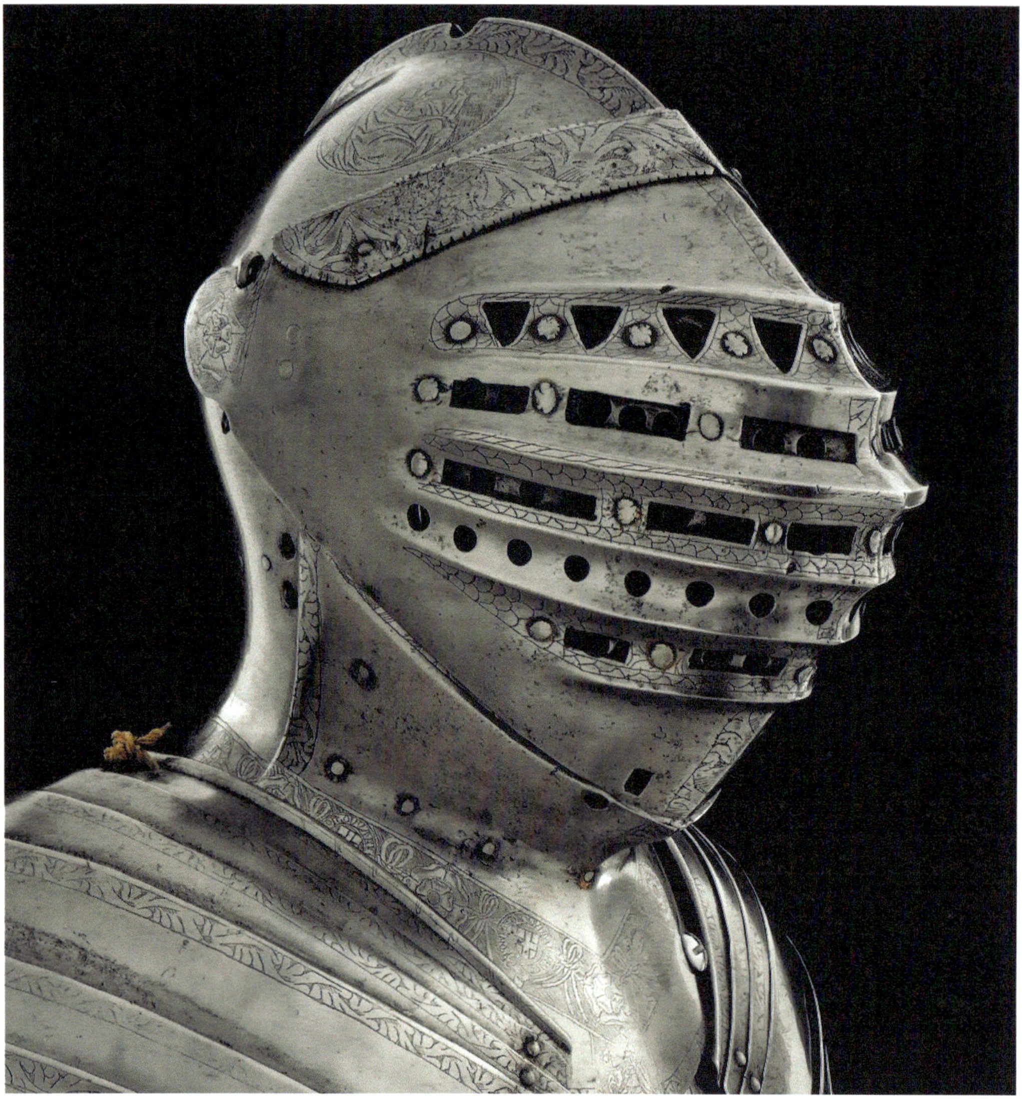

Figure 26 Decoration on the bascinet varies widely from the finely etched collar of the Order of the Garter and the roundels on the skull to the less accomplished scale motif around the visor slots. © Royal Armouries (II.7).

(E.Cl. 1618) (Figure 24) and on the painting of the *Coronation of David* at the Musée de Cluny in Paris (inv. no. Cl. 822b, circle of the Picard 'Master of Antoine Clabault') may simply point to a shared north-western European/Flemish style.[72] In fact, it is possible that some – or indeed all – of the elements of the armour could have been made at Southwark prior to 1520 by either Henry's Flemish or Italian armourers. At the very least they appear to have been responsible for the fabrication of the pauldrons and tonlet, both of which compare poorly to the quality of the work evident on II.6 and the slightly later armour of Maréchal Fleuranges at Paris. Although the records are incomplete, particularly regarding the Milanese armourers at Henry's court, it is certainly known that both Peter Fevers and Jacob 'Copyn' de Watte began to make armours for Henry shortly after their arrival in England in 1511.[73]

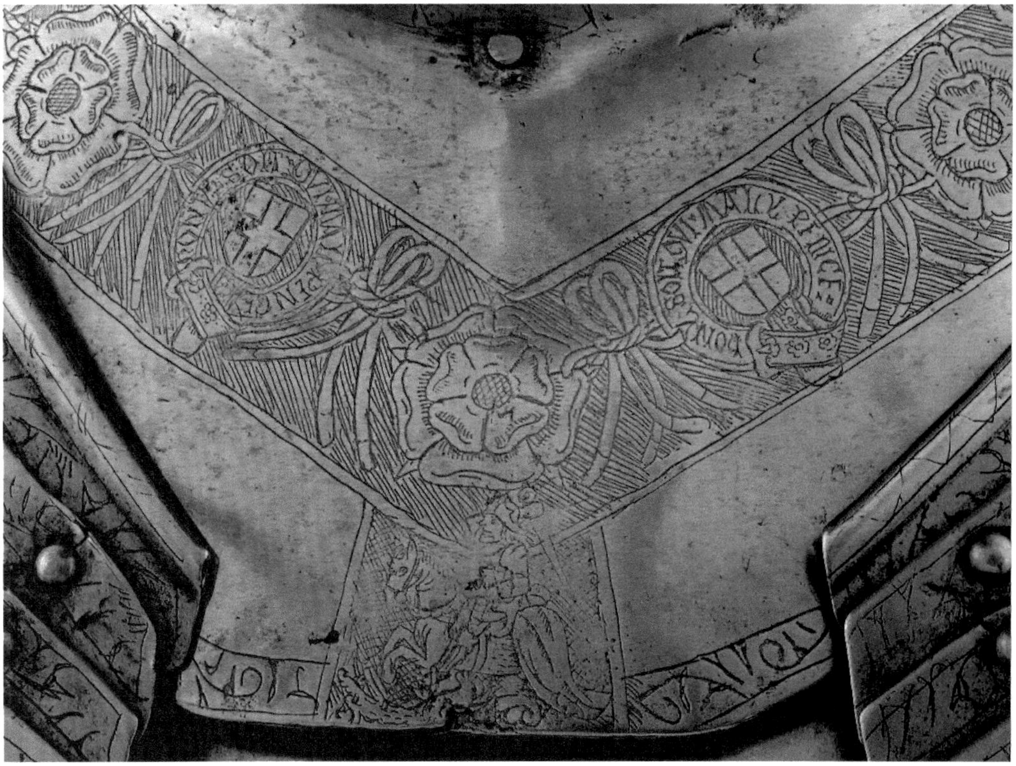

Figure 27 Detail of the collar of the Order of the Garter showing the difference in background etching between the Great George and the collar itself. © Royal Armouries (II.7).

Given the heterogeneity of the armour, the decoration was key to uniting the various parts and, importantly, glorifying Henry as king of England. However, the variable quality of the etching is indicative of the time constraints faced by Henry's armourers and is not of a standard expected on princely armour. The overall Italianate scheme is comparable to the decoration found on various Italian and Flemish armours, particularly that of Philip the Handsome.[74] Therefore it is highly likely that the embellishment of Henry's tonlet armour was undertaken by his Italian or Flemish armourers – or perhaps both – as careful examination shows that more than one individual was responsible.

The decorative scheme on the tonlet (Figure 25) consists of etched and formerly gilt foliate motifs and scale work together with Tudor or 'union' roses arranged alternately with blank panels in a chequerboard pattern (further roses can be found on the visor arms).[75] The various motifs are arranged in bands of three on the front of the tonlet with scrolling tendrils and scales occupying the uppermost bands, coarse foliage in the middle and Tudor roses at the base. In contrast, the back of the tonlet only incorporates the latter two motifs. Close observation of the etched roses reveals at least six designs, some more skilfully executed than others: the majority of those on the front of the tonlet, for example, are encircled by rings whereas those on the back are of various forms. The rose was, of course, the pre-eminent badge of the Tudors and was widely used by Henry to assert his dynasty and right to rule. Its prominence on the armour is therefore not surprising, particularly in the context of an Anglo-French tournament.[76] As has previously been observed, the scrolling foliate designs on the tonlet are particularly crude although the gilding must have detracted from this. The upper three bands featuring scrolling tendrils and scale work are of a better quality, although quite why

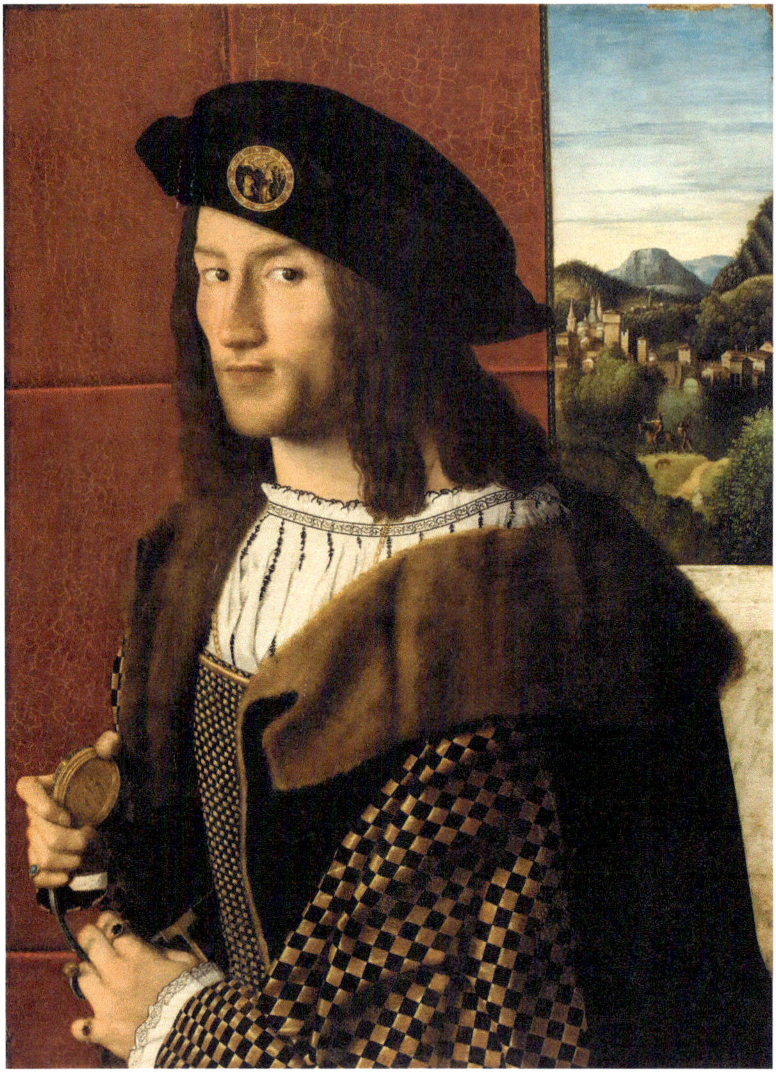

Figure 28 'Clothes maketh the man'. Although fulfilling a practical function, Henry's tonlet armour was also a visual representation of his taste, wealth and power. © Gallerie Nazionali di Arte Antica, Roma (MIBACT) – Bibliotheca Hertziana, Istituto Max Planck per la storia dell'arte/Enrico Fontolan.

this scheme was not continued on the back of the tonlet is unclear, but it may have been omitted to save time.

The bascinet has been very finely etched with the collar of the Order of the Garter, the garter itself appearing below the knee on the left greave (Figure 26). The ends of the visor arms have also been etched with roses. Scale work is present around the visor apertures, while the reinforcing band, the medial ridge and the bascinet's borders have all been decorated with foliate motifs. The top of the bascinet skull has also been etched with two roundels: the first containing an image of the Virgin and Child and the second St George; scenes that are also repeated on the lowest lames of the pauldrons. The cuirass, vambraces and legs have also been etched with foliate bands and scale work. As a genuinely pious individual with a keen interest in religion, it is not surprising that the warrior patron saint of England, St

George, should feature on Henry's armour (the saint had previously appeared on Henry's 'Silver and Engraved Armour' of c.1515). The decision to include the image of the Virgin and Child may be less obvious, however, Mary was the focus of considerable devotion by Henry which continued following the break with Rome in 1533.[77]

As noted above, the quality of the etching varies across the armour. The collar and garter are the finest and are clearly the work of the same person who, as Blair observed, may also have been responsible for the decoration on a magnificent gilt buckler, probably made for Henry VIII, in the Musée de l'Armée (I 6).[78] He must also have been responsible for the pendant Great George as the saint's head overlaps the collar (Figure 27). If so, stylistic similarities also indicate that he was responsible for the religious scenes on the helmet and pauldrons although he may not have added the cross-hatched ground as it differs completely from that on the collar and garter. Admittedly this view contradicts Blair, who saw the collar and garter as being by a different person.[79] Given the differences in the quality of the scale work, foliate motifs and tendrils, and the roses on the tonlet, it is highly likely that two other less-talented hands were involved. Clearly time was short as the hurriedness of the work is evidenced by a section on the tonlet that was etched in error then left incomplete. Furthermore, some of the roses on the front right of the tonlet are so sketchy that they could only have been added at the last minute. This lack of concentration can also be seen on the breastplate, where the scale motif at the bottom of the central flutes has been carried over onto the ribs.

Unfortunately, no evidence remains of any surface treatment on the plain undecorated areas of the armour. Nonetheless, it is possible that these parts were blued or blackened rather than being left as bare silver-grey metal.[80] In addition to covering up any mistakes, this would have created a striking blue/black and gold appearance that would have been particularly eye-catching, especially on the tonlet. Black, an expensive and highly fashionable colour for men's clothing favoured by Henry VIII, worked particularly well when paired with precious metals and jewels.[81] An impression of this can be seen on the painting of an *Unknown Gentleman*, c.1512, by Bartolomeo Veneto, which shows the sitter wearing a chequered black and gold gown.[82] This type of ostentatious display was ideally suited to the Field of Cloth of Gold where all the participants sought to outdo each other with the richness of their clothing. Although, unfinished, we may reasonably assume that had the original foot combat armour been completed, it would have been similarly embellished with striking designs personal to Henry and his dynasty (Figure 28).

Conclusions

The two foot combat armours made for the Field of Cloth of Gold are among the earliest surviving products of Henry's royal workshops at Greenwich. They were produced by a talented team of Continental armourers under the direction of the Fleming, Martin van Royne. Charged with creating a field armour with reinforcing pieces, Henry's armourers built upon preexisting knowledge to overcome the shortcomings of earlier designs and produce a technically sophisticated armour and a triumph of the armourer's craft. Although fulfilling a practical defensive function and ensuring the king's safety, the real motivation for its creation must have been the desire to inspire awe and wonder in all who saw it. In the eyes of many, England was still an insular medieval country: Henry's ambition was to make it a modern Renaissance state at the forefront of art and technical innovation.[83] The Field of Cloth of Gold was the ideal stage on which to demonstrate that his court armourers could compete with the best in Europe. However, it was not to be: Sir Richard Wingfield's letter of 16 March put paid to Henry's hopes. Yet, the creation of the tonlet armour under difficult circumstances reflected the consummate skill and ingenuity of his newly established court workshop. With resources and manpower stretched, Henry's armourers were forced to utilise as many pre-existing pieces as possible to ensure that the armour would be completed on time. In order to unify the numerous disparate pieces from which it was assembled, the armour was richly embellished with etched and gilt designs incorporating Tudor roses, Christian iconography and England's premier order of chivalry, the Order of the Garter. Having lost much of its original finish it is hard now to appreciate the dazzling effect it must have had and its positive role in shaping Henry's identity on an international stage. It also fulfilled its practical function as according to the chronicler Edward Hall, despite the combat being so vigorous 'that the fyre sprang out of their armure', Henry avoided injury.[84]

The two armours of Henry VIII therefore provide a fascinating insight into armour technology and design in the early 16th century and the importance of armour in expressing the identity power, and ambition of the wearer. They show that despite England being on the periphery of Europe, it was outward-looking and fully engaged with the latest developments in European armour technology. Doubtless further examination of both II.6 and II.7 will reveal more fascinating details.

Acknowledgements

My sincere thanks to Dr Tobias Capwell, Dr Stefan Krause, Dr Jonathan Tavares, Dr Ralph Moffat and Ms Marina Viallon for their much-valued assistance and advice throughout this project and for commenting on earlier drafts of this article.

Notes and references

1. For a detailed study of the foot combat at the Field of Cloth of Gold see Dowen, K. (forthcoming) 'The foot combat at the Field of Cloth of Gold', in K. Dowen and S. Hurst (eds), *The Field of Cloth of Gold* (provisional title). Woodbridge: Boydell and Brewer.
2. Blair, C. (1995) 'King Henry VIII's tonlet armour', *Journal of the Arms and Armour Society* 15(3): 85–108; Rimer, G., Richardson, T. and Cooper, J.P.D. (eds) (2009) *Henry VIII Arms and the Man 1509–2009*. Leeds: Royal Armouries; Richardson, T. (2013) 'The king and the astronaut', *Arms and Armour* 10(1): 3–13.
3. Lescaille, J. (1520) *L'ordonnance et ordre du tourney, ioustes, & combat a pied, & a cheval*. Paris.
4. Lescaille 1520 (cited in note 3), ff. 4r–4v. My thanks to Marina Viallon for her translation and commenting on the meaning of 'armet'.
5. Jusserand, J.J. (1901) *Les sports et jeux d'exercise dans l'ancienne France*, p. 138; Prost, B. (ed.) (1878) *Traicté de la forme et devis comme on faict les tournois*. Paris: A. Barraud, pp. 226 and 235. See also Fallows, N. (2010) *Jousting in Medieval and Renaissance Iberia*. Woodbridge: Boydell Press, pp. 76 and 94.
6. My thanks to Dr Tobias Capwell and Marina Viallon for their thoughts on this matter.
7. Williams, A. and de Reuck, A. (1996) *The Royal Armoury at Greenwich 1515–1649: A History of its Technology*. Leeds: Royal Armouries, p. 29; Ian Eaves, personal communication, June 2020.
8. Terjanian, P. (2009) 'The king and the armourers of Flanders', in G. Rimer, T. Richardson and J.P.D. Cooper (eds), *Henry VIII: Arms and the Man 1509–2009*. Leeds: Royal Armouries, pp. 155–9, esp. p. 155. See also Blair, C. (2002) 'The will of Martin van Royne and some lists of Almain armourers at Greenwich', *Journal of the Arms and Armour Society* 17(4): 199–216.
9. Cripps-Day, F.H. (1934) *Fragmenta Armamentaria, Volume I, Part II*. Frome: printed by Butler and Tanner Ltd for the author, pp. 20–22; Richardson, T. (2009) 'The Royal Armour Workshops and Greenwich', in G. Rimer, T. Richardson and J.P.D. Cooper (eds), *Henry VIII: Arms and the Man 1509–2009*. Leeds: Royal Armouries, pp. 148–54, esp. pp. 149–50; Reitzenstein, A. (1954) 'Die Landshuter Plattner Wolfgang und Franz Grosschedel', *Munchner Jahrbuch der bildenden Kunst* 3(5): 142–53, esp. pp. 145–6. For a list of all the Almains employed in 1520 see Blair 2002 (cited in note 8), pp. 199–216, esp. pp. 205–6.
10. Baltasar Bullato was described as armourer to the king in 1532: see Cripps-Day 1934 (cited in note 9), pp. 19–20.
11. Cripps-Day 1934 (cited in note 9), pp. 15–20; Richardson 2013 (cited in note 2), pp. 35 and 56.
12. Richardson 2013 (cited in note 2), pp. 38–42; Russell, J.G. (1969) *The Field of Cloth of Gold*. London: Routledge and Kegan Paul, p. 22; Caley, J. (ed.) (1825) 'Two papers relating to the interview between Henry the Eighth of England and Francis the First of France', *Archaeologia* 21(2): 175–91.
13. See Dowen forthcoming (cited in note 1).
14. Stefani, F. *et al.* (eds) (1890) *I Diarii di Marino Sanuto*, vol. XXIX. Venice, p. 247.
15. Cited in Dillon, H.A. (1904) 'Barriers and foot combats', *Archaeological Journal* 61: 276–308, esp. pp. 283 and 293.
16. 'Item, the armour for the body is like a cuirass or like the foot harness that one calls a tonlet', in Bennet, E. (trans.) (1997) *Rene of Anjou, King Rene's*

Tournament Book: Traictie de la forme et devis d'ung tournoy. Available at: https://www.princeton.edu/~ezb/rene/renebook.html (accessed 12 March 2021).

17. Blair 1995 (cited in note 2), p. 89. For other styles of armour worn at the foot combat during the 15th century see Dillon 1904 (cited in note 15), pp. 288–93.
18. New York Pierpont Morgan Library, ff. 122v–123r. My thanks to Dr Ralph Moffat for allowing me to use his transcription of the text. See also: Réne of Anjou, 'A Treatise on the Form and Organization of a Tournament'. Available at: https://www.princeton.edu/~ezb/rene/renebook.html (accessed 22 February 2021).
19. See, for example, an armour of c.1525 depicted in the Thun-Hohenstein Sketchbook, Uměleckoprůmyslové Museum, Prague, GK-11.571-A f. n51r; the armour of Charles V at the Real Armería, Madrid (A 93); the armour of Emperor Maximilian (B 71) and the 'Eagle Garniture' (*Adlergarnitur*) of Ferdinand II, 1547 (A 638) at the Kunsthistorisches Museum Vienna (A 638); and G 181 and G 182, both mid-16th century, Musée de l'Armée, Paris. In order to facilitate movement and allow the individual to fight both on foot and in the saddle, the lames of the skirt were joined together with internal leather straps which allowed them to collapse and extend. However, this was not without its limitations: see Capwell, T. (2015) *Armour of the English Knight 1400–1450.* London: Thomas del Mar, pp. 127–30.
20. Blair 1995 (cited in note 2), p. 92.
21. See, for example, an inventory of an armourer's shop in Tours of 1512: 'ung harnois complet pour combater à pié, reste les avant bras et gathelez' cited in Giraud, J.P. (ed.) (1894) *Documents pour server à l'histoire de l'armament au moyen âge et a la Renaissance* 1. Lyon, p. 174.
22. Gay, V. (1928) *Glossaire Archéologique du Moyen Âge et de la Renaissance*, vol. 2. Paris: Éditions Auguste Picard, p. 407; Comptes de Francois d'Angouleme, 1514, Archives Nationales, KK240 ff. 132v–133; Kerr, R. (ed.) (1932) *Acts of Lords of Council in Public Affairs, 1501–1554.* Edinburgh, p. 59; Dillon, H.A. (1888) 'Arms and armour at Westminster, the Tower, and Greenwich, 1547', *Archaeologia* 51: 219–80, esp. p. 279. My thanks to Marina Viallon for providing me with the reference to Francis and Dr Ralph Moffat for the reference to James IV.
23. Terjanian, P. (2014) 'Zwei Entwürfe zu Harnischen im Auftrag von Kurfürst Christian I. von Sachsen', *Dresdener Kunstblätter* 1: 38–47, esp. p. 43. My thanks to Dr Stefan Krause for drawing my attention to this article.
24. KHM A 474. See also Jeremais Schemel's 'Reit- und Turnierbuch' of 1570, Kunsthistorisches Museum Vienna, MSKK5247 and the Thun-Hohenstein Sketchbook, Uměleckoprůmyslové Museum, Prague, GK-11.571-A, f. 31r for late 16th-century examples.
25. For further examples of armours with laminated defences see Chapter 2 in this volume.
26. The earliest depiction of which I am aware is located on the tomb of Pierre d'Évreux-Navarre, Comte de Mortain at the Musée du Louvre, Paris (c.1412–15) which shows laminated defences on the back of the legs. The use of laminated defences had a far older history: Xenophon, for example, described cavalry of the 4th/5th century BC wearing 'Cheira' or articulated armguards. See Bishop, M.C. (2002) *Lorica Segmentata. Volume 1. A Handbook of Articulated Roman Plate Armour.* Braemar: Armatura Press, p. 18.
27. My thanks to Dr Tobias Capwell for drawing my attention to Rogier van der Weyden's depiction of St George.
28. BNF MS Fr. 2695 f. 27v.
29. See also Chapter 2 in this volume.
30. Terjanian, P. (2019) *The Last Knight: The Art, Armor and Ambition of Maximilian I.* New York: Metropolitan Museum of Art, p. 27; Breiding, D. (2005) 'Horse armour in medieval and Renaissance Europe: an overview', in S.W. Pyhrr, D.J. La Rocca and D. Breiding (eds) (2005) *The Armoured Horse in Europe 1480–1620.* New York: Metropolitan Museum of Art, pp. 8–18, esp. p. 14. The incomplete horse armour of Frederick III is now at the Kunsthistorisches Museum Vienna (A 69). Pier Innocenzo da Faerno of Milan had already experimented with an enclosed plate crinet as can be seen on the famous armour of c.1450 at the Historisches Museum der Stadt, Vienna (127.151–.153, 127.157–.159). For illustrations of Maximilian's horse armour see: Thun-Hohenstein Sketchbook, Uměleckoprůmyslové Museum, Prague, GK-11.571 -B. f. 1rv Praze, and KHM GG 2375, a copy of a lost painting by Pierre Coustain.
31. Thun-Hohenstein Sketchbook, Uměleckoprůmyslové Museum, Prague, GK-11.571-B ff. n27v–n28r. Lorenz's son Kolman continued to push the limits of technology with the production of the spectacular armour of Wilhem von Rogendorf, dated 1523 in the Kunsthistoriches Museum Vienna (A 374). See Krause, S. (2017) *Fashion in Steel: The Landsknecht Armour of Wilhelm von Rogendorf.* London: Yale University Press. See also Chapter 2 in this volume.
32. Terjanian 2019 (cited in note 30), pp. 32–7; Oberhammer, V. (ed.) (1954) *Die Innsbrucker Plattnerkunst, Katalog.* Innsbruck: Tiroler Landesmuseum Ferdinandeum, pp. 20–21. For a field armour with laminated defences over the inner bend of the elbow attributed to Conrad Seusenhofer, see Wallace Collection A28. See

also Bookman, H., Burschel, P. and Quaas, G. (1992) *Eisenkleider. Plattnerrbeiten aus drei Jahrhunderten.* Berlin: Deutsches Historisches Museum, p. 49.
33. Balfour Paul, J. (ed.) (1901) *Accounts of the Lord High Treasurer of Scotland* 3. Edinburgh: General Register House, p. 363; Caldwell, D.H. (1981) 'Royal patronage of arms and armour making in fifteenth- and sixteenth-century Scotland', in D.H. Caldwell (ed.), *Scottish Weapons and Fortifications, 1100–1800.* Edinburgh: John Donald Publishers Ltd, pp. 73–93. My thanks to Dr Ralph Moffat for the reference to the Moncurs of Dundee.
34. Thomas, B. (1977) 'Konrad Seusenhofer Studien. Zu seinen Spätwerken zwischen 1511 und 1517', in *Gesammelte Schriften zur Historischen Waffenkunde* 1. Graz: Akademische Druck -u. Verlagsanstalt, pp. 543–76, esp. pp. 572–4.
35. The use of the pollaxe in the foot combat was probably one of the factors which led to the development of enclosed armour. However, they were not used at the Field of Cloth of Gold probably because they were deemed too dangerous.
36. Terjanian 2019 (cited in note 30), pp. 17 and 227.
37. Multiple articulating narrow lames continued to be used on both field and tournament armours until the late 17th century. See, for example, the 1668 armour for Louis XIV by Francesco da Garabgnate at the Musée de l'Armée, Paris (G 125).
38. My thanks to Dr Stefan Krause at the Kunsthistorisches Museum Vienna for allowing me to study the inventory files and photographs of E1 and to Olivier Renaudeau at the Musée de l'Armée, Paris and Julie Rohou at Musée national de la Renaissance, Château de Écouen for permission to study the armours in the collection. The Silva armour is the only surviving armour of its type to have been made by an Italian, albeit for a foreign customer: see Rimer *et al.* 2009 (cited in note 2), p. 116.
39. For a detailed description of the armour's construction see Richardson 2013 (cited in note 2), pp. 3–13.
40. 1761. *London and its Environs Described*, vol. VI. London: R. and J. Dodsley, p. 183.
41. See Krause, S. (forthcoming) 'The ceremonial armor for the young Emperor Charles V in Vienna: the German Empire, Burgundy, and a failed plan for a Habsburg-Tudor alliance', in E. Gertsma (ed.), *Collectors, Commissioners, Curators: Studies in Medieval Art Inspired by Stephen Fliegel.* Berlin: De Gruyter/Medieval Institute Press. See also Chapter 2 in this volume.
42. The locking gauntlet on the armour of Claude de Vaudrey at Kunsthistorisches Museum Vienna is associated.
43. Musée de l'Armée, Paris G 46.
44. Anglo, S. (2000) *The Martial Arts of Renaissance Europe.* London: Yale University Press, pp. 168–71; Young, A. (1987) *Tudor and Jacobean Tournaments.* London: George Philip, pp. 81–2. For an overview of the introduction and early use of the barrier, see Dowen forthcoming (cited in note 1).
45. Dillon 1888 (cited in note 22), pp. 119–280, esp. p. 280.
46. The three armours in Paris, for example, feature simpler designs and so leave gaps when the legs are extended. Rather than separate laminated defences secured to the main body of the armour, the armour of Louis II features a highly complex system of articulating plates and sliding rivets to provide both full protection and movement.
47. SP.1, 20, ff. 44–47 and see SP.1, 19, ff. 229–230 for the date of this letter.
48. National Archives SP.1, 20, ff. 44–47. This is clearly the document referred to in a letter of 16 March (SP.1, 19, ff. 229–230).
49. See for example, Richardson, G. (2013) *The Field of Cloth of Gold.* London: Yale University Press, p. 125.
50. It is worth noting that there is no evidence that the armour was changed because of the inclusion of the two-handed swords or a change to fighting over the barrier. Indeed, it would have been extremely unusual, not to mention dangerous, if the foot combat had taken place within an enclosure without a dividing barrier. Contra Watts, K. (2020) 'The Field of Cloth of Gold: arms, armour and the sporting prowess of King Henry VIII and King Francis I', in A.V. Murray and K. Watts (eds), *The Medieval Tournament as Spectacle: Tourneys, Jousts and Pas d'Armes 1100–1600.* Woodbridge: Boydell and Brewer, pp. 208–37, esp. p. 217.
51. Richardson 2013 (cited in note 49), pp. 35–7, 75, 105–6.
52. 'Expenses of Sir Edw. Guldeforde, master of the Armory, and others, for horses and stuff provided for the King's armoury, by John Frognall': see Brewer, J.S. (ed.) (1867) *Letters and Papers Foreign and Domestic, Henry VIII. Volume 3 1519–1523,* p. 413.
53. Revels Accounts: see Brewer 1867 (cited in note 52), pp. 181, 1555–6.
54. 11 March 1520: To Hen. Smith, for finishing the armory house, the millhouse, the bridge at Greenwich, and a closet at Windsor 300l. To Thos. Forster, for the armoury house, mill and conduit at Greenwich, 300l: Brewer 1867 (cited in note 52), p. 1539.
55. For a detailed description of Henry's tonlet armour see Blair 1995 (cited in note 2), pp. 85–109. Compared to II.6, Henry's 'Tonlet Armour' is considerably lighter at 64 lb (29 kg).

56. Blair 1995 (cited in note 2), p. 103.
57. Dillon 1888 (cited in note 22), p. 280.
58. The suggestion by Claude Blair and Ian Eaves that it was originally designed for the tourney is a possibility. However, distinguishing them from those made for the foot combat is certainly not always possible: see Blair 1995 (cited in note 2), p. 103; Eaves, I. 1993. 'The tournament armours of King Henry VIII of England', *Livrustkammaren*, pp. 3–46, esp. p. 16.
59. The legs were acquired by the Tower Armouries in 1971.
60. Eaves 1993 (cited in note 58), pp. 16 and 18.
61. The inclusion of tonlet armours is a surprising and puzzling aspect to this painting. I am not aware that this style of armour is seen in any other portrayal of a historical battle. See Zygulski, Z. (1979) 'The Battle of Orsha', in R. Held (ed.), *Art, Arms and Armour: An International Anthology.* Chiasso: Acquafresca Editrice, pp. 108–43, esp. p. 126. The armour of Maximilian I in Vienna (Kunsthistorisches Museum) is somewhat old-fashioned and reflects the conservative style of an armourer long experienced in producing certain pieces: see Terjanian 2019 (cited in note 30), pp. 94–7.
62. Blair 1995 (cited in note 2), p. 103.
63. The crude fleur-de-lys incised into the upper left front of the tonlet remains unidentified.
64. Berchet, G. (ed.) (1883) *I Diarii di Marino Sanuto. Tom X.* Venice: F. Visentini, p. 313; Le Glay, A.J.G. (ed.) (1838) *Analectes historiques ou documents inédits.* Paris, p. 182. See also Bolland, C. (2011) *Italian Material Culture at the Tudor Court*, unpublished PhD thesis, University of London.
65. Blair 1995 (cited in note 2), p. 104; Blair, C. (1965) *The Silvered Armour of Henry VIII in the Tower of London.* Oxford: Society of Antiquaries, pp. 8–14, Rimer *et al.* 2009 (cited in note 2), p. 124; Von Klapsia, H. and Thomas, B. (1977) 'Harnischtudien. Stilgeschichte des Deutschen Harnisches von 1500 bis 1530', in Thomas, B (ed.), *Gesammelte Schriften zur Historischen Waffenkunde* 1. Graz: Akademische Druck -u. Verlagsanstallt, pp. 619–638.
66. Boccia, L.G. and Coelho, E.T. (1967) *L'arte dell'armatura in Italia*. Milan: Bramante Editrice, pp. 194–7.
67. The right vambrace features two ribs on the upper cannon beneath the turning joint and a single rib beneath the couter on the lower cannon. The left vambrace only has a single rib on the lower cannon.
68. Other *'alla Tedesca'* armours with scalloped lames include the anime armour of Gian Giacomo de Medici in Kunsthistorisches Museum Vienna (A 404) and the main plate of a pauldron at the Higgens Armory Museum, Worcester, MA (928.12).
69. Blair 1995 (cited in note 2), p. 92.
70. Boccia, L.G. (1982) *Le armature di S Maria della Grazie di Curtatone di Mantova*. Milan: Bramante Editrice, p. 265 and black-and-white plate 402.
71. My thanks to Dr Stefan Krause for his thoughts on the cuisses and poleyns.
72. Blair 1995 (cited in note 2), p. 104. The central spine down the centre of the sabatons is also found on the remains of a Flemish or English sabaton from the Church of St Lawrence, Hatfield (Yorks) on loan to the Royal Armouries, Leeds (AL.288).
73. Richardson 2009 (cited in note 9), p. 149.
74. Eaves 1993 (cited in note 58), p. 18.
75. Under Henry VIII the rose had become the pre-eminent badge of the house of Tudor and was widely used by him to promulgate the dynasty. See Fox-Davies, A.C. (1907) *Heraldic Badges.* London: John Lane, pp. 51–2; Anglo, S. (1992) *Images of Tudor Kingship.* London: Seaby. pp. 74–93.
76. Fox-Davies 1907 (cited in note 75), pp. 512; Anglo 1992 (cited in note 75), pp. 74–93.
77. See Rex, R. (2014) 'The religion of Henry VIII', *The Historical Journal* 57(1): 1–32, esp. pp. 12–13.
78. Edwards, I and Blair, C. (1982) 'Welsh bucklers', *The Antiquaries Journal* 62(1): 74–115, esp. p. 89.
79. Blair 1995 (cited in note 2), pp. 94–98.
80. Rimer *et al.* 2009 (cited in note 2), p. 127.
81. Hayward, M. 2007. *Dress at the Court of King Henry VIII.* Leeds: Maney Publishing, pp. 120–21.
82. Richardson 2009 (cited in note 9), Galleria Nazionale d'Arte Antica, Rome. My thanks to Dr Tobias Capwell for drawing my attention to this portrait.
83. See, for example, the painting of the *Legend of Saint Ursula: The Meeting of the Betrothed Couple and the Departure of the Pilgrims* by Vittore Carpaccio, 1495, Venice. For commentary on this see Giry-Deloison, C. (2012) *1520 Le Camp du Drap D'Or/The Field of Cloth of Gold.* Paris: Somogy Editions d'Art, pp. 25–7 and generally, Starkey, D. (1991) *Henry VIII: A European Court in England.* London: Collins & Brown.
84. Ellis, H. *et al.* (eds) (1809) *Hall's Chronicle: containing the history of England, during the reign of Henry the Fourth, and the succeeding monarchs, to the end of the reign of Henry the Eighth, in which are particularly described the manners and customs of those periods.* London: J. Johnson, p. 618.

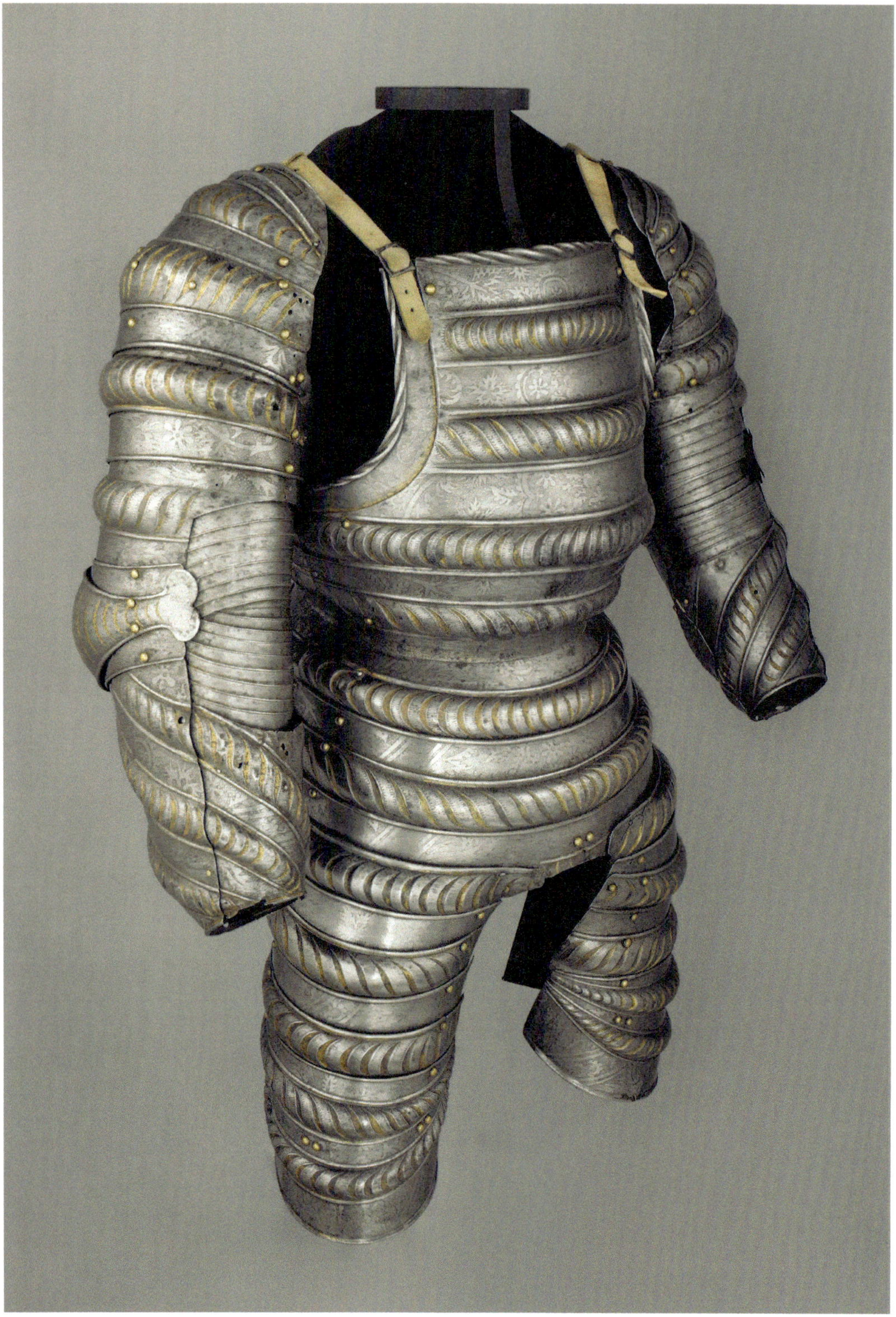

Figure 1 Elements of a Landsknecht armour, Innsbruck, c.1515. The Wallace Collection, inv. no. A28. © Trustees of the Wallace Collection.

2

Cloth of steel: elements of a Landsknecht armour in the Wallace Collection

Tobias Capwell

Introduction

Many superlative examples of Renaissance armour-making survive today in the Wallace Collection. Of these, some have been the focus of intense study and publication over many years, with the result that their origins, dating, typologies and functions are well understood and the facts generally agreed upon. Others, however, remain somewhat mysterious or have yet to attract focused investigation. One such armour is inv. no. A28 (Figure 1), historically catalogued by Mann as a 'puffed and slashed' armour made in Germany, possibly in Augsburg, with the etching ascribed to Daniel Hopfer.[1] Slightly later, Gamber credited the making of the armour to Kolman Helmschmid of Augsburg.[2] Norman accepted both of these attributions, adding the suggestion that the etching could have been the work of either Daniel Hopfer or Hans Burgkmair.[3]

The attribution of this armour to Augsburg is understandable given the fame of the Helmschmids and of their work in this style in particular – Kolman was almost certainly responsible for the most celebrated of puffed and slashed armours, that of Wilhelm Freiherr von Rogendorf (1481–1541). Made in 1523, the majority of this armour (Figure 2) is now in the Imperial Armoury, Vienna.[4] Although the creators of the Rogendorf and Wallace Collection armours had similar technical and artistic intentions, their individual realisations are in fact very different in both construction and decoration.

This chapter proposes a new attribution for this treasure of the Wallace Collection, while also exploring its original extent and function, possibly as a 'Landsknecht garniture' – a multi-purpose armour system, the design of which was

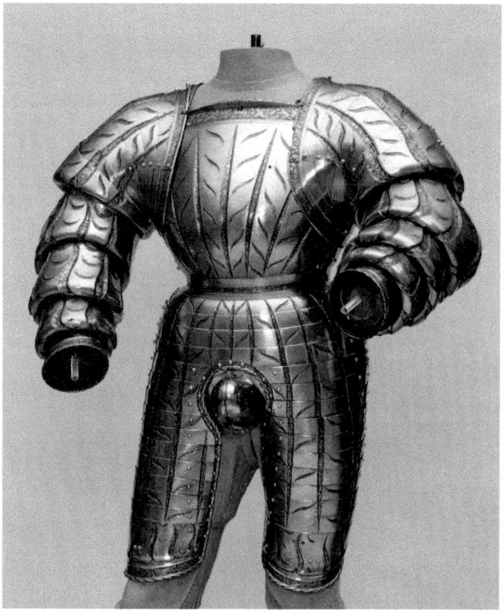

Figure 2 Elements of a Landsknecht garniture of Wilhelm Freiherr von Rogendorf, made by Kolman Helmschmid, Augsburg, dated 1523. Imperial Armoury, Kunsthistorisches Museum, Vienna, inv. no. A374. KHM–Museumsverband.

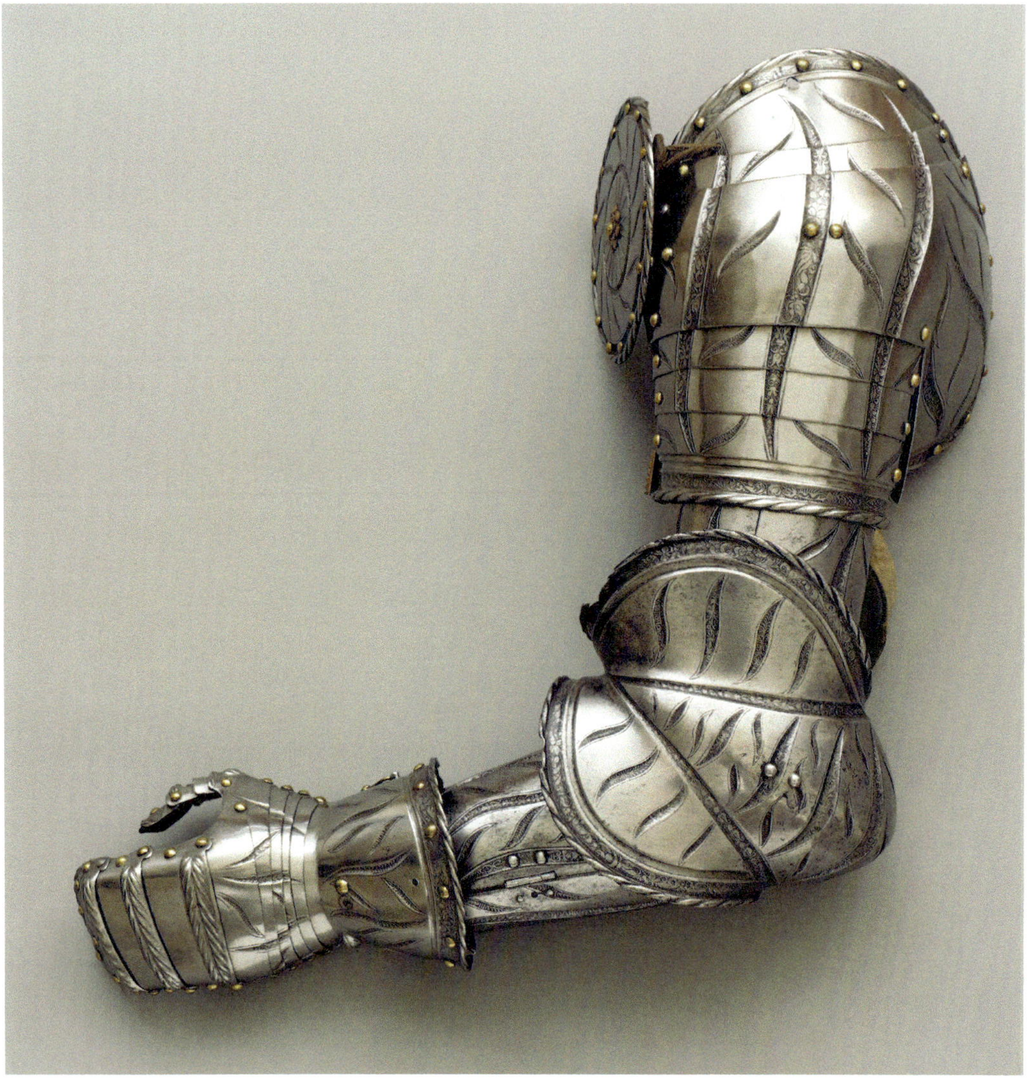

Figure 3 Hand, arm and shoulder defences for the field, from the Rogendorf Landsknecht garniture, 1523. Here reconstructed from parts preserved in the Imperial Armoury, Vienna (pauldron, besagew, lower cannon and gauntlet) and the Wallace Collection (upper cannon and couter). KHM–Museumsverband.

determined by the distinctive modes of warfare practised by German and Swiss mercenaries during the first half of the 16th century.

The Landsknecht garniture

The existence of this very specific type of field garniture, developed in the second decade of the 16th century, was first postulated by Stefan Krause, director of the Imperial Armoury at Vienna, in discussions with the present author, during preparation of his book on the Rogendorf armour.[5] The Rogendorf armour is clearly a garniture of some kind since, in addition to the exuberant costume vambraces designed to simulate the billowing sleeves of a Landsknecht doublet, it also had another set of simpler shoulder, arm, hand and leg defences for the field (Figure 3). Of these, the pauldrons, besagews, right lower cannon,

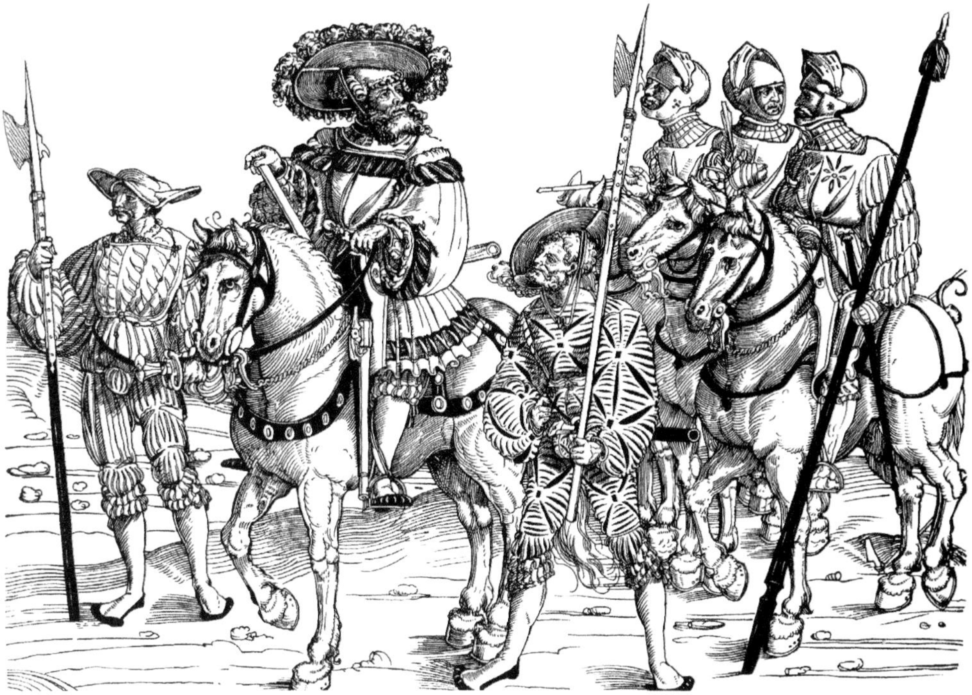

Figure 4 A Master of Ordinance and his Landsknecht guard, including demi-lancers, by Erhard Schön, Nuremberg, c.1525–30.

gauntlets and short cuisses with poleyns are also preserved in Vienna, while the right upper cannon and left couter are now in the Wallace Collection;[6] the left upper cannon, right couter and left lower cannon had already been lost by the late 16th century.

There is no evidence that the Rogendorf armour ever included any other pieces of exchange beyond the surviving elements and the missing pair mates to the extant couter and cannons. No trace of an original helmet remains, although the garniture must have included one.[7] As it is however, the Rogendorf garniture serves the needs of a well-armed infantryman or light cavalryman. The presence of a lance rest, cleverly hidden in the turned and roped edge of the right side of the breastplate, reveals the intention that the armour should also serve the requirements of a medium cavalryman or demi-lancer (Figure 4); it is strong enough to support a spear or demi-lance, but far too delicate to endure the much greater impact of the heavy cavalry lance.[8] The garniture therefore must have had a burgonet and falling buffe to complete the requirements of these fighting roles. Crucially, the Rogendorf armour does not appear to have ever had any pieces for heavy cavalry deployment or any parts for formal or courtly combats such as the joust or tourney. The costume sleeves would have been worn in parades and perhaps also in other courtly contexts,[9] therefore it was designed to serve exclusively within the military culture of the Landsknechts, who fought as infantry, sometimes with the support of light and medium cavalry. Landsknecht commanders such as the Freiherr von Rogendorf, despite being knights with the right to deploy as heavy cavalry, sometimes chose instead to advance into battle on foot with their pikemen, halberdiers and handgunners, or, remaining in the saddle, to arm themselves as light cavalry or demi-lancers.[10] It is for these leaders of Landsknecht forces that the South German Landsknecht garniture was devised.

'Costume' armour

A more complete understanding of this genre of armour design requires that a clear distinction be drawn between the intended function of an armour and its style of ornament. Sculpted decoration as ostentatious as simulated puffing and slashing immediately draws the eye, and can easily drown out other essential qualities of the work. The term 'costume armour' – often applied to any harness decorated in this manner – can also mislead, since it suggests typological homogeneity while also implying that such armours were somehow non-functional or purely ceremonial. It is perhaps easy to dismiss armour that simulates textile clothing as inherently less practical, to interpret it as Blair once did, as 'symptomatic of the beginning of the decline of armour as a thing of practical use ... many of these armours were designed chiefly for parade, not for protection in battle.'[11] In fairness to Blair, in 1958 he did not have access to decades of later research on Renaissance armour and garnitures in particular. Nevertheless, it is not uncommon for such armours still to be viewed in this way.

In reality, cloth simulation appears on armour of many different types intended for diverse purposes of which parade was just one. Indeed, the earliest hint of the idea is found on a standard field armour made for Philip I 'the Handsome', King of Castile and the son of Emperor Maximilian. This armour, now in the Real Armería in Madrid, was made not in Germany but in Brussels in 1500, and incorporates a helmet sculpted into the form of a fashionable hat.[12] The earliest surviving armour referring to puffed and slashed clothing was made c.1511–12 in Innsbruck, for Philip's son, the 11-year-old future Emperor Charles V (Figure 5). Although this armour, the work of Hans Rabeiler (first recorded 1501, died 1519), was never finished, the work progressed far enough to reveal that it was intended as a full knightly field armour optimised for heavy cavalry use.[13] Although the young owner would not have been expected to fight on the battlefield, such equipment was essential to his ongoing martial training. The 'costume' decoration – realised as narrow puffed and slashed bands placed horizontally on the cuirass, pauldrons, vambraces and cuisses – is simply a luxurious addition to what is otherwise perfectly functional fighting equipment.

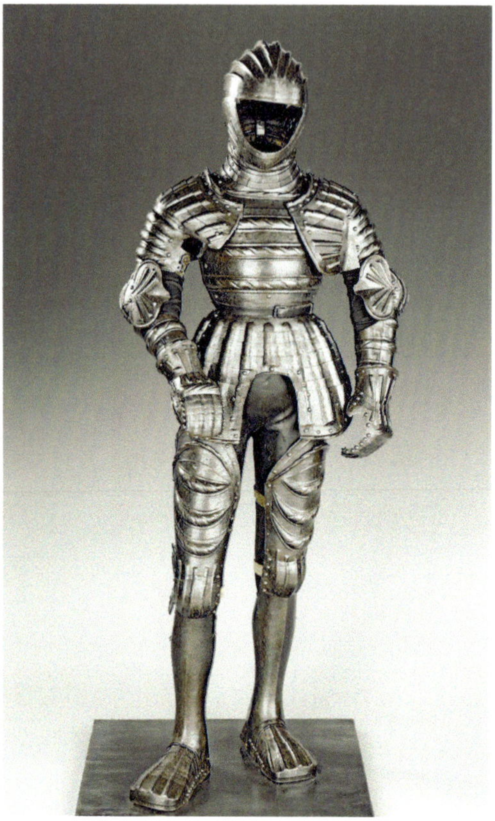

Figure 5 Unfinished child's armour of the future Emperor Charles V, made by Hans Rabeiler, Innsbruck, c.1511–12. Imperial Armoury, Kunsthistorisches Museum, Vienna, inv. no. A186. KHM–Museumsverband.

The second earliest armour is another made slightly later in Innsbruck for Charles when he was about 13 (Figure 6) by Maximilian's court armourer Konrad Seusenhofer between 1512 and 1514.[14] Although it is technically quite different from the Wallace Collection armour, aspects of the decoration of both harnesses are closely comparable (see below). This armour exhibits more obvious parade characteristics, foremost the steel base simulating a pleated textile skirt, and yet it is still a highly functional armour; as intended on the earlier unfinished boy's armour, this second armour even once included a lance rest.[15]

The fashion for armour decoration simulating puffs and slashes does not seem to have spread

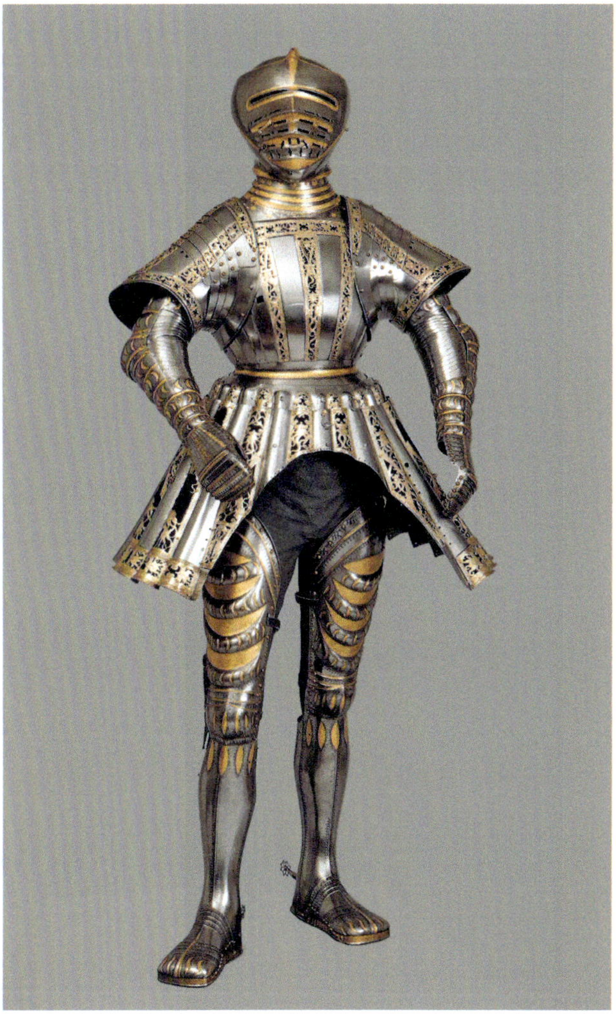

Figure 6 Child's armour of the future Emperor Charles V, made by Konrad Seusenhofer, Innsbruck, c.1512–14. Imperial Armoury, Kunsthistorisches Museum, Vienna, inv. no. A109. KHM–Museumsverband.

beyond Innsbruck until the 1520s, when work in this style started to be produced in Augsburg and Nuremberg. The Rogendorf armour is dated 1523; although it may not have been the first Augsburg armour decorated in this way, it is the earliest instance currently known. Fashions could spread very quickly in the 16th century and the Seusenhofer boy's armour may have been a carrier, since it was sent to Augsburg for specialist decoration.[16] Of course, Augsburg craftsmen were frequently employed by the Habsburg court in many capacities, and transference of technical and decorative ideas back and forth between Innsbruck and Augsburg armourers would have occurred quite naturally via their personal contacts, exposure to each other's work and through the requests of shared clients.[17]

Nevertheless, the surviving material evidence suggests that the idea for the style originated in Innsbruck in the 1510s, with masters in Augsburg creating their own interpretations from c.1520. The place of the Wallace Collection armour in this story is thus highly significant since it appears to incorporate substantial parts of the only known Landsknecht garniture made in Innsbruck to the same (or closely similar) technical specifications as the Rogendorf armour. Crucially, however, the Wallace Collection armour was made years earlier, perhaps c.1515. It is therefore one of the first realisations of the concept of such a specialised infantry (and perhaps also light cavalry) garniture closely associated with the Landsknecht sub-culture.

The Wallace Collection armour

The homogeneous garniture elements in the Wallace Collection comprise breast- and backplates, the former with a skirt of four plates and thigh-length tassets – the latter including a rump defence – and fully-enclosed vambraces with integral spaudlers. There is a hole located centrally near the bottom edge of the skirt, over the area of the groin, provided for the attachment of a codpiece, now lost. Since the 19th century, these parts have been associated with a close helmet with puffed and slashed decoration of a similar date, but which does not belong to the other parts of the armour.[18] At some point, long after the armour's working lifetime almost certainly in the 19th century, a pair of top neck lames from a contemporary gorget were combined with recycled pieces of late medieval mail (installed inside out) to fill the gaps left in the armour by the missing gorget, and to provide the lower half of the turning joint required to associate the close helmet with the rest of this armour (Figure 7). However, this rather improvised system of assembly lacks a historical basis.

Although it realises the idea of a foot combat armour in the Landsknecht style in quite a different way, there are important technical points of comparison between the Wallace Collection armour and the Rogendorf garniture in Vienna (Figure 8). Krause has noted that the fully enclosed costume vambraces and rump defence of the latter were probably intended for parade use, as well as, possibly, a type of formal combat on foot forming part of a courtly spectacle.[19] This idea is also supported by the form of the (removable) front skirt and tassets of the Rogendorf

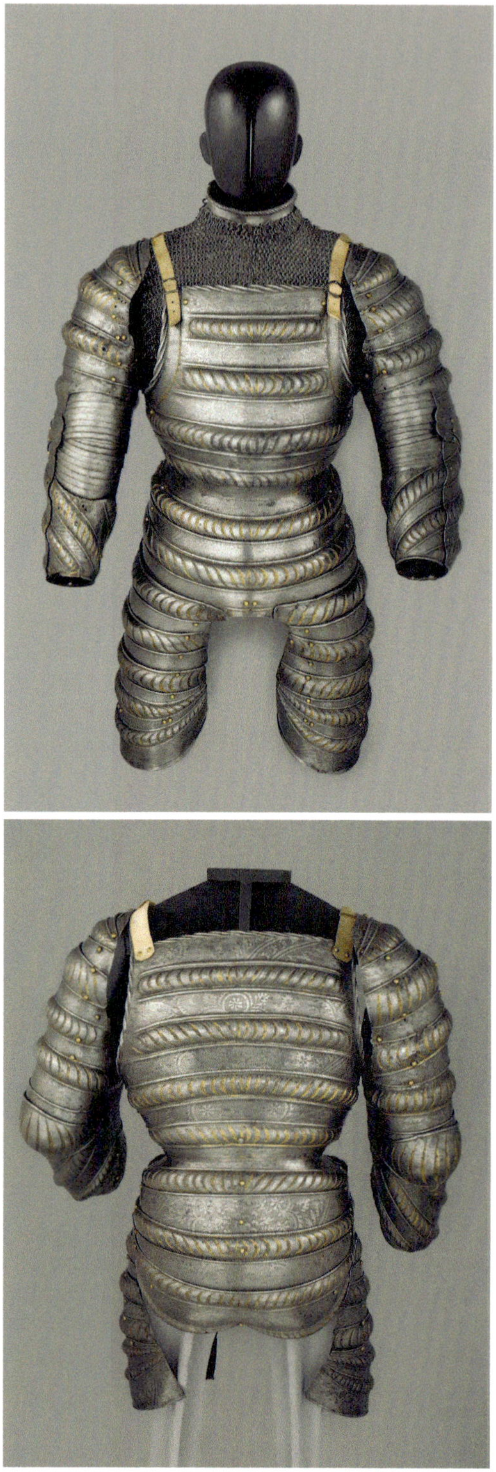

Figure 7 Elements of a Landsknecht armour, Innsbruck, c.1515. The Wallace Collection, inv. no. A28. © Trustees of the Wallace Collection.

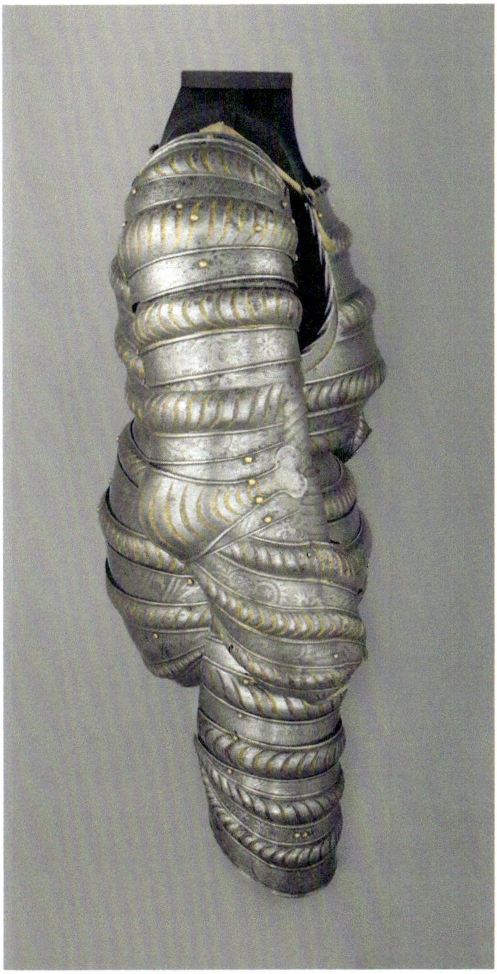

Figure 8 Elements of a Landsknecht armour, Innsbruck, c.1515. The Wallace Collection, inv. no. A28. © Trustees of the Wallace Collection.

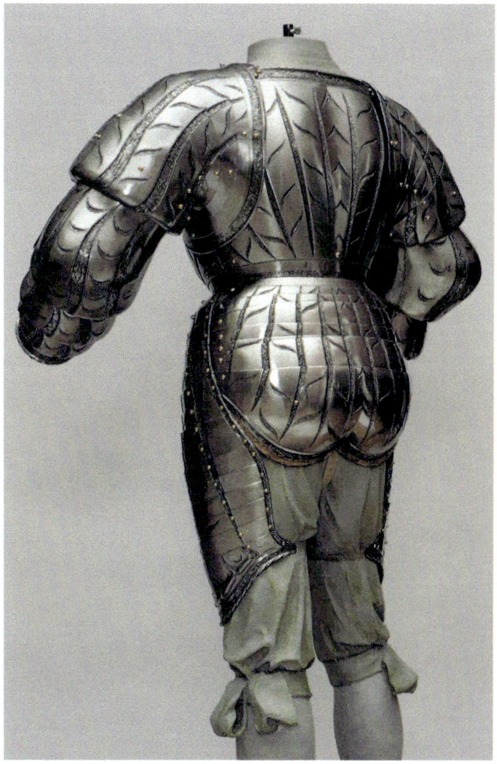

Figure 9 Elements of a Landsknecht garniture of Wilhelm Freiherr von Rogendorf, made by Kolman Helmschmid, Augsburg, dated 1523. Imperial Armoury, Kunsthistorisches Museum, Vienna, inv. no. A374. KHM–Museumsverband.

armour, designed to be worn with the rump plate, which are fitted to the body very closely and incorporate a codpiece. Notably, the tassets wrap around the lateral surfaces of the thighs to give more protection to the back of the leg with the trailing edge of each even drawn out into the bend of the knee, like the wing of a standard poleyn (Figure 9). Such extensive coverage of the rear of the lower body and legs was at this date only found on specialised armours for formal foot combats.[20] It is important also to mention that another Landsknecht garniture, made slightly later by Kolman Helmschmid (Figure 10), possibly for Jerzy Herkules Radziwill (1480–1541) in

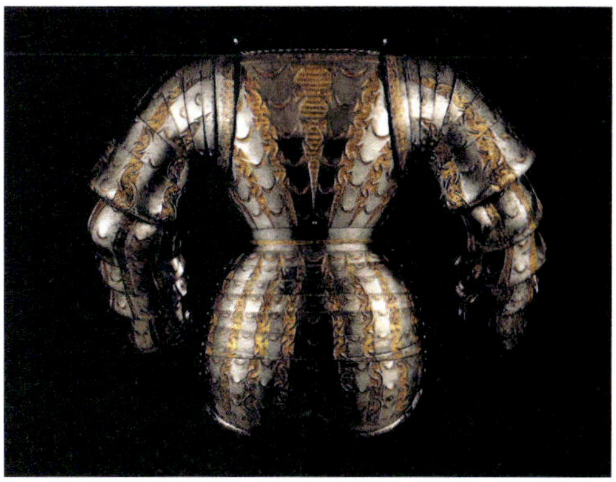

Figure 10 Elements of a Landsknecht garniture, possibly of Jerzy Herkules Radziwill, made by Kolman Helmschmid, Augsburg, c.1525. The Metropolitan Museum of Art, New York, inv. nos. 24.179, 26.188.1-2, 2.29.158.363a-b; Backplate and rump: Gift of Bashford Dean, 1924; vambraces: Mrs. Stephen V. Harkness Fund, 1926; top lames of vambraces: Bashford Dean Memorial Collection, Funds from various donors, 1929.

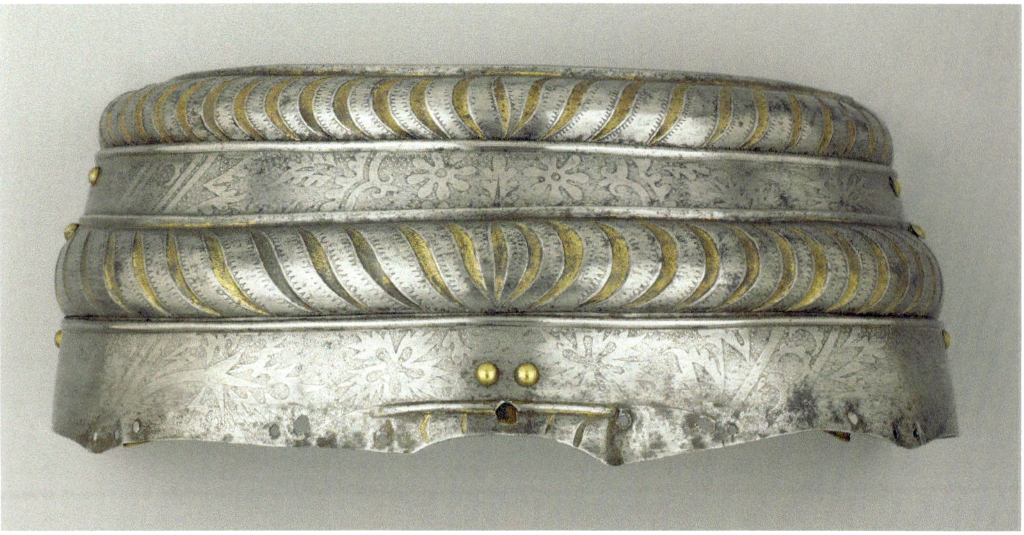

Figure 11 Front skirt of a Landsknecht armour, Innsbruck, c.1515. The Wallace Collection, inv. no. A28. © Trustees of the Wallace Collection.

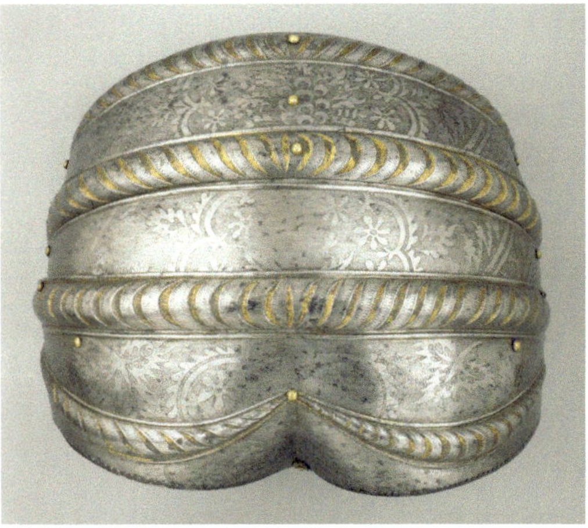

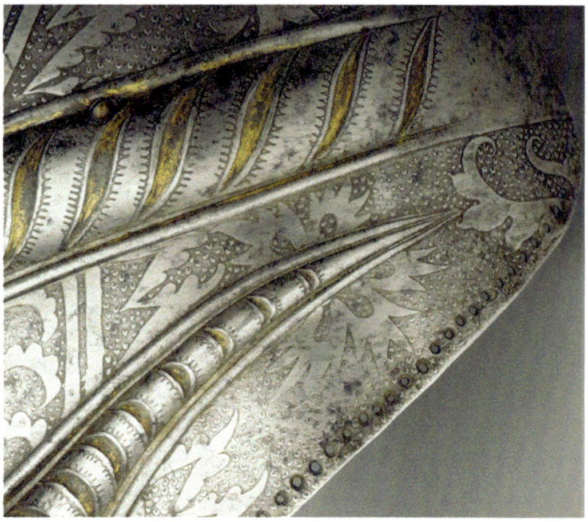

Figure 12 Hoguine of a Landsknecht armour, Innsbruck, c.1515. The Wallace Collection, inv. no. A28. © Trustees of the Wallace Collection.

c.1525,[21] shows evidence of once having included fully enclosed cuisses as a more advanced stage of the concept.[22] When the Rogendorf garniture was configured for field use, the whole front skirt and integral foot combat tassets would have been removed, along with the rump defence, and replaced with a rear skirt lame, shorter front skirt and tassets for the field (which have not survived) and the short cuisses and poleyns (which remain in Vienna).

The technical configuration of the lower cuirass of the Wallace Collection is almost identical in concept, if not execution. The front skirt (Figure 11) is intended to support a codpiece (now lost), and the tassets also wrap around the thighs of the wearer to a greater extent than those of a typical field armour. The lower edge of the rump defence of the Wallace armour (Figure 12) is set with a row of small, closely spaced holes for a decorative leather or textile edging, and provided with

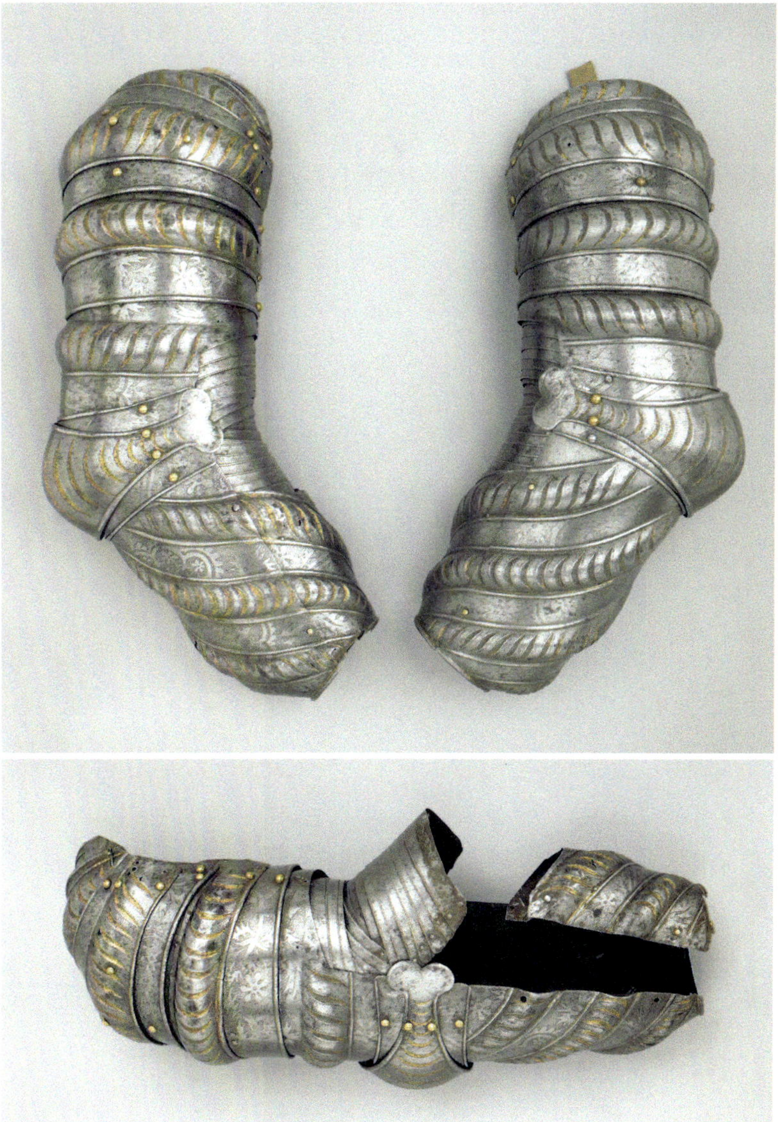

Figure 13 Arm defences of a Landsknecht armour, Innsbruck, c.1515. The Wallace Collection, inv. no. A28. © Trustees of the Wallace Collection.

the attachment point for a groin strap attaching the rump plate to the codpiece between the legs. Both of these features survive on the Rogendorf armour.

The Wallace Collection and Rogendorf armours both also include voluminous, fully enclosed 'costume' vambraces as well as integral shoulder plates in these defences. Interestingly, they achieve this desired result in quite different ways. The Wallace Collection arm defences are constructed in essence like a standard vambrace, with a main elbow plate fitted with articulation lames, which in turn are attached to full upper and lower cannons (Figure 13). The anterior surfaces of the elbows, usually left open on most styles of vambrace, are here protected with 'gussets of plate', skilfully engineered sets of 16 lames each, designed to contract and expand as

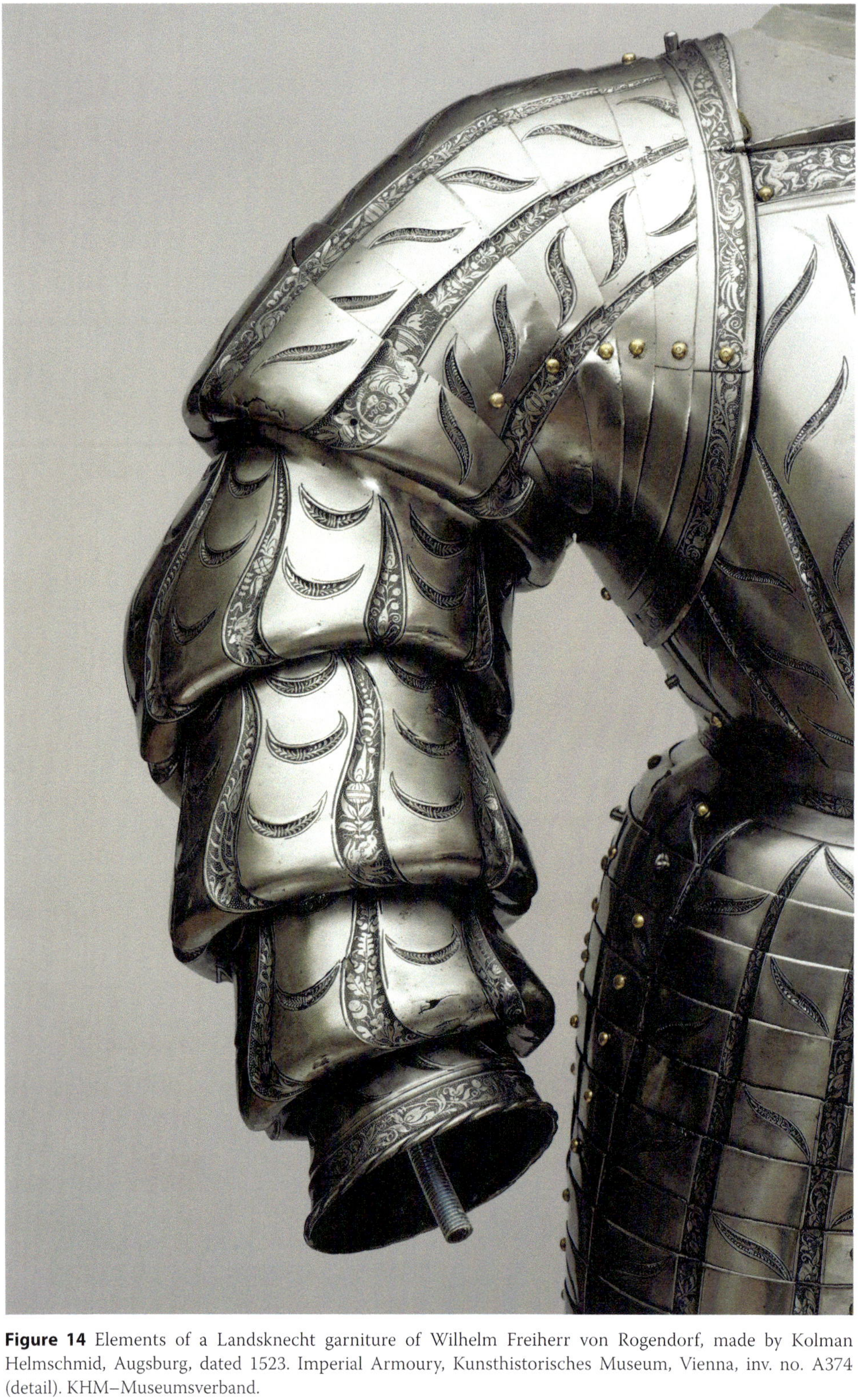

Figure 14 Elements of a Landsknecht garniture of Wilhelm Freiherr von Rogendorf, made by Kolman Helmschmid, Augsburg, dated 1523. Imperial Armoury, Kunsthistorisches Museum, Vienna, inv. no. A374 (detail). KHM–Museumsverband.

the elbow joint flexed and extended. The upper cannons have been extended upwards with two articulation lames covering the top of the arm, with the top plate joined to the shoulder assembly by means of a slotted turning joint. This system seems to have inspired a restorer in the 19th century who added an additional lame at the top of each spaudler so that the shoulder defences could be hung from the shoulder straps of the cuirass. On each side this added plate, crudely etched in an attempt to match it to the rest of the armour, is cut with a long front-to-back slot along which runs a rivet affixing a leather loop that passes through the shoulder strap to attach the arm to the body. Like the makeshift mail and plate neck piece, this shoulder system has no historical precedent: rather it is a rough improvisation, almost certainly installed at the same time as the neck defence was fabricated to facilitate association of the close helmet. On each spaulder, what is now the second lame from the top is the original top lame; this plate has a finished outer edge and a central hole that was once part of the original shoulder fastening system. Most likely, the lost gorget was mounted with short vertical posts which passed through these holes from underneath to hold the shoulder defence in place in common with the most typical shoulder fastening on South German armour of this period.

The Rogendorf garniture is famed for its slightly more original and radical approach to the fully enclosed arm and shoulder defences for parade and foot combat (Figure 14). The Helmschmids had already perfected fully enclosed shoulder defences, which even cover the whole area of the underarms (left open on the Wallace Collection armour). Developed from the early 15th-century technology of gusset plates or *croissants de plate*, designed to reduce the gaps in plate protection under the arms,[23] the Helmschmids were incorporating such enclosed shoulder defences on armours made for Maximilian, Archduke of Austria (later Emperor Maximilian I) by the mid-1480s.[24] From this point into the early 16th century, the Helmschmids appear to have provided enclosed shoulder defences for use in both the *Kolbenturnier* and for formal foot combat.[25] In the 16th century they began to be included on 'costume' armours, first on the second child's armour made for the future Charles V in Innsbruck and then on both of the extant Landsknecht armours made by Kolman Helmschmid.[26]

On the Rogendorf armour the shoulder defences have then been extended down the arms in a series of three socketed, hoop-like segments, with each piece reducing slightly in circumference so as to fit inside the one above it. Each metal 'sleeve' is completed with a wrist- or cuff-plate at the bottom. This innovative and essentially unprecedented construction granted an enhanced verisimilitude to the conceit of textile transformed into steel. Although the creator of the Wallace Collection armour was a highly skilled metalworker, able to achieve a wonderful sense of flowing cloth in the way he rendered the twists in the forearm and the sagging just above the wrist, he was ultimately limited by his adherence to standard vambrace construction methods. Even in advance of the consideration of their differing approaches to the decoration, two completely different artist's 'voices' are evident in the builds of these armours.

Parts of a Landsknecht garniture in the Royal Armouries

The original extent of the Wallace Collection garniture is brought into clearer focus by a group of armour parts preserved in the Royal Armouries in Leeds (Figure 15). Comprising a gorget, a rear skirt lame, a pair of tassets and a pair of vambraces with integral spaudlers, these pieces are so similar in style to the Wallace Collection armour that they appear, at first glance, to belong to it; together they would form almost a complete Landsknecht garniture closely akin (in functional terms) to the Rogendorf armour. Taken as a single assemblage, there is no duplication or repetition: each of the Royal Armouries' parts represents a piece missing from the Wallace Collection armour. The gorget is perhaps the most obvious loss from the latter, which necessitated the fabrication of the aforementioned (and inaccurate) mail and plate neck piece in the 19th century. Intriguingly, the skirt lame for the backplate would have been required if and when the rump defence was not in use; to configure the armour for war, this would have

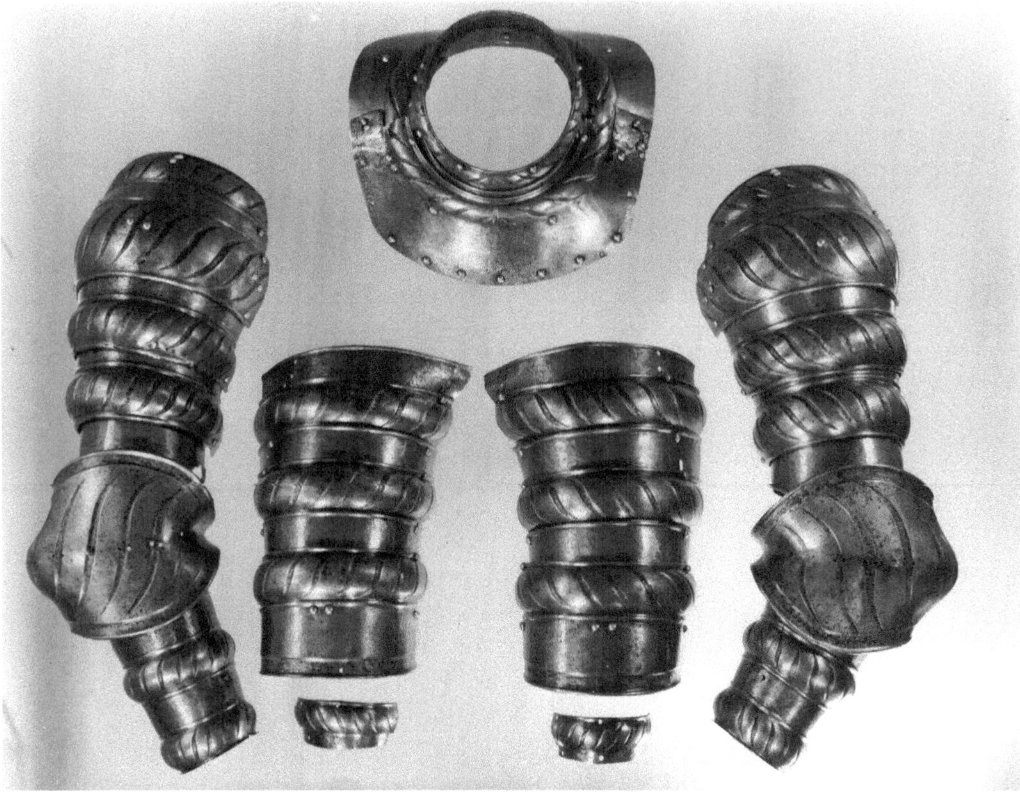

Figure 15 Elements of an armour in the Landsknecht style, Innsbruck, probably by Konrad Seusenhofer, c.1515. Royal Armouries, Leeds, inv. no. II.10. The waist lame has not been included. © Royal Armouries.

needed to be removed and replaced with a skirt lame, which undoubtedly would have incorporated a finished edge in the form of a final puffed and slashed band, almost exactly like the one preserved in Leeds. The Leeds tassets might seem redundant, given that the Wallace Collection armour has a pair, but they would not have been had the Wallace and Leeds parts come from the same garniture. The Royal Armouries tassets are of the standard thigh-length type designed for the field, while the Wallace Collection tassets are those, wrapping further around the legs, designed to accompany the full vambraces and rump defence to compose the parade and foot combat armour.

Finally, the Leeds vambraces are also those for routine field use (Figure 16). The construction is lighter, more flexible and generally simpler, as would be expected of the armour for war. The spaudlers are each made up of five plates, the articulation cleverly hidden within the puffed and slashed bands. The vambraces are of the typical three-part construction comprising a partial upper cannon (protecting the lateral surface of the upper arm and leaving most of the medial face uncovered), a one-piece 'floating' couter and a two-piece enclosed lower cannon. The concept is essentially the same as the field vambraces of the Rogendorf armour, although again the sculptural form, the hand of the maker, is different. The Leeds couters include a cosmetic nod to costume fashion found nowhere else – pairs of unnecessary and non-functional holes have been placed on either side of each of the long recessed, etched and gilt slashes, so that they could be pointed 'closed' as they would have been if made in textile. The presence of such playfulness in the decoration of a war armour is a significant demonstration that clothing simulation was never restricted purely to armour for parade and ritual use away from the battlefield.

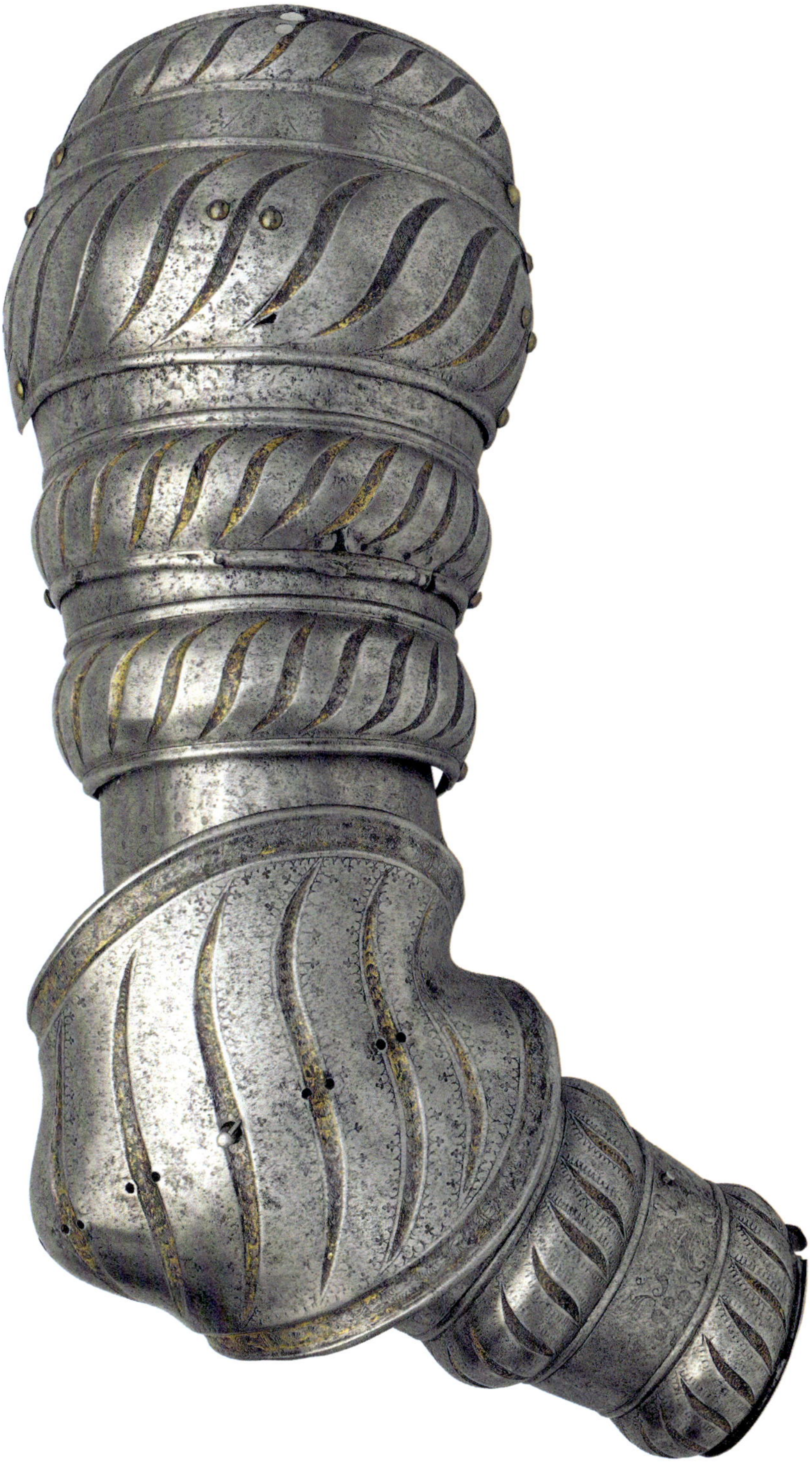

Figure 16 Right arm defence from an armour in the Landsknecht style, Innsbruck, probably by Konrad Seusenhofer, c.1515. Royal Armouries, Leeds, inv. no. II.10. © Royal Armouries.

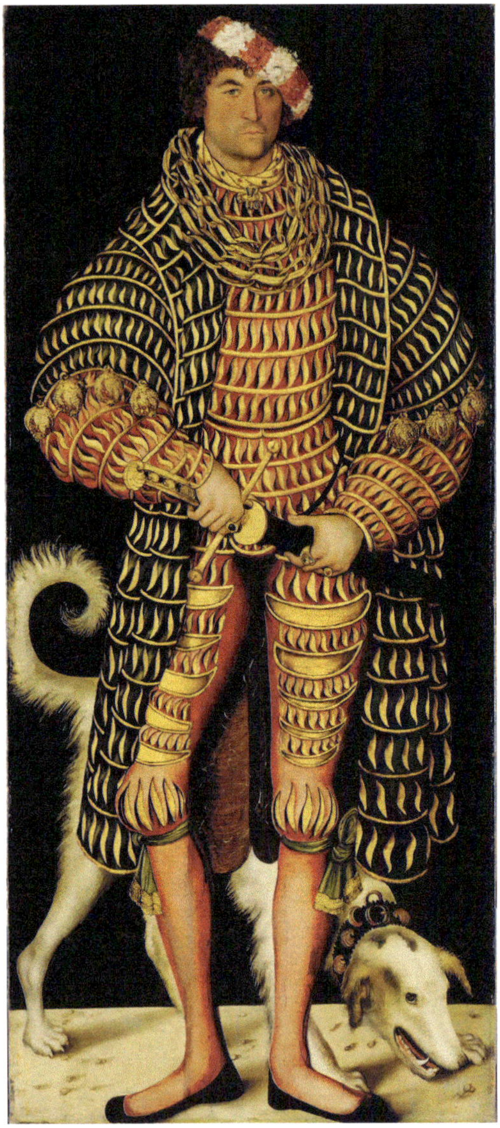

Figure 17 Lucas Cranach the Elder, *Henry the Pious, Duke of Saxony*, c.1514. Gemäldegalerie, Dresden, inv. no. 1906 G. © bpk / Staatliche Kunstsammlungen Dresden / Elke Estel / Hans-Peter Klut.

Taken together, the Wallace and Leeds elements form an almost complete Landsknecht garniture precisely comparable to the concept and extent of the Rogendorf armour. However, one fact demolishes the idea that this is a single garniture now spread across two locations. The etched decoration at Leeds does not match that on the Wallace armour – therefore the Royal Armouries parts cannot have come from the same garniture (see below). Indeed, while most of the Leeds parts match, the gorget carries an etched border containing a pattern of undulating clouds, found on no other piece. Therefore, it would seem that the Leeds group encompasses parts from at least two different costume armours.

For analytical purposes, the ornamental aspects of the Wallace Collection armour must be split into two discussions, of the sculptural forms hammered into the plates and of the acid-etched decoration. Independently, each presents a solid justification for attribution of the work to Innsbruck. Taken together, the case becomes even stronger.

Decorative form and sculpted ornament

The method employed on the Wallace Collection armour to simulate puffed and slashed textile in steel is very specific. Clothing of this fashion was constructed in a number of ways, of which two methods are most relevant for present purposes. The first method, represented on the Wallace Collection armour, involved the loose outer facing material being gathered horizontally at intervals causing the bands of material between the gatherings to bulge outwards. These bands were then slashed with short vertical, diagonal or crescent-shaped cuts, allowing the contrasting interlining to show through. Sometimes two slash types were employed in alternation, making the garment seem even more flamboyant or, to tone down the slashing while also showing off the rich facing to a greater extent, the puffed and slashed bands were alternated with plain strips without slashing. Perhaps the most famous depiction of this 'horizontal' style is the earlier of Lucas Cranach's two portraits of Henry the Pious, Duke of Saxony (c.1514) (Figure 17).[27] The second method involved the garment being made even more loose-fitting in the sleeves and cut instead with long, vertical or diagonal slashes to reveal the interlining. A number of illustrations of clothing in this style are found in the 'Book of Clothes' kept over many years by the Augsburg accountant Matthäus Schwarz,[28] especially in the images dating from the 1520s and early 1530s (Figure 18).[29] In cases where

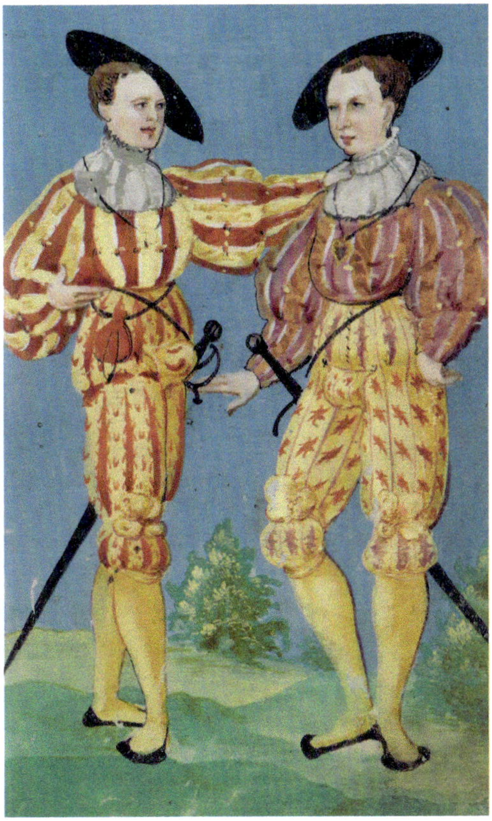

Figure 18 '1530, to please King Ferdinand during the Imperial Diet, when Emperor Carolus was here.' From the Book of Clothes of Matthäus Schwarz, Augsburg, c.1530–40. Herzog Anton Ulrich-Museum, Brunswick, inv. no. Hs 27 N. 67a, ff. 102, 103r. © bpk / Herzog Anton Ulrich-Museum.

the slashes in the facing material were especially long, their edges were sometimes tied together at intervals, a fashion depicted in the later of the two portraits of Henry the Pious by Cranach (c.1526) (see Figure 20 below),[30] and on the couters of the field vambraces preserved in Leeds.

Interestingly, all of the extant armour in the Landsknecht style confirmed as having been made in Augsburg, and the Helmschmids in particular, simulates this second 'vertical' style of Landsknecht fashion – the slashed decoration on both the Rogendorf and Radziwill armours is arranged vertically with the challenge of simulating the distinctively capacious sleeves having required Kolman Helmschmid to invent a totally new way of building plate defences for the arms.

In contrast, puffed and slashed armour of the Innsbruck school normally followed the first 'horizontal' style. The earliest surviving costume armour of the German Renaissance, the unfinished child's armour begun by Hans Rabeiler in Innsbruck, works according to this fashion, with narrow horizontal puffed and (diagonally) slashed bands embossed and chased into the breastplate, backplate, shoulders, vambraces and cuisses. Here the puffed and slashed bands are spaced quite far apart with plain, smooth areas between; these would almost certainly have been decorated with etching and gilding had the armour been completed. Another essential ingredient of the Innsbruck horizontal style was the way in which the puffed and slashed bands were framed above and below with embossed welts having a rounded cross-section, resembling a turned edge but generally passing through the main surface of the plate rather than finishing the edge (Figure 19). These cord-like welted borders simulate the piping used to frame the slashed bands on textile clothing in this style, which was often made of the same material used for the interlining, as shown in the earlier Cranach portrait of Henry the Pious in which the slashed bands are bordered with piping made, like the interlining, of cloth-of-gold. Interestingly, all the narrow welted borders on the Wallace Collection armour were once gilded like the interior of the simulated slashes; only traces of gold now remain on them.

The first Henry the Pious portrait also shows the way in which the horizontal style, when applied to the hosen, often included longer, wider slashes on the thighs, set between the puffed bands so that they bagged elegantly. Alternatively, the bands themselves were tapered towards their ends on either side, with the swag-like form actually built into the garment. These treatments of the slashed hosen are represented on the child's armours of the future Charles V: the bands spaced between long yawning slashes are rendered with clever conceit on the second armour of 1512–14: the metal seems to sag under the influence of gravity. Meanwhile, the tapered swag-bands were selected for simulation on the slightly earlier, unfinished armour of 1511–12.

A third variation in the treatment of the hosen within the horizontal style was to allow the puffed

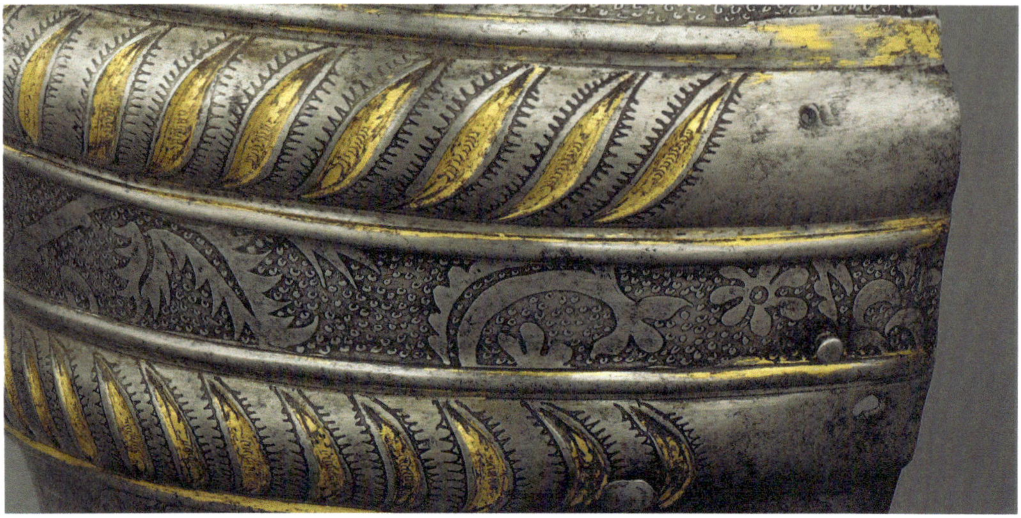

Figure 19 Backplate of a Landsknecht armour, Innsbruck, c.1515. The Wallace Collection, inv. no. A28 (detail). © Trustees of the Wallace Collection.

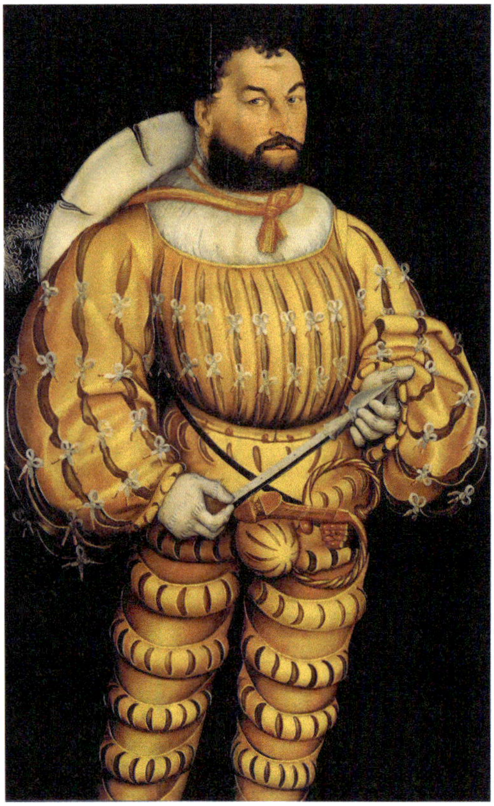

Figure 20 Lucas Cranach the Elder, *Henry the Pious, Duke of Saxony*, c.1526. Gemäldegalerie, Dresden, inv. no. H 75. © bpk / Staatliche Kunstsammlungen Dresden / Elke Estel / Hans-Peter Klut.

bands to maintain their level orientation without additional slashing, and with only more subtle tapering of the bands at the outer edges, yielding a gentler upwards curvature at the sides. Hosen of this form are illustrated in the later Henry the Pious portrait (worn in combination with a 'vertical' style doublet with billowing sleeves). This is also the construction simulated on the tassets of the Wallace Collection armour, where the upper three bands are placed horizontally while the bottom two are narrower, tapered and drawn up at the sides, the lowermost to a greater extreme than the one above. This rendition succeeds in giving some sense of the way the material would hang were it textile, although the effect is subtler than on the second boy's armour of Charles V.

Another important sculptural expression of the maker's hand is the treatment of the roped edges. The edges on Landsknecht armours made by the Helmschmids tend to be quite fine, with the twists usually being quite tight. The larger turns on the top and sides of the breastplate never stray far from their technical function as stop-ribs. In fact, on the Rogendorf armour, the top edge has been finished with a quite practical stop-rib of narrow oblong cross-section with only a nod to roped decoration running along the forward face; the upper and lower sides have been left plain and smooth. Also, as is typical of

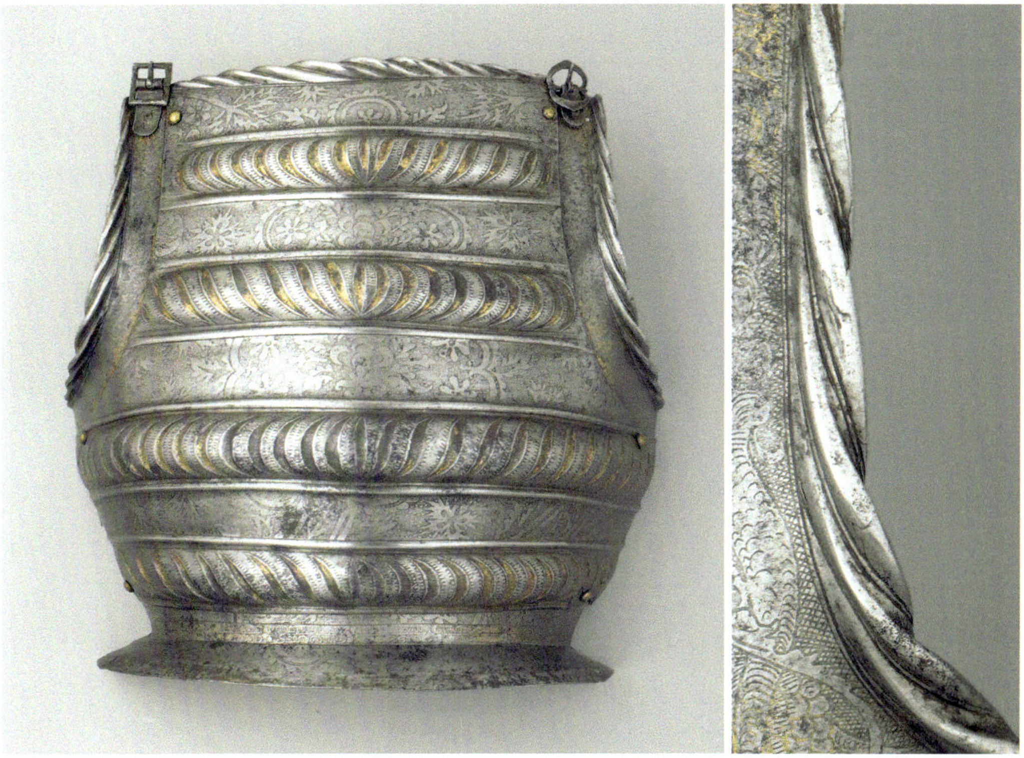

Figure 21 Breastplate of a Landsknecht armour, Innsbruck, c.1515. The Wallace Collection, inv. no. A28. © Trustees of the Wallace Collection.

Augsburg work, the twist direction changes at the midline, with the roping running inwards on both sides to converge on a pair of vertical grooves placed in the middle just below the base of the wearer's throat. This converging style of roping is also employed on the narrower turn of the backplate. On the Wallace Collection armour, however, the top and side edges of the breast- and backplates have been thickly turned and roped in a different way (Figure 21). The twisting cords are very strongly modelled with the troughs accentuated with a pair of fine incised lines. The twists are also noticeably looser and more drawn out than most Augsburg examples. Roping closely comparable to that found on the Wallace Collection armour is found on numerous surviving examples of Innsbruck armourers' work of the same period, including the unfinished boy's armour of Charles V, an exceptional armet made in the Maximilian style, probably by Konrad Seusenhofer in c.1510–15 and now in the Royal Armouries,[31] and the field armour of Matthäus Lang von Wellenburg, made by Konrad Seusenhofer c.1511 (see Figure 40 below).[32]

The Glasgow vambrace

Roped edges of this same form appear again on a right vambrace from an otherwise lost child's armour, c.1515–20, now in the collection of Glasgow Museums (Figure 22).[33] Although this piece has never been specifically attributed to Innsbruck, it exhibits several key characteristics which suggest that origin. Beyond the roped edges (at the wrists), the puffing and slashing has been rendered in the horizontal style with welted borders comparable to the unfinished boy's armour of Charles V, the Wallace Collection armour and the Leeds pieces. However, it differs from those examples in that the short slashes are arranged vertically, rather than diagonally or formed as crescents. The couter is decorated with longer lenticular slashes,

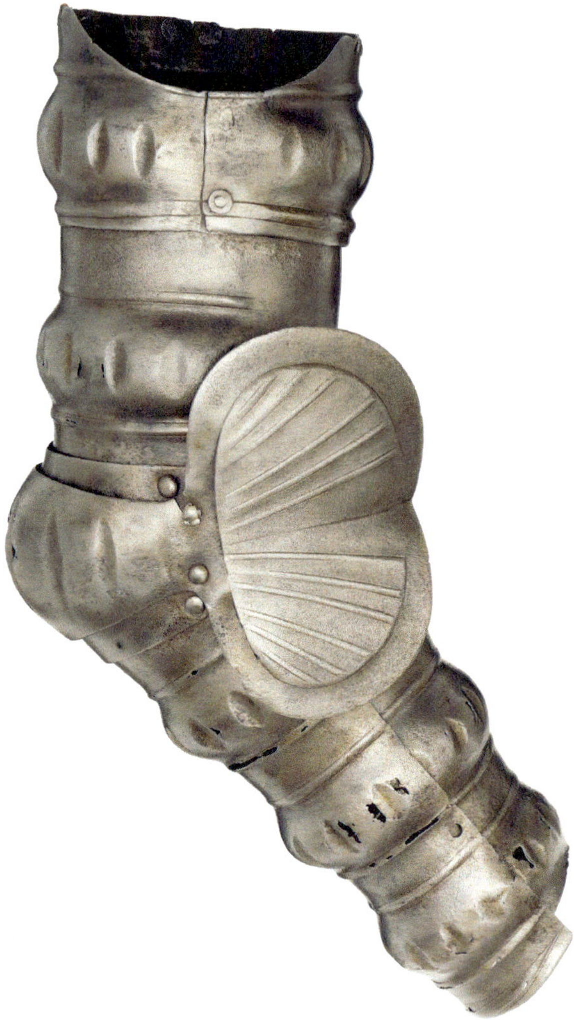

Figure 22 Right vambrace from a child's armour in the Landsknecht style, probably made in Innsbruck, possibly by Konrad Seusenhofer, c.1515. Glasgow Museums, inv. no. E. 1939.65.ce.

also aligned vertically. Lenticular slashes are also a key decorative motif on the skirted field armour of Frederick II, Duke of Legnica (1480–1547), also known as Friedrich von Liegnitz, made by Konrad Seusenhofer c.1515 (see Figure 41 below).[34] The Glasgow vambrace may have been left unfinished, as suggested by Laking,[35] although its somewhat bedraggled surface condition may instead be the result of 'cleaning' with acid in the 19th century. Regardless, no trace of any etched decoration remains, if it ever existed. Another connection to other examples of the Innsbruck School is provided by the couter wing. Its rounded but almost rectangular shape is typical of Innsbruck work of the first quarter of the 16th century, as is the slightly sunk outer border within which the surface is embossed and incised with fluting in the 'Maximilian' style. Notably, the leading edge of the wing has only been slightly pushed down into the bend of the elbow. Comparable wings are found on numerous pieces of armour made in Innsbruck around this time. Examples include two pairs of complete leg armour made c.1510–15, one set now in Vienna and the other in the Wallace Collection,[36] and the Wellenburg armour in Vienna. The Glasgow vambrace thus represents an important stylistic bridge between 'costume' armours made in Innsbruck and standard field armours produced in the same place at around the same time.

Mi-parti design

Slightly later, in the 1520s, the Helmschmids in Augsburg created some of the most breathtaking and creatively inspired fusions of field and costume armour, proving again how the two styles validated and flattered each other: field armour could be worn on parade and costume armour might well go into combat. Several remarkable illustrations are preserved in the Thun-Hohenstein albums: these may simply have been Helmschmid concepts that were never realised or which were made but have since been lost. In one illustration, of a standing figure clad in a complete armour of polished steel, a cuirass, gorget and visored armet made in the typical, densely fluted 'Maximilian' style are combined with puffed and slashed pauldrons, vambraces, gauntlets, leg armour, and sabatons.[37] Here the Landsknecht styling is quite ostentatious and yet in no way does it compromise the practical functionality of the harness. An even more advanced integration of the two styles is found elsewhere in the same album, split between two pages of illustrations showing the disassembled elements of another field armour displaying puffed and slashed decoration.[38] Here however, the entire harness is decorated mi-parti, wherein the body is divided vertically straight down the midline, with the right side decorated in one way and the left in another (Figure 23). Mi-parti design was a cornerstone of the Landsknecht style where it was routinely used to heighten the flamboyance of the wearer's appearance, increase the number of bright colours and patterns displayed on the body, and express a distinctive brand of exuberant grandiosity. In the case of the mi-parti armour recorded in the Thun-Hohenstein album, cunningly, the side-to-side contrast is technical as well as decorative. The left leg, pauldron, vambrace and gauntlet are embossed and chased with

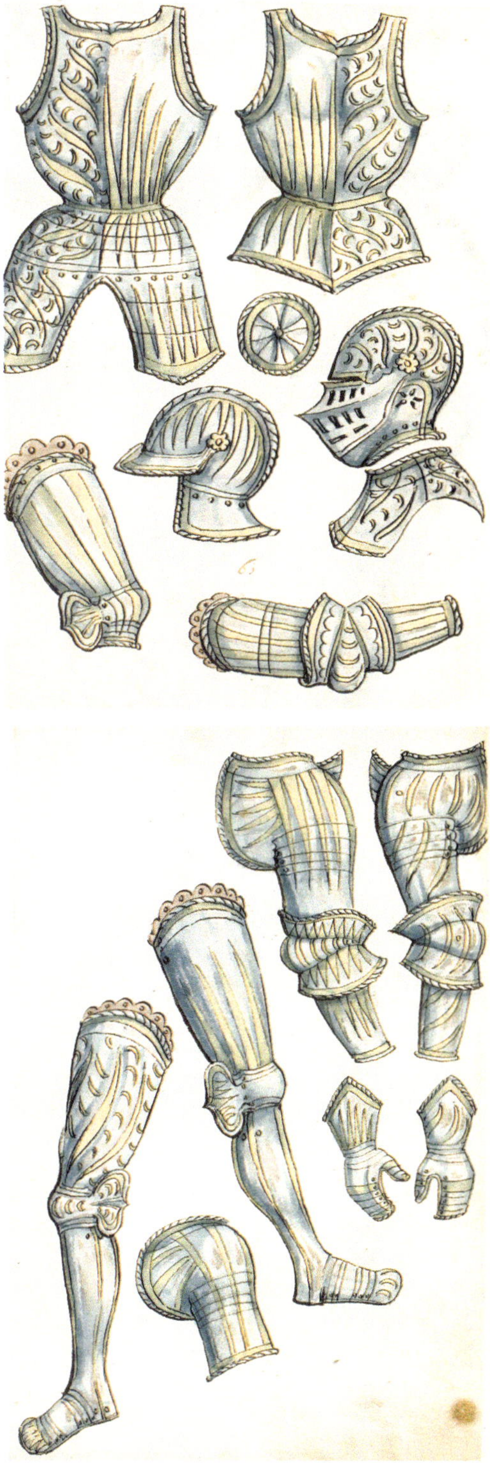

Figure 23 A disassembled field armour decorated mi-parti, Augsburg, c.1525. Museum of Decorative Arts, Prague, inv. no. GK 11.572-B, ff. 52v, 71r.

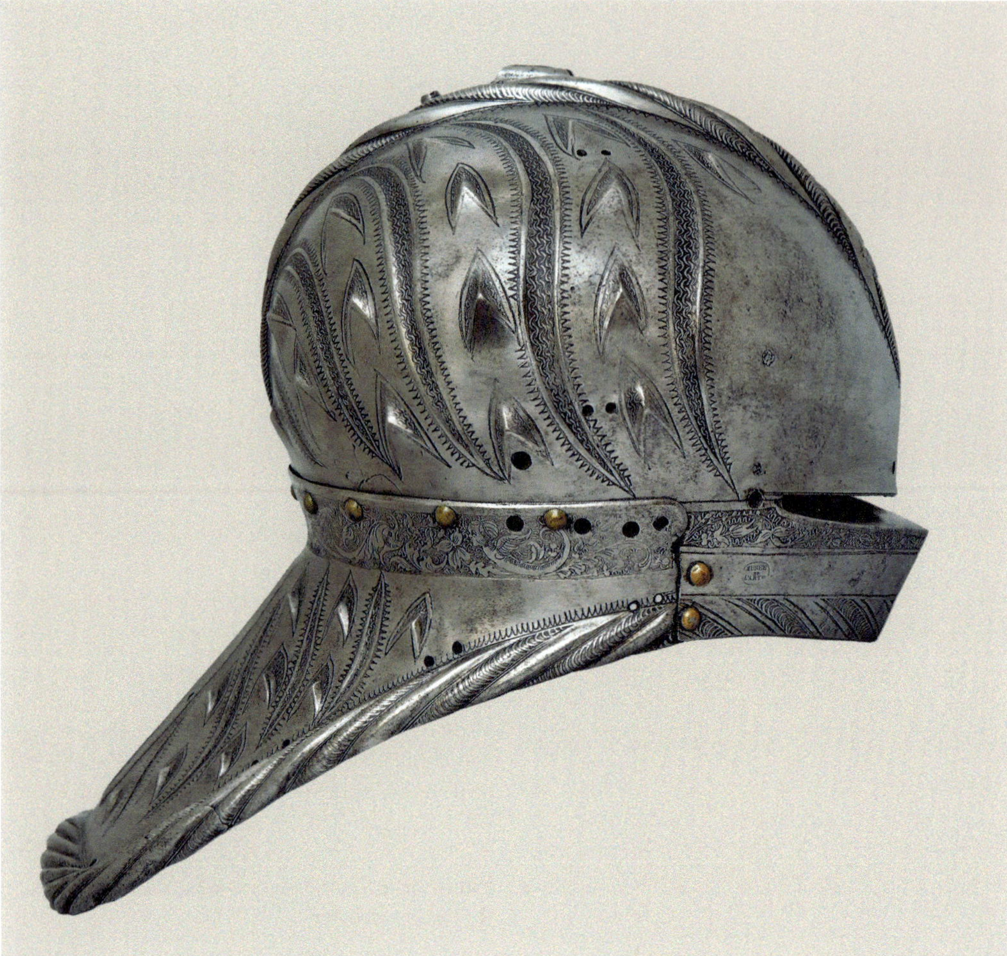

Figure 24 Sallet for the Rennen, made by Konrad Seusenhofer, Innsbruck, c.1515. State Historical Museum, Moscow, inv. no. 82001-2295. © State Historical Museum, Moscow.

puffing and slashing, with a cloth-of-gold interlining simulated with fire gilding within the slashes. Their mates on the right, in strong contrast, are fashioned as plain, smooth field armour decorated with solar rays created through selective gilding but not rendered in relief. Interestingly, the patterns are reversed on the cuirass, with the puffing and slashing placed on the right side and the smooth surface on the left.

Regardless of whether or not these armours were actually made, proof of the Helmschmids' implementation in steel of mi-parti design exists in the form of a single breastplate, probably made by Kolman, now preserved in the Army Museum, Warsaw.[39] Here the left side of the breastplate has been decorated with diagonally slashed vertical bands separated by groups of four longitudinal flutes. The treatment of the slashes is very similar to that used on the Rogendorf armour. The right side contrasts strongly with a predominantly horizontal scheme, but without slashing; the main horizontal bands are formed of recessed bands and flutes, while the areas between the bands are bridged with rows of very narrow recessed and etched vertical bars.

However, as with the essential idea of rendering puffed and slashed clothing in steel, which appeared in Innsbruck work before that made

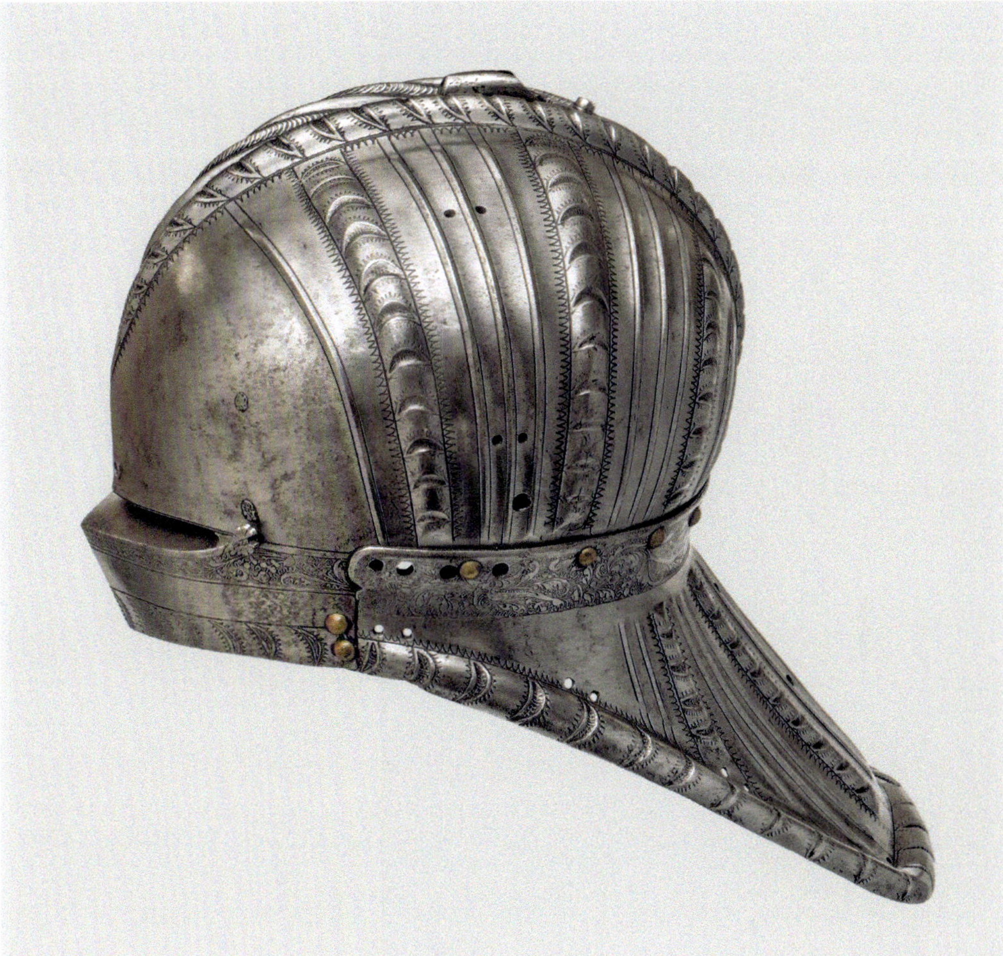

in Augsburg, mi-parti decoration in particular occurred first in the workshop of Konrad Seusenhofer. It is found on an extraordinary sallet for mock jousts of war or Rennen, made in the Seusenhofer workshop c.1515 (Figure 24), formerly in the Musée de l'Armée and now, like the Legnica armour, in the State Historical Museum, Moscow.[40] Here the left side of the helmet is embossed, chased and etched with puffed and slashed bands of virtually identical workmanship to those on the Wallace Collection armour, combined with vertical fluting of the 'Maximilian' style. The right side of the helmet – in an expression of the sheer daring and flair of the maker – is decorated with long, recurving slashes, between which embossed and etched swallowtail cut-outs have been placed in vertical rows. Within both the long, gently curving slashes and the swallowtails, the metal has been etched – not with a simulation of some rich textile, but instead, remarkably, with the unmistakable interlocking pattern of mail. The right dilge from the same Rennen armour also survives in Moscow and is entirely decorated like the right side of the helmet. We may assume that the lost left dilge was decorated with the same puffed and slashed scheme found on the left side of the helmet, and that the Rennzeug was completed with a mi-parti cuirass, bevor and vamplate all featuring some ingenious combination of the two patterns.

Notably, the Wallace Collection armour also displays mi-parti decoration of a sort, although it is much subtler: the puffed and slashed bands alternate between diagonal and crescent-shaped

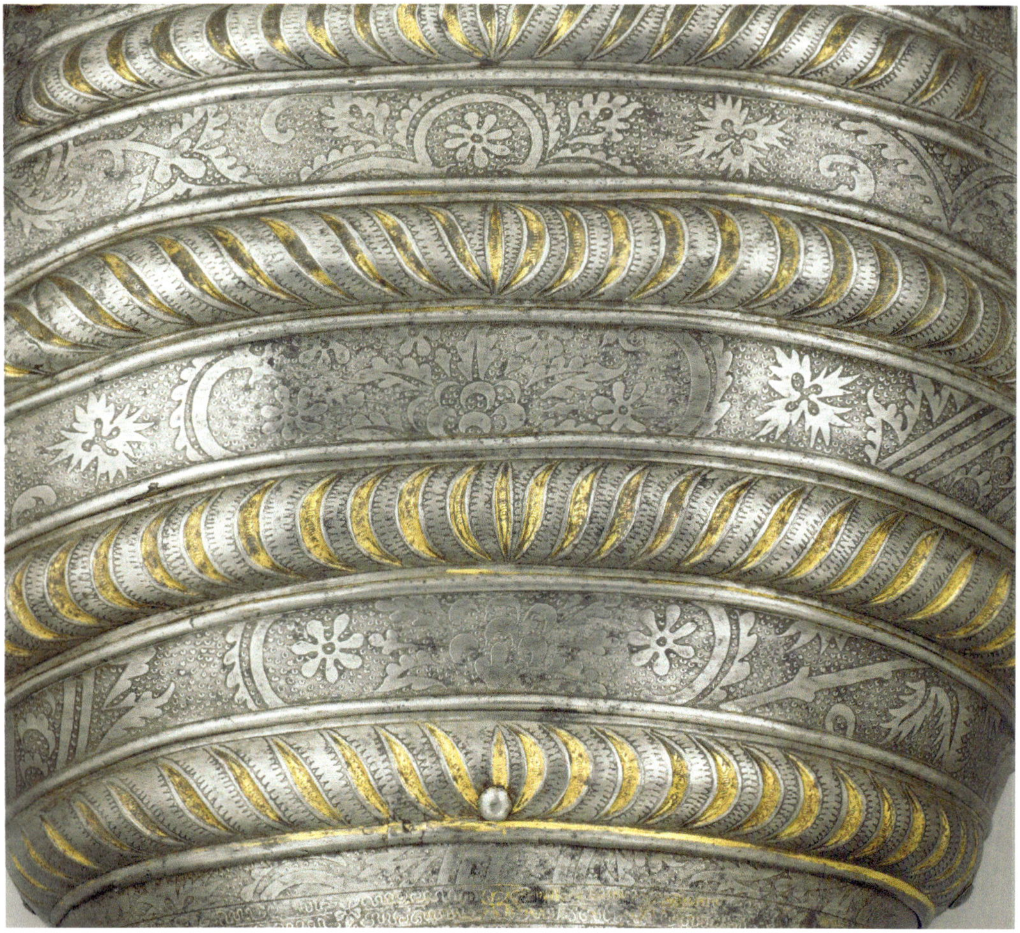

Figure 25 Backplate of a Landsknecht armour, Innsbruck, c.1515. The Wallace Collection, inv. no. A28. © Trustees of the Wallace Collection.

slashes. On the arm defences and tassets, this means that a diagonally slashed band is placed between bands of crescents above and below, and vice versa. On the cuirass however, the slashes alternate not only on the bands above and below, but also within the same horizontal band either side of the midline of the body (Figure 25). This means that although the slashing on the arms matches, that on the tassets differs left to right since the uppermost puffed band on the left is diagonally slashed while that on the right is 'cut' with crescents (Figure 26).

The use of contrasting slashing styles on either side of the midline is (to my knowledge) found on no other costume armour. Also, the slashing pattern does not normally change or alternate on horizontally oriented puffed and slashed bands as it does on the Wallace Collection armour. In fact, the only other instances of this kind of alternation are found on products of the Seusenhofer workshop – the second boy's armour of the future Charles V in Vienna and two fragments from an otherwise lost boy's armour, commissioned by King Ferdinand of Croatia, Hungary and Bohemia (later Emperor Ferdinand I) for his 5-year-old nephew, the future King Philip II of Spain, and made by Hans Seusenhofer in 1532 (Figure 27). In both of these instances, the slashing alternation does not involve the slashing type, but rather its direction: all the slashes are of the crescent variety, but they face alternately left and right, almost like rows of oversized mail links.

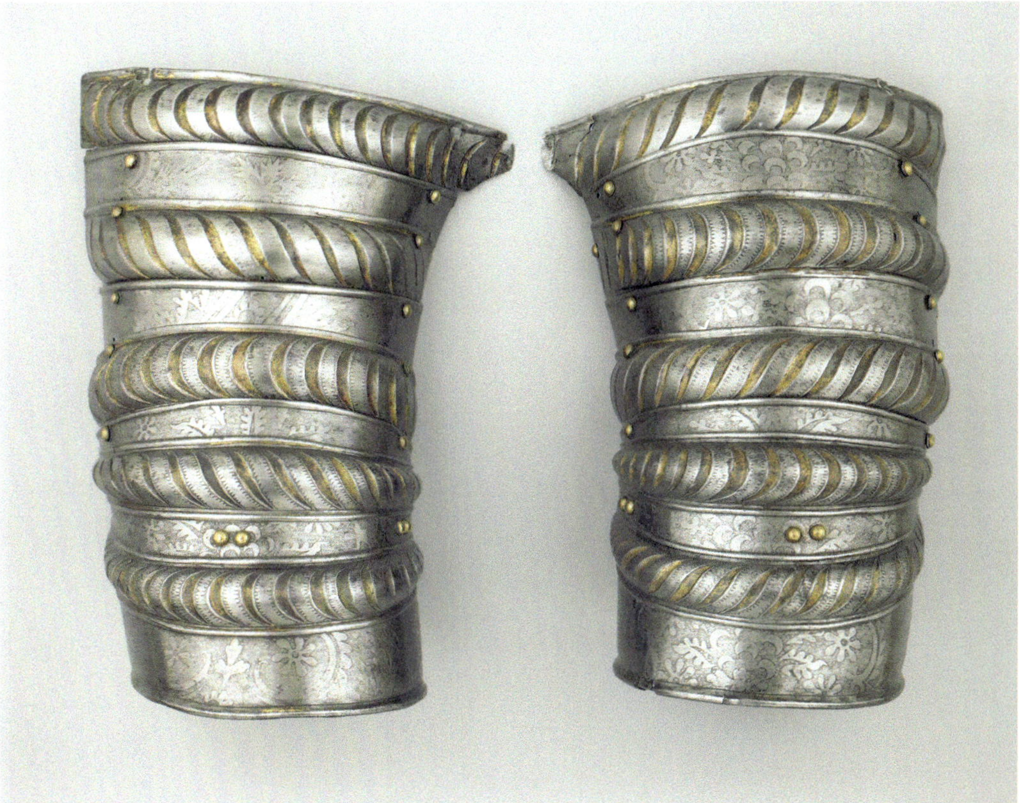

Figure 26 Tassets of a Landsknecht armour, Innsbruck, c.1515. The Wallace Collection, inv. no. A28. © Trustees of the Wallace Collection.

a

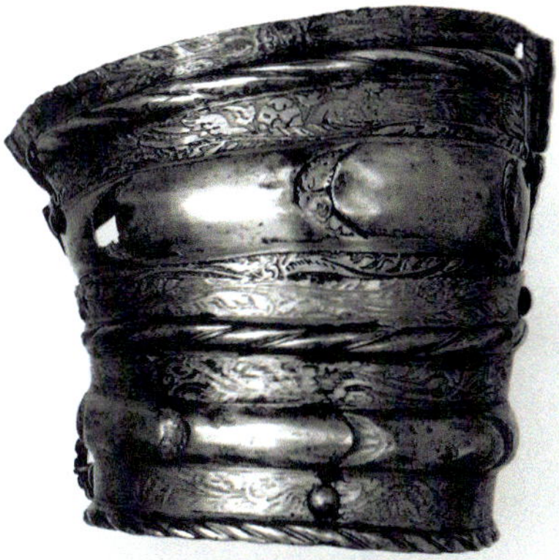

b

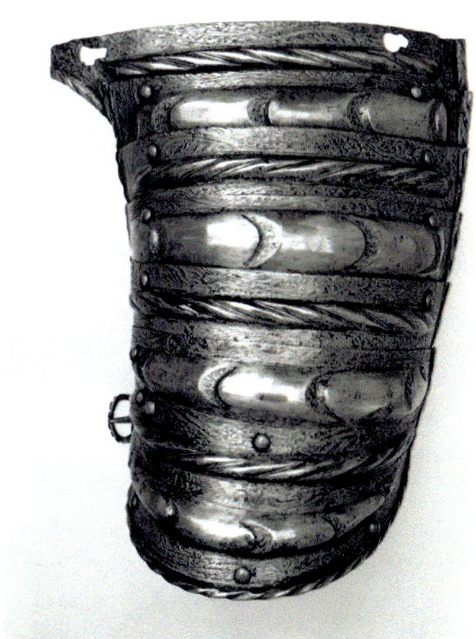

Figure 27 Left shoulder plates (a) and left tasset (b) from a child's armour of the future King Philip II of Spain, made by Hans Seusenhofer, Innsbruck, 1532. The Metropolitan Museum of Art, New York, inv. no. 29.158.121, .304; Bashford Dean Memorial Collection, funds from various donors, 1929.

Etched decoration

The acid etching found on the Wallace Collection armour is very extensive, covering almost all of its surface area and following a complex overall scheme designed both to meet the contemporary expectations for a rich armour and to contribute to the simulation of luxurious textile (Figure 28). For the purposes of discussion, the etcher's work on this armour can be separated into five distinct parts:

- The etched pattern fully covering the otherwise plain areas interspersed between the puffed and slashed bands.
- The wide etched bands covering the whole surface of the sliding gusset plates fitted to the sides of the breastplate.
- The etching used to suggest the interlining behind the slashes.
- The borders of the slashes.
- The unique pattern filling the etched and gilt waistband of the cuirass.

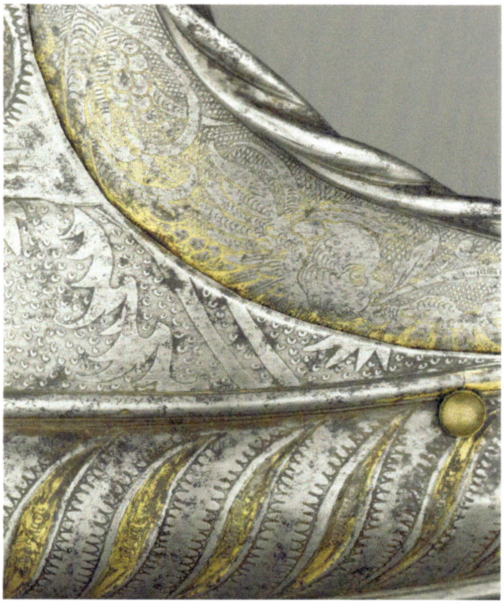

Figure 28 Breastplate of a Landsknecht armour, Innsbruck, c.1515. The Wallace Collection, inv. no. A28 (detail). © Trustees of the Wallace Collection.

The narrow horizontal panels spaced between the puffed and slashed bands have been fully etched to resemble silk velvet (probably Italian) interwoven with gold or silver gilt looped threads to create a pattern of precious metal leaves and flowers laid against the velvet ground (Figure 29). To extend the illusion, the ground has been filled with minute crescent-shaped granulations instead of the round stipples or cross-hatching more typically used to texture the ground in etched decoration of this period (see Figure 31 below). Here, the tiny etched crescents are intended to suggest the small loops of gold or silver wire pushed through the velvet areas of the textile to make the cloth sparkle (see Figure 32 below). Indeed, the Italian Renaissance term for this technique was *allucciolato*, from the Italian *luccioli*, meaning 'glowing in the night sky'.[41] Numerous examples of such textiles survive; an especially useful comparison to the Wallace Collection armour is a large piece of red silk velvet woven with the popular 'pomegranate' pattern in silver-gilt threads made in Florence in the late 15th century and now in the Victoria and Albert Museum in London.[42] Taken on its own and without context, the etched flowers and foliage on the Wallace Collection armour might seem somewhat rudimentary – of secondary quality even – however, greater appreciation follows an informed reading as an imitation of a very specific type of textile, characterised by floral decoration rendered in a distinctive way.

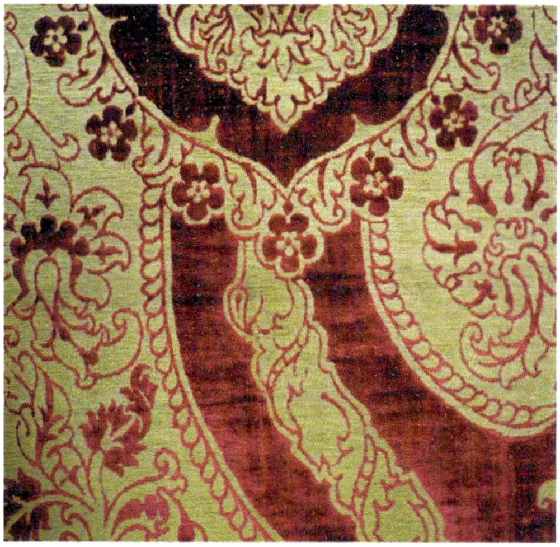

Figure 29 Silk velvet, Florence, c.1475. Victoria and Albert Museum, London, inv. no. 81A-1892.

The representation of patterned textile by means of etching is not unique to the Wallace Collection armour: closely similar patterns

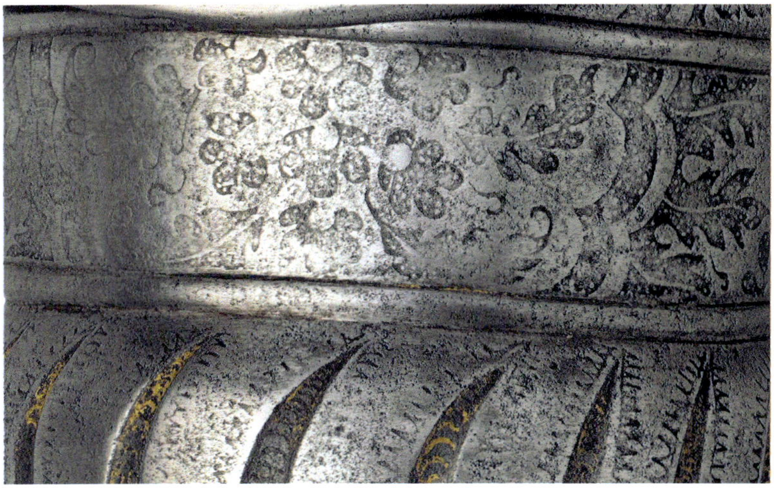

Figure 30 Arm defence from an armour in the Landsknecht style, Innsbruck, probably by Konrad Seusenhofer, c.1515. Royal Armouries, Leeds, inv. no. II.10 (detail). © Royal Armouries.

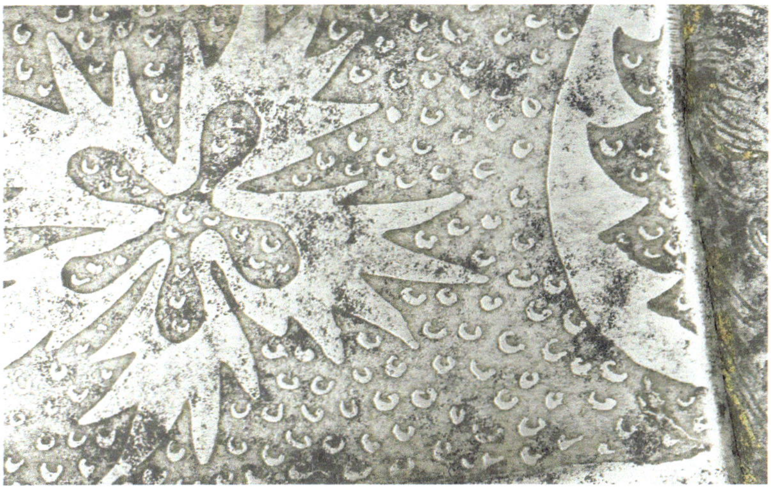

Figure 31 Breastplate of a Landsknecht armour, Innsbruck, c.1515. The Wallace Collection, inv. no. A28 (detail). © Trustees of the Wallace Collection.

are found on the armour parts in the Royal Armouries, which almost certainly come from the same workshop as the Wallace Collection armour. The etcher's hand might even be the same, although the scheme here is sufficiently different in detail to indicate that the Leeds pieces come from a different armour, albeit one made in the same style, by the same armourers, at the same time. At Leeds, the flowers, leaves, fronds etc. are the areas eaten away by the acid, while the background has been left proud and polished – the reverse of the treatment on the Wallace Collection armour (Figure 30). Also, at Leeds standard stippling has been used to accent the etched areas with no signs of etched allucciolato loops.

Simulated loops do appear on at least two surviving works almost certainly made in the Helmschmid workshop: a close helmet with mask visor of c.1515 and a set of exchange pieces

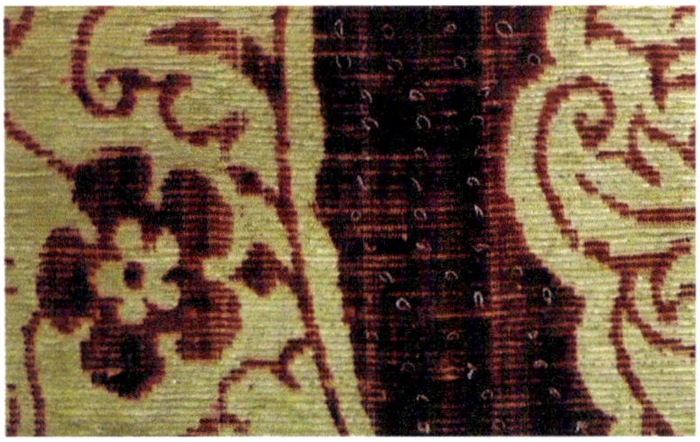

Figure 32 Silk velvet, Florence, c.1475. Victoria and Albert Museum, London, inv. no. 81A-1892 (detail).

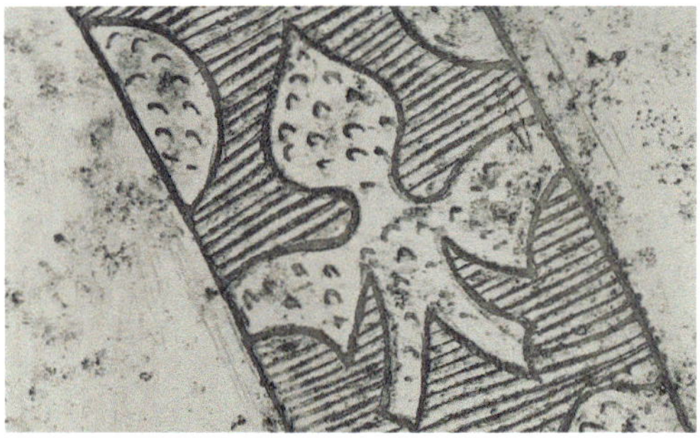

Figure 33 Gardbrace for the Feld Rennen and Freiturnier, made by Kolman Helmschmid, Augsburg, c.1510–20. The Metropolitan Museum of Art, New York, inv. no. 29.150.6; Bashford Dean Memorial Collection, Bequest of Bashford Dean, 1928.

for the German Feld Rennen ('field joust') and Freiturnier ('free tourney'), c.1510–15 (Figure 33) in the Metropolitan Museum of Art, New York.[43] In both of these cases however, the style of the floral pattern is very different both in its forms and in the way it has been realised. The Augsburg examples have been accomplished purely through the use of the 'basic' etching technique, wherein all aspects of the design have been scratched away from a surface fully coated in the resist medium. The areas of the pattern which would have been worked in precious metal thread are here delineated with shading comprised of densely packed and parallel etched lines, while the loops, etched into the surface, fill the contrasting zones between. A very different approach has been taken by the etcher responsible for the decoration of the Wallace armour: he has instead employed the 'raised' etching technique – painting the flowers, foliage and even all of the allucciolato loops onto the surface of the metal with the resist. Although sharing the basics of a design idea, when compared to the Augsburg examples and the Wallace armour, they are quite different both in drawing style and technique.

The sliding gusset plates on the sides of the breastplate of the Wallace Collection armour are fully etched and were originally entirely gilt. The design takes a predominantly aquatic theme, with the main vertical area on each side occupied by a long-bodied, sinuous mermaid (Figure 34). Below the mermaid on the right side is a coiled sea serpent with a dolphin-like head, and on the left, a winged head. Instead of arms, long elegant fronds resembling tendrils of seaweed extend from the mermaids' shoulders to twist around their tails, while foliage sprouts above their heads; all is set against a densely cross-hatched ground. Although it involves the more typical theme of terrestrial vegetation, rather than marine subjects, the etching on an armet made in the Seusenhofer workshop c.1510–15 (Figure 35) is similar to that at the Wallace Collection. Now in Vienna,[44] it is decorated with criss-crossing recessed bands filled with a characteristic mixture of foliage, flowers and human heads placed against a hatched ground, the treatment of the foliage being especially comparable. All has been rendered the same general style as the Wallace Collection armour but in an inferior hand, probably that of an assistant. Interestingly, the cuirass belonging to the same otherwise lost 'lattice-work' armour is preserved at the museum in Krems, Austria.[45] Here a net of recessed intersecting bands run across the main surface of the breastplate and skirt, while on the waist plate they are etched but not recessed.

The sliding gusset plates on the breastplate of the Rogendorf armour are also fully etched but in a completely different style, by the hand of Daniel Hopfer – perhaps the greatest etcher of the period (Figure 36). Hopfer decorated many of the armours made in Augsburg by Kolman Helmschmid but had nothing to do with the Wallace Collection armour, despite the various attributions of it to him made in the 20th century.[46] Hopfer's style is exceptionally

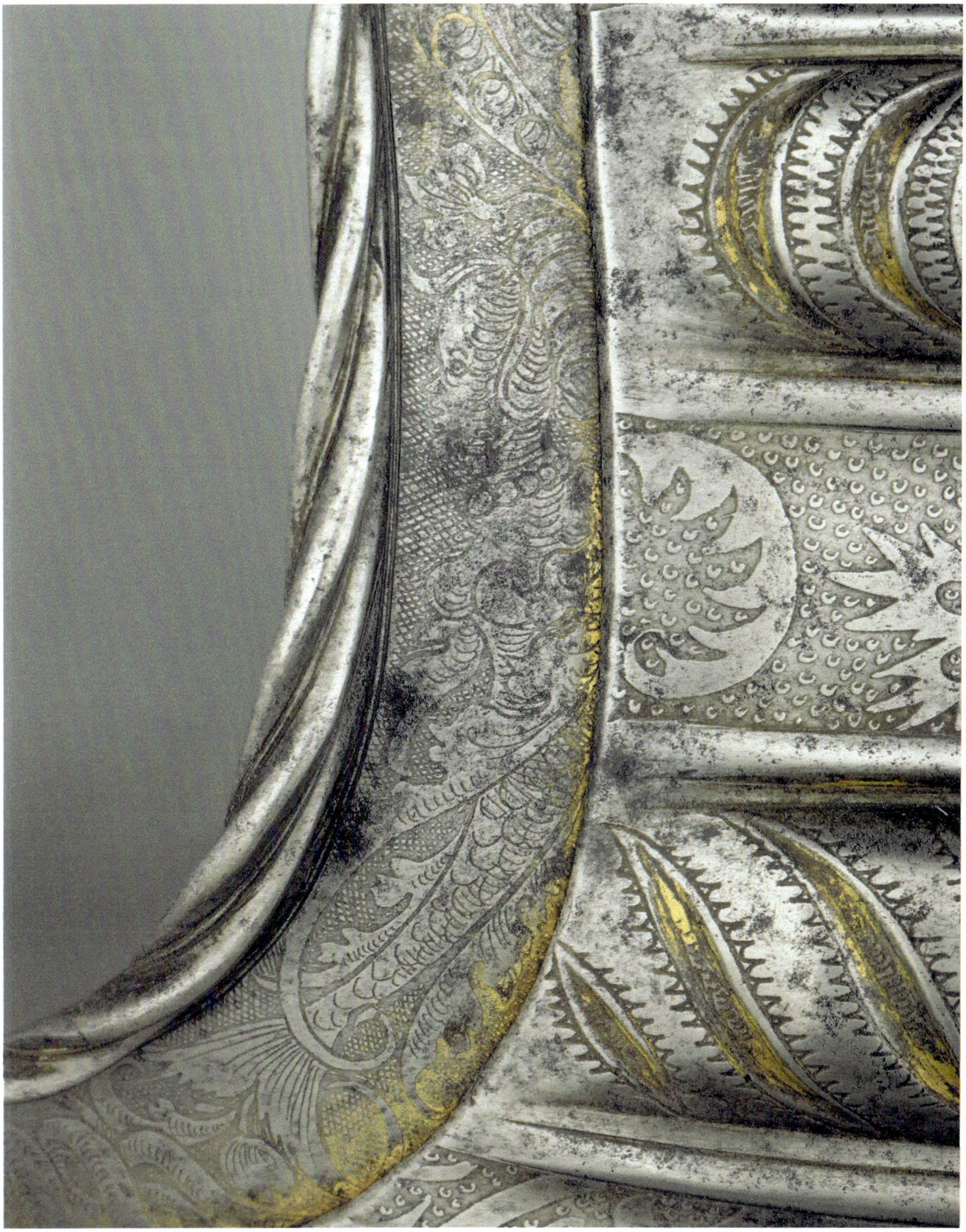

Figure 34 Breastplate of a Landsknecht armour, Innsbruck, c.1515. The Wallace Collection, inv. no. A28 (detail). © Trustees of the Wallace Collection.

precise, delicate and sharply defined, and usually involves quite small-scale figures, wings, heads, rectangular cartouches and scrolling foliage rendered in raised etching and placed against very fine granulated backgrounds created through the painstaking application of thousands upon thousands of tiny spots of resist. Again, when full figures are included – humans, animals or

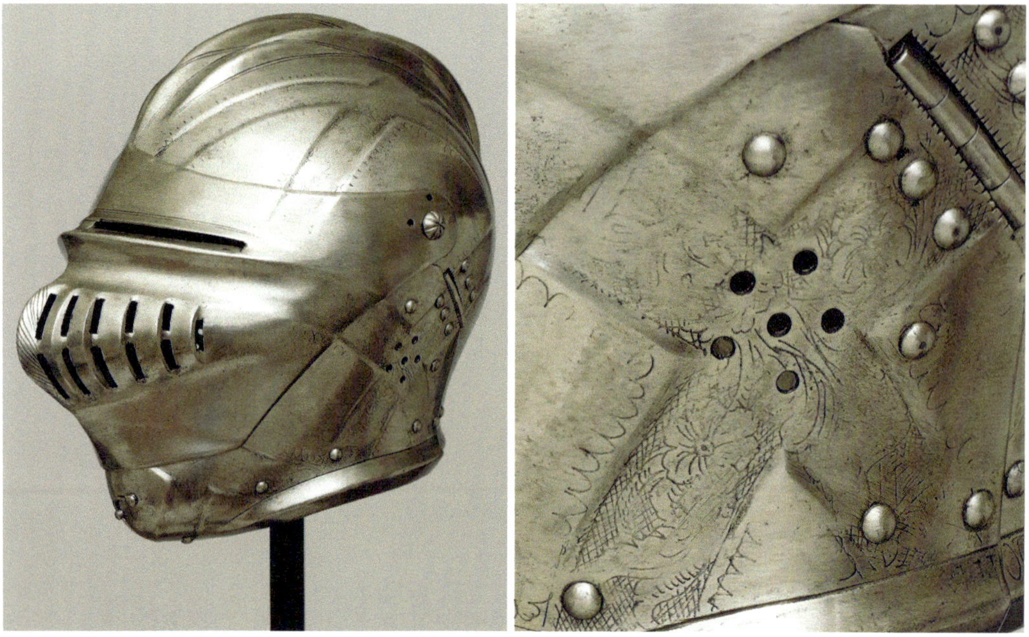

Figure 35 Armet, made in the Seusenhofer workshop, Innsbruck, c.1510–15. Imperial Armoury, Kunsthistorisches Museum, Vienna, inv. no. A295. KHM–Museumsverband.

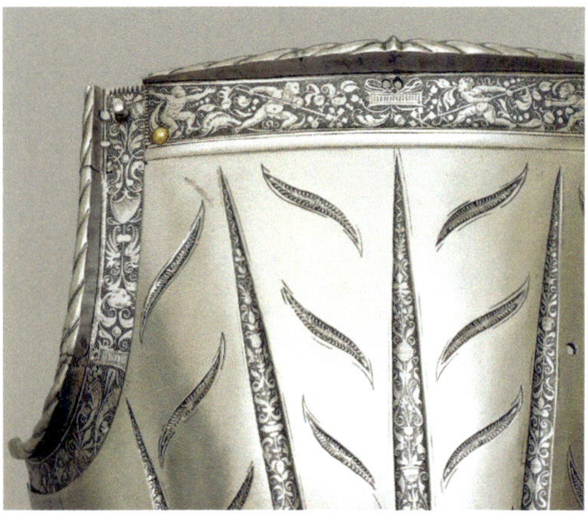

Figure 36 Breastplate from the Landsknecht garniture of Wilhelm Freiherr von Rogendorf, made by Kolman Helmschmid, Augsburg, dated 1523. Imperial Armoury, Kunsthistorisches Museum, Vienna, inv. no. A374 (detail). KHM–Museumsverband.

mythical beasts – they are generally kept quite small in proportion to the piece of armour on which they have been placed. The figurative etching on the Wallace Collection armour is thus dissimilar in every respect: the larger scale mermaids fill most of the available space, they have been set against a hatched ground, and execution has been restricted to the basic etching technique. All of these qualities amount to a style of etching which is entirely unlike anything produced in Augsburg. At the same time, it is a close match to the etched decoration found on numerous examples made in Innsbruck in the mid-1510s.

The interior of the recessed slashes on the Wallace Collection armour, etched and gilt to suggest a cloth-of-gold interlining showing through the cuts in the outer facing, also exhibit etched ornament typical of Innsbruck work in the second decade of the 16th century (Figure 37). It features two patterns – dense cross-hatching and foliage consisting of long, spear-like leaves reaching up and down towards the ends of the slash. These leaves have flame-like edges, while their interior is detailed with rows of minute strokes – single scratches of the etcher's needle-like graving tool – bridging the width of the leaf. Both patterns appear on others of the surviving pieces already mentioned. The slashes on the gorget of the Leeds group carry the cross-hatched pattern;

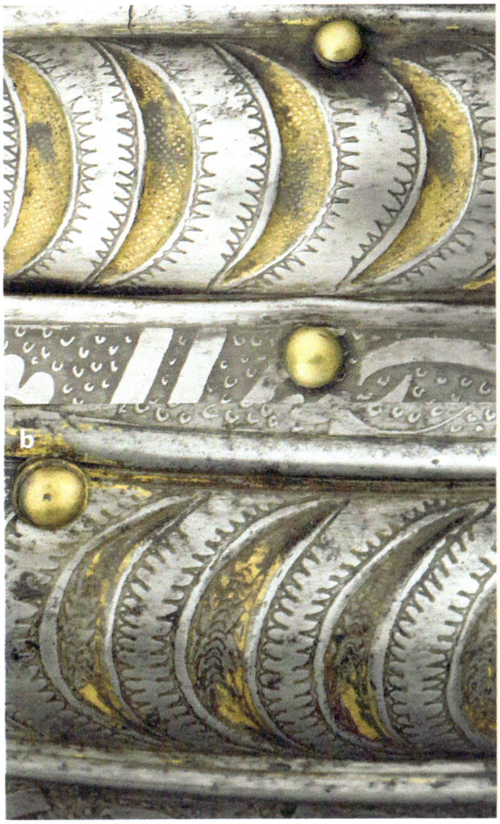

Figure 37 Hoguine (a) and breastplate (b) of a Landsknecht armour, Innsbruck, c.1515. The Wallace Collection, inv. no. A28 (details). © Trustees of the Wallace Collection.

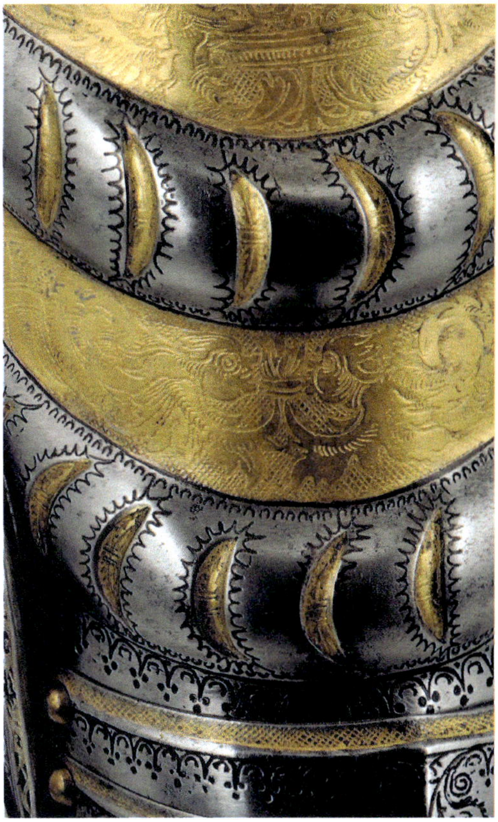

Figure 38 Right cuisse of a child's armour of the future Emperor Charles V, made by Konrad Seusenhofer, Innsbruck, c.1512–14. Imperial Armoury, Kunsthistorisches Museum, Vienna, inv. no. A109 (detail).

interestingly, all the other pieces at Leeds exhibit a different pattern within their slashes of flower blossoms laid down like a row of overlapping scales. The same cross-hatching is also found on the steel-skirted boy's armour of Charles V, where it is used not within the slashes but rather to accent the recessed edges of the poleyn lames and articulation plates at the top of the cuisses (Figure 38). Cross-hatched slashes are also present on a right pauldron in Vienna, the only surviving element of a field armour made c.1515 by Konrad Seusenhofer (Figure 39).[47] This otherwise lost armour combined the dense fluting of the 'Maximilian' style with puffed and slashed decoration – another example of how aspects of the 'costume' style could be integrated into even the most practical war harness. The presence of a row of crescent-shaped slashes, filled with cross-hatching and framed above and below by narrow welted borders, reveals this tantalising fragment to be a close contemporary relative of the Wallace Collection armour.

The second leafy pattern employed within some of the slashes on the Wallace Collection armour is very similar to that used inside all the small, crescent-shaped slashes on the skirted boy's armour of Charles V, except in that case they are girded in the middle by a small ring, with the leafy fronds extending above and below its upper and lower edges.

Perhaps the most eye-catching and consistent aspect of the etching on Innsbruck costume armour in general is the treatment of the external borders of the slashes. In the 1510s, the

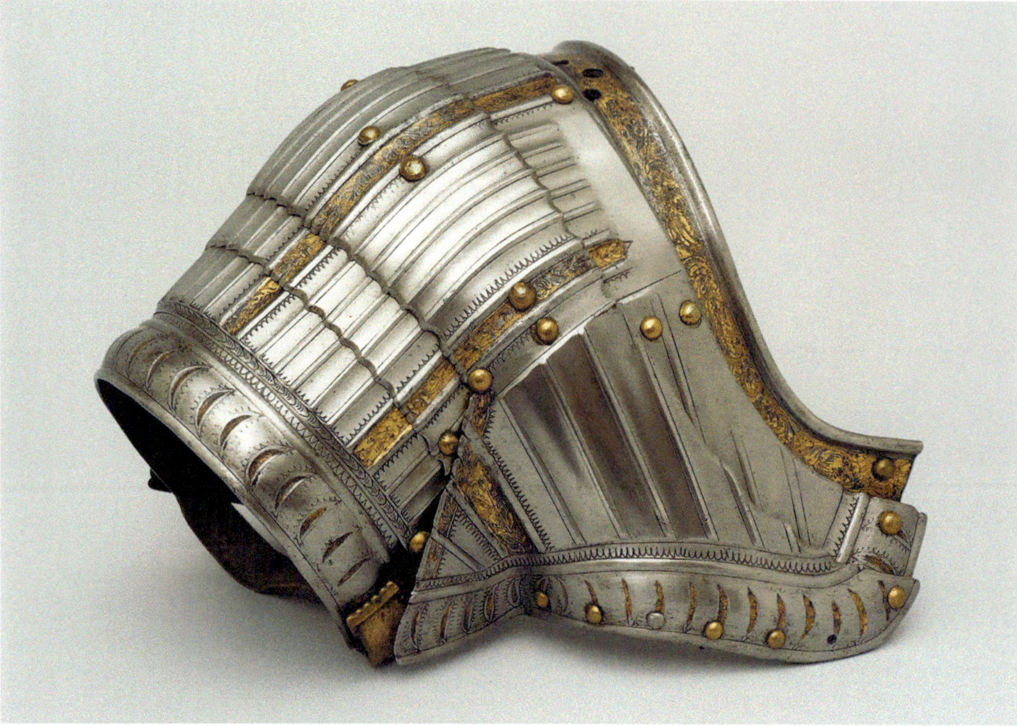

Figure 39 Right pauldron from a field armour associated with King Ludwig II of Hungary, made by Konrad Seusenhofer, Innsbruck, c.1515. Imperial Armoury, Kunsthistorisches Museum, Vienna, inv. no. A179. KHM–Museumsverband.

products of the Seusenhofer workshop often included fine engrailed lines etched around each slash. Beyond its presence everywhere on the Wallace Collection armour, fine engrailing is employed around the slashes of the Charles V boy's armour, all of the Leeds pieces, the Rennhut in Moscow and the 'King Ludwig' pauldron. Beyond the emphasis of crescent-shaped slashes in particular, these same engrailed lines are often used elsewhere in Innsbruck decorative schemes (but not on the Wallace Collection armour) to strengthen the visual impact of key features, such as the edges of the long curving slashes on the right side of the Moscow sallet and right dilge; various borders and only some of the slashes on the Legnica armour; framing the sprays of flutes and along borders on the Wellenburg armour (Figure 40); along the welted borders of the King Ludwig pauldron; and to enhance the recessed bands on the armet in Vienna and breastplate at Krems, from the otherwise lost lattice-work armour.

An elaboration of the engrailed border found on many surviving Innsbruck pieces of this period involves a trio of dots being added to the engrailed peaks, with the troughs sometimes developed with the addition of small crescent-shaped marks (Figure 42). This is the main form of etched border employed on the second boy's armour of Charles V in addition to the plain engrailing etched around the slashes and on parts of the Legnica armour (Figure 41). A simpler version, with only every third peak given a decorated terminal (here a solid form resembling a three-leaf clover) is found on the couters of the Leeds group (Figures 42 and 43), and on the dilge in Moscow. Neither of these variants are present on the Wallace Collection armour, which employs plain engrailing exclusively. Yet its presence emphasises the close workshop family relationship between all of the pieces that do carry it, each of which is stylistically associated with the Wallace armour in other ways.

Finally, the presence of an etched and gilt waistband is another key trait of fine armours

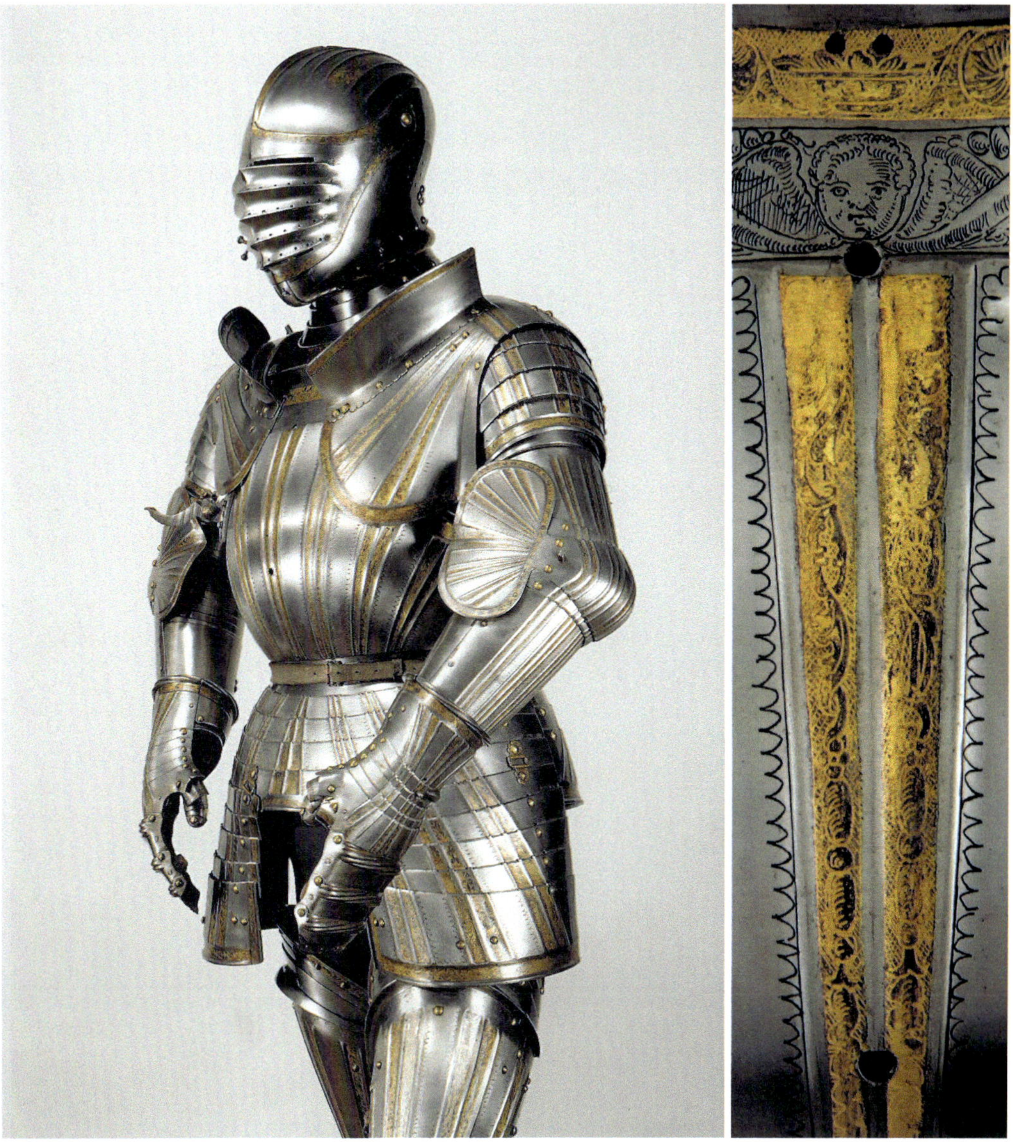

Figure 40 Field armour of Matthäus Lang von Wellenburg, made by Konrad Seusenhofer, Innsbruck, c.1511. Imperial Armoury, Kunsthistorisches Museum, Vienna, inv. no. A244. KHM–Museumsverband.

made in the Seusenhofer workshop during the second decade of the 16th century. The cuirass of the Legnica armour displays a band of flowering foliage framed above and below with a simple etched riband, while the Krems breastplate exhibits a row of closely set, circular blossoms with very fine petals – a pattern found (in other locations) on the Wellenburg armour and the Charles V boy's armour. The Wallace Collection armour also includes such an etched and gilt waistband, containing especially fine ornament involving a double row of very narrow horizontally oriented crawling vines, undulating very tightly and broken at intervals by small flower blossoms (Figure 44). To my knowledge, this pattern appears in only one other instance: on the skirted boy's armour of Charles V (Figure 45).

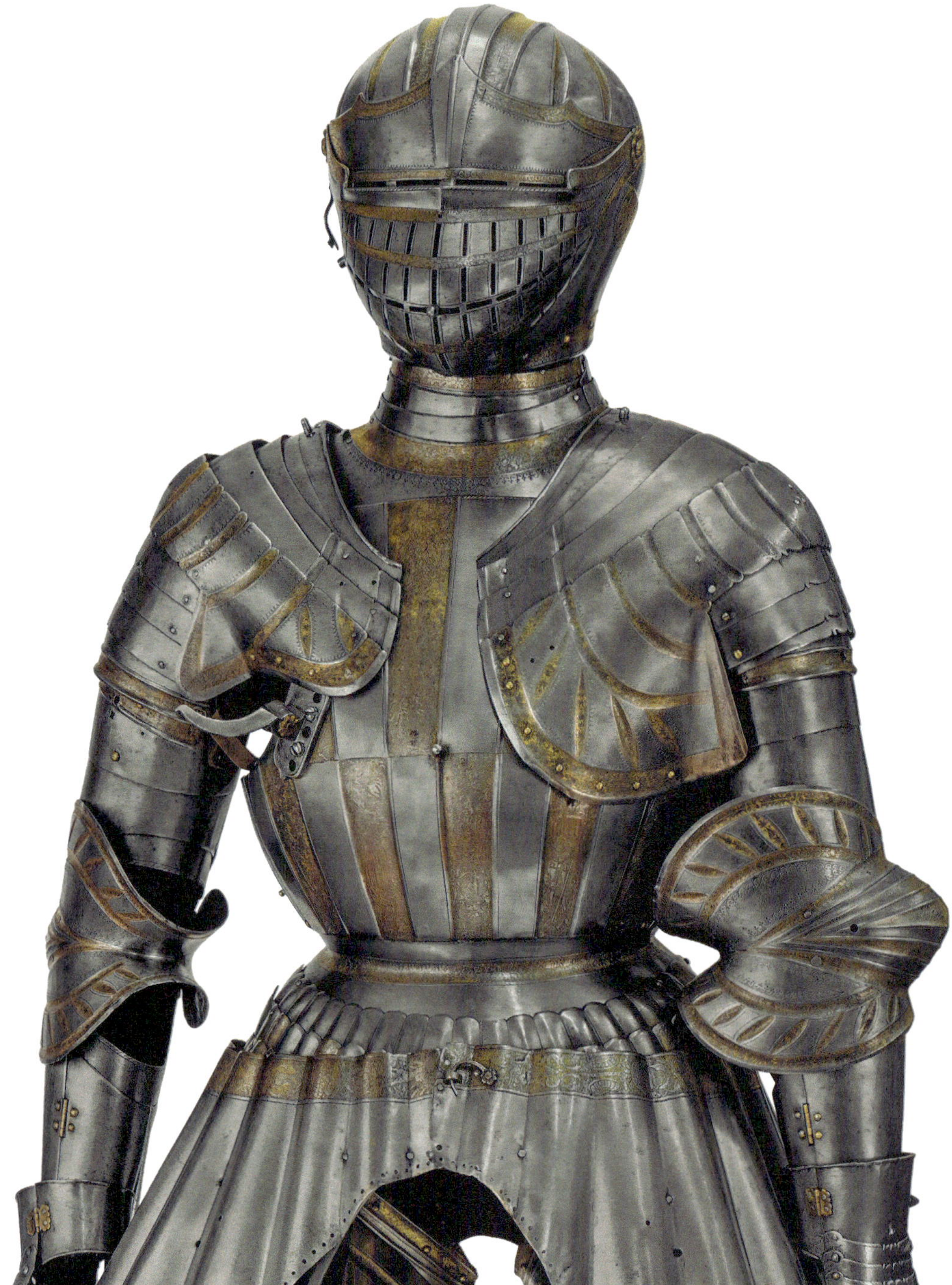

Figure 41 Armour of Frederick II, Duke of Legnica, made by Konrad Seusenhofer, Innsbruck, c.1515. State Historical Museum, Moscow, inv. nos. 2150-1, 2209-22. © State Historical Museum, Moscow.

Figure 42 Right cuisse of the child's armour of the future Emperor Charles V, made by Konrad Seusenhofer, Innsbruck, c.1512–14. Imperial Armoury, Kunsthistorisches Museum, Vienna, inv. no. A109 (detail).

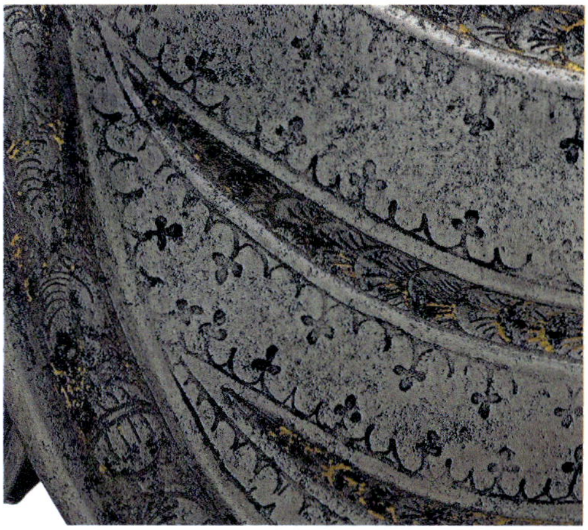

Figure 43 Arm defence from an armour in the Landsknecht style, Innsbruck, probably by Konrad Seusenhofer, c.1515. Royal Armouries, Leeds, inv. no. II.10 (detail). © Royal Armouries.

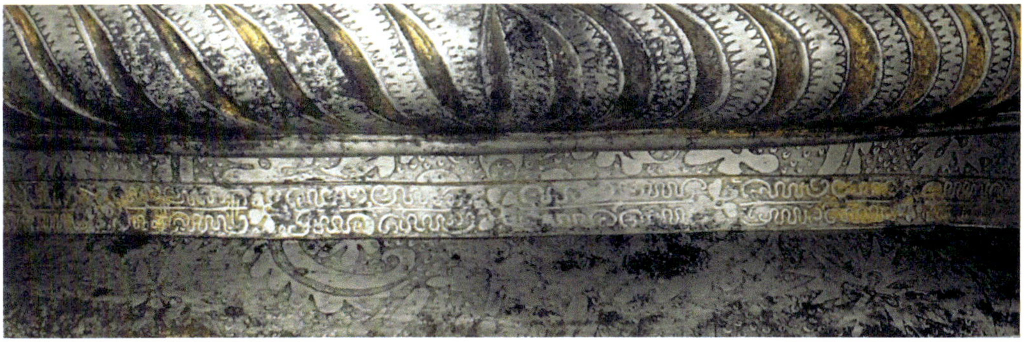

Figure 44 Breastplate of a Landsknecht armour, Innsbruck, c.1515. The Wallace Collection, inv. no. A28 (detail). © Trustees of the Wallace Collection.

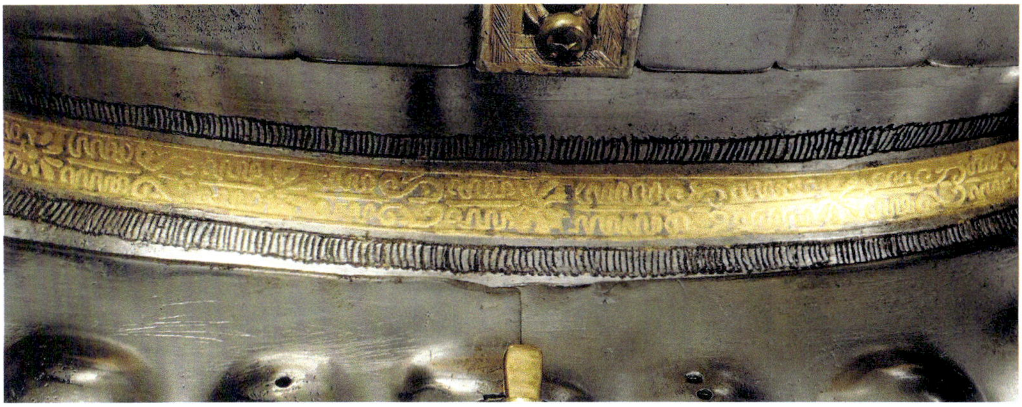

Figure 45 Breastplate of a child's armour of the future Emperor Charles V, made by Konrad Seusenhofer, Innsbruck, c.1512–14. Imperial Armoury, Kunsthistorisches Museum, Vienna, inv. no. A109 (detail). KHM–Museumsverband.

Metallurgy

In his seminal work on the metallurgy of late medieval and Renaissance armour, Alan Williams included an examination of the Wallace Collection armour in his chapter on the work of Augsburg armourers since that was the attribution current at the time – repeatedly published and not yet questioned.[48] However, the results of its metallographic analysis are also consistent with an Innsbruck origin, being congruent with the metallurgy of other surviving Innsbruck pieces, especially products of the Seusenhofer workshop.[49] The Wallace Collection armour is made of high-quality medium-carbon steel, hardened to an average of 305 VPH by means of fully quenching in its many parts and tempered to produce a very fine-grained pearlite and martensite microstructure with, crucially, very little slag. Very similar results, reached by means of the same heat-treatment processes, are revealed in other works by Konrad Seusenhofer, including an armet c.1511–14 in the Wallace Collection (a medium-carbon steel, 0.5% C, hardened by fully quenching and tempering);[50] the Wellenburg armour (a medium-carbon steel, 0.5% C, hardened by fully quenching and tempering to 468 VPH);[51] the mask helmet of Henry VIII at the Royal Armouries (a medium-carbon steel, 0.6–0.7% C, the skull hardened by near-full quenching and tempering, the visor and cheeks made of a similar steel but only air-cooled while the horns were made of an entirely different metal);[52] a shaffron of the Emperor Maximilian in Vienna (a medium-carbon steel, 0.6% C, having had an accelerated cooling to 300 VPH);[53] and the King Ludwig pauldron (medium-carbon steel, 0.5% C, hardened by fully quenching and tempering to 400 VPH).[54] Interestingly, the earliest examples of the puffed and slashed style forged in steel – the Innsbruck child's armours of the future Charles V – are both made out of a lower carbon steel, and the hardness of the second, skirted armour (244–366 VPH) is still comparable to the Wallace Collection armour, although the heat treatment for hardening may have been modified by the extensive gilding.[55]

The best Augsburg work is also usually made of hardened and tempered medium-carbon steel, although the overall hardness is sometimes somewhat lower than that of the best Innsbruck armour. Moreover, the microstructure of the Rogendorf armour differs notably,[56] being fine pearlite and ferrite of 280 VPH – a medium-carbon steel that was not fully quenched. The upper cannon and couter from the Rogendorf armour in the Wallace Collection were never hardened at all.[57]

Conclusions

Traces of fastenings on the shoulders and at the groin verify that the Wallace Collection armour once had a gorget and codpiece, now lost. Beyond those two parts, there is no definite proof of the former existence of any other pieces. It is not a very great leap of imagination to postulate that the armour was originally provided with a helmet and hand protection of some kind, but beyond that, we rapidly sink deeper into the realm of supposition. It is possible that Wallace Collection A28 was made exclusively for parade and for use in formal, courtly combats on foot, without any parts for war. This study has shown that it almost certainly dates from c.1515, and not from the 1520s as previously assumed. Also, it was clearly made in Innsbruck, not Augsburg, probably in the workshop of Konrad Seusenhofer.

This origin, together with the new dating, relocates this fascinating work to an earlier chapter in the story of costume armour of the German Renaissance. It was a near contemporary of the boy's armours of the future Charles V and may even have been on the maker's workbench at the same time as the later of those two diminutive masterpieces. Taking account of its now earlier dating, and the lack of any proof for the original inclusion of exchange pieces for war, we must allow for the possibility that the conception of the Landsknecht garniture was not yet complete when the Wallace Collection armour was made. Regardless, its creation could have contributed some of the critical inventive components necessary for the realisation of the fully fledged system after c.1520.

Similarly, there is almost no direct evidence that the lost costume armour, to which all but one of the Leeds pieces belonged, was anything other than a standard field armour, albeit quite a flamboyant one. That group is comprised almost entirely of parts for the field, with the tantalising exception of the rear skirt lame. The presence of this piece implies a garniture of some kind, since it would only have been required to fill the gap left at the lower edge of the backplate when some other interchangeable gluteal defence was removed. The existence of the rear skirt lame does not, however, prove that a rump defence once formed part of the armour to which the Leeds group belonged.

Although the front skirt and rump defence of the Wallace Collection armour could originally have been removable, there is no definite physical evidence on the pieces themselves that confirms that they were: there are no turn-pins or keyhole slots. Currently the parts are held together with modern nuts and bolts. The armour could once have had a fastening system which permitted the cuirass to be easily reconfigured through the use of exchange pieces, but equally, the defences for the lower body could have simply been riveted onto the front and rear waist flanges.

The material evidence certainly does show that all the technological and artistic ingredients necessary to construct a Landsknecht garniture for the field, parade and formal foot combat had been developed in Innsbruck by c.1515. The Wallace Collection armour reveals the evolution at that point of specialised foot combat armour in the Landsknecht style, while the Leeds pieces confirm the existence of parts designed for infantry use in the field (and perhaps also light–medium cavalry service) but decorated in the same horizontal variant of the puffed and slashed style. However, we cannot at this point be certain that all of these innovations were actually ever combined into a single en suite armour system before the creation of the Helmschmid garnitures of the 1520s; but it is a possibility.

It seems a highly improbable coincidence that the Leeds pieces encompass almost all of the parts which would have been required by the Wallace Collection armour had it actually been one of the first Landsknecht garnitures – a technological innovation subsequently developed further in the Rogendorf and Radziwill garnitures. The forging and sculpting techniques observed at the Wallace Collection and Leeds are identical, and the decoration is very close indeed. The presence of a rear waist lame in Leeds is especially provocative. It is difficult therefore to avoid the assumption that the Wallace and Leeds pieces must have some kind of unknown shared history beyond their date and place of manufacture. However, at present little is known about their provenances. The Wallace Collection armour originates from the collection of Sir Samuel Rush Meyrick, who probably acquired it after 1825 but before 1830, when it was published in Joseph Skelton's *Engraved Illustrations of Ancient Armour* (1830) (Figure 46).[58] Unfortunately, nothing is known of this armour before it appeared in the Meyrick Collection. A similar dead end currently exists in the provenance of the Leeds pieces, which cannot be traced before their sale at auction in London in 1841.[59] Could they all come from a single, older source? At one time, might they have been assembled as one group, containing parts of several fragmented garnitures, once preserved in some German, Austrian or East European dynastic collection dispersed in the late 18th or early 19th century? Or is their complementary technical compatibility really just a flukish coincidence?

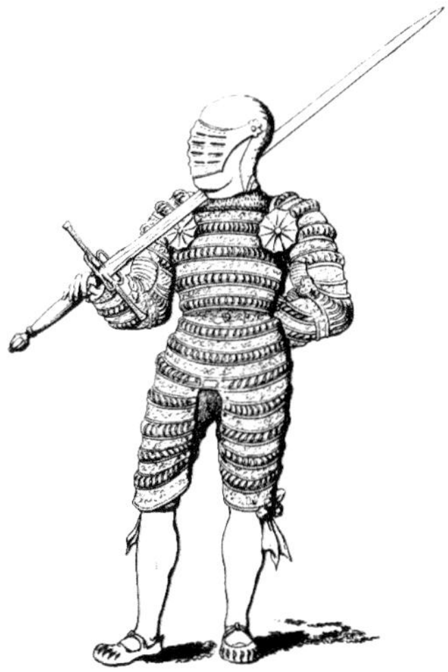

Figure 46 Joseph Skelton, *Engraved Illustrations of Ancient Armour from the Collection of Sir S.R. Meyrick at Goodrich Court, Herefordshire*, 1830, Plate 19. Wallace Collection Library and Archive.

Regardless, the Innsbruck School of armour making is now even more strongly represented at the Wallace Collection, which can now present a more detailed record of the incredible talent, creativity and originality of the artist-craftsmen active at the Imperial Court workshop under the mastery of Konrad Seusenhofer.

Notes and references

1. Mann, J.G. (1962) *Wallace Collection Catalogues: European Arms and Armour*, 2 vols, vol. 1. London: The Wallace Collection, pp. 26–7. Mann's source for the Hopfer attribution was Bruno Thomas, who was the first to notice a resemblance between the etching of A28 and that of the equestrian armour of Frederick II, Duke of Legnica, then in the Museum für Deutsche Geschichte, Berlin. The Legnica armour had been identified (correctly) as the work of Konrad Seusenhofer of Innsbruck by Paul Post in 1928. However, Post (incorrectly) attributed the etching to Daniel Hopfer of Augsburg: see Post, P. (1928) 'Ein Frührenaissanceharnisch von Konrad Seusenhofer mit Ätzungen von Daniel Hopfer im Berliner Zeughaus', *Jahrbuch der Preuszischen Kunstsammlungen* 49(3): 167–86. Interestingly, Mann accepted the attribution of the etching and applied it to the Wallace Collection armour as well, while also ignoring Post's identification of the armourer, a trend that continued in the scholarship for many years.
2. Gamber, O. (1975) 'Kolman Helmschmid, Ferdinand I und das Thun'sche Skizzenbuch', *Jahrbuch der Kunsthistorischen Sammlungen in Wien* 71: 30–31.
3. Norman, A.V.B. (1986) *Wallace Collection Catalogues: European Arms and Armour Supplement*. London: The Wallace Collection, p. 28.
4. Gamber, O. and Thomas, B. (1976) *Katalog der Leibrustkammer: Der Zeitraum von 500 bis 1530*, vol. 1. Vienna: Kunsthistorisches Museum/Verlag Anton Schroll, pp. 227–8.
5. Krause, S. (2017) *Fashion in Steel: The Landsknecht Armour of Wilhelm von Rogendorf*. Vienna: Kunsthistorisches Museum.
6. For the field pieces in Vienna, see Krause 2017 (cited in note 5), pp. 62–6; for the parts in the Wallace Collection (inv. no. A245) see Thomas, B. (1950) 'Portions of the Rogendorf armour in the Wallace Collection', *The Burlington Magazine* 92(587): 173–4; Mann 1962 (cited in note 1), pp. 174–5; Gamber 1975 (cited in note 2), p. 32; Norman 1986 (cited in note 3), p. 82.
7. The close helmet with which the armour was displayed until very recently was first associated with it in the later 16th century, and does not match, either in its typology or decoration.
8. Krause 2017 (cited in note 5), p. 64. Light lance rests secreted into the construction of light and medium cavalry breastplates by means of ingenious spring mechanisms were unique to the Helmschmid workshop – an idea introduced by Kolman and developed by his son and successor as master, Desiderius. As well as the one on the Rogendorf armour, at least two were built into armours made for Charles V by Desiderius – the three-quarter demi-lancer cuirass of the 'Cleves' garniture (1543, now in Vienna, inv. no. A546) and the anime cuirass of the 'Mühlberg' garniture (1544, Real Armería, Madrid). I am grateful to Álvaro Soler del Campo for allowing me to examine this armour and for revealing the hidden lance rest mechanism. Practical experiments have shown that impacts delivered with the heavy cavalry lance typical of the late 15th and early 16th century, by an armoured rider using a lance rest and seated in a war saddle, can yield energy transferences ranging between around 100 and 300 joules, i.e., blows up to around four times harder than those typically struck with the strength of the arms alone. The experiments also suggested that even higher impacts were probably possible, up to perhaps 600 joules. Conversely, blows struck with lighter spears without full armour ranged between around 60 and 150 joules: see Williams, A., Edge, D. and Capwell, T. (2016) 'An experimental investigation of late medieval combat with the couched lance', *Journal of the Arms and Armour Society* 22(1): 1–16.
9. Archduke (later Emperor) Maximilian may have sparked a wider trend of noblemen marching as Landsknechts when he joined his infantry on foot in his formal entry into the City of Ghent in 1485; for this and later instances of Landsknecht involvement in parades and formal display, see Krause 2017 (cited in note 5), p. 58. A courtly mock combat fought with pikes between teams of pikemen (or noblemen dressed as pikemen) is illustrated in *Freydal*, one of three pseudo-historical and partially autobiographical works on the life of Maximilian (Kunsthistorisches Museum, Vienna, inv. no. KK 5073) produced between 1512 and 1515; see Krause, S. (2019) *Freydal, Medieval Games: The Book of Tournaments of Emperor Maximilian I*. Cologne: Taschen, pp. 226–7 (fol. 116). See also Krause 2017 (cited in note 5), pp. 56–7.
10. Landsknecht officers are often shown mounted and armed as light cavalry in German and Swiss prints of c.1520-40; see Breünner-Enkevoerth, A.J. (1883) *Römisch Kaiserliche Majestät Kriegsvölker im Zeitalter der Landsknechte*. Vienna: C.J. Wawra. See also the hand-tinted series *Warhafftige beschreibung des andern Zugs in Osterreich wider den Turcken*, illustrated by Michael Ostendorfer and printed by Hieronymus Formschneyder in 1539 (Albertina, Vienna, inv. no. DG1959/276) and in the *Kriegsbüchlein* of Reinhard zu Solms (c.1540; reproduced in Breünner-Enkevoerth 1883).
11. Blair, C. (1958) *European Armour circa 1066 to circa 1700*. London: Batsford, p. 116.
12. Patrimonio Nacional, Madrid, Real Armería, inv. no. A11. See Valencia de Don Juan, Conde (1898) *Catálogo Histórico-Descriptivo de la Real Armería*

de Madrid. Madrid: Hauser y Menet, pp. 8–10; Terjanian, P. (2019) *The Last Knight: The Art, Armor and Ambition of Maximilian I*, exh. cat. New York: Metropolitan Museum of Art, cat. no. 76, pp. 162–4.

13. The heavy cavalry typology of this armour is indicated by the armet set on a turning-joint at the neck, the asymmetrical pauldrons (with the right cut away for the couched lance), and the short tassets and high cuisses. A lance rest was intended but never installed; see Terjanian 2019 (cited in note 12), pp. 170–72.
14. Imperial Armoury, Kunsthistorisches Museum, Vienna, inv. no. A109. See Thomas, B. (1949) 'Konrad Seusenhofer – Studien zu seinen Spätwerken zwischen 1511 und 1517', *Konsthistorisk Tidskrift* 18: 37–70; Gamber and Thomas 1976 (cited in note 4), pp. 216–17 and plate 110; Terjanian 2019 (cited in note 12), cat. no. 82, pp. 172–5. For other examples see also Chapter 1 in this volume.
15. Interestingly, the Anglo-Flemish silvered and engraved armour of King Henry VIII (1515), which also includes a highly decorated steel base inspired by the work of Innsbruck craftsmen, is comprised of pieces for heavy cavalry use, but its unusually light breastplate is not pierced for a lance rest, suggesting that it was only ever intended for parade use: see Blair, C. (1965) 'The Emperor Maximilian's gift of armour to King Henry VIII and the silvered and engraved armour at the Tower of London', *Archaeologia* 99: 1–56.
16. Terjanian 2019 (cited in note 12), p. 175. The contribution of Augsburg craftsmen to this project was probably restricted to the pierced copper alloy panels applied to the cuirass, base and shoulder defences (and also probably the original helmet); the etching was almost certainly executed in Innsbruck (see below).
17. See also Chapter 1 in this volume.
18. Inv. no. A162. See Mann 1962 (cited in note 1), p. 134; Norman 1986 (cited in note 3), p. 60. This helmet is of a different typology to the armour with which it is associated, having been designed for medium and heavy cavalry service. The decoration is also different, although the rendering of the slashes and the etching are consistent with an Innsbruck origin. Further research on this helmet is needed.
19. Krause 2017 (cited in note 5), pp. 55–7.
20. For further discussion of the design of foot combat armour see Chapter 1 in this volume.
21. This armour was preserved from c.1533 in the dynastic armoury of the Radziwill family at Nesvizh Castle in Poland until c.1924. The backplate, rump plate and parade-tournament arm defences are now in the Metropolitan Museum of Art in New York (inv. nos. 24.179; 26.188.1, .2; 29.158.363a, b), while the breastplate, right tasset, left cuisse plate and three-part vambraces for the field with integral spaudlers are now in the Musée de l'Armée, Paris (inv. no. G. 45): see Dean, B. (1926) 'Puffed and slashed armor of 1525', *The Metropolitan Museum of Art Bulletin* (November): 260–64; Grancsay, S.V. (1931) 'The mutual influence of costume and armor: a study of specimens in the Metropolitan Museum of Art', *Metropolitan Museum Studies* (June): 194–208; Gamber 1975 (cited in note 2), p. 33; Reverseau, J.-P. (1982) *Musée de l'Armée Paris: Les armes et la vie*. Paris: Dargaud, p. 132; Krause 2017 (cited in note 5), pp. 46–9.
22. Krause 2017 (cited in note 5), p. 58.
23. For early examples of gusset plates see Chapter 1 in this volume.
24. For a further discussion on this topic see Chapter 1 in this volume.
25. An important surviving pair are preserved in the Imperial Armoury in Vienna, inv. no. A79; see Gamber and Thomas 1976 (cited in note 4), p. 102 and fig. 39. Enclosed shoulder defences are illustrated as part of the equipment both of the club tourney and foot combat in the first Thun album; see Terjanian, P. (2011–12) 'The art of the armorer in late medieval and Renaissance Augsburg: the rediscovery of the *Thun* sketchbooks', *Jahrbuch des Kunsthistorischen Museums Wien* 13/14: 382–3, image 105 and pp. 340–41, image 27. See also Chapter 1 in this volume.
26. Enclosed shoulder defences are also illustrated as part of another Helmschmid costume armour that was lost or never actually made, illustrated in the first Thun album; see Terjanian 2011–12 (cited in note 25), pp. 378–9, image 97.
27. Staat Kunstsammlung, Dresden, inv. no. 1906 G.
28. Herzog Anton Ulrich-Museum, Brunswick, inv. no. Hs 27 N. 67a. Schwarz kept this diligent pictorial record (begun in descriptive form in 1510 and illustrated from 1520) of his many sets of fine clothes year by year from his birth in 1497 to 1560. A second, much shorter book was kept by his son Veit Konrad c.1561 (inv. no. Hs 27 N. 67b): see Hayward, M. and Rublack, U. (eds) (2015) *The First Book of Fashion: The Book of Clothes of Matthäus and Veit Konrad Schwarz of Augsburg*. London: Bloomsbury.
29. See for example Hayward and Rublack 2015 (cited in note 28), pp. 100, 111, 116, 134.
30. Staat Kunstsammlung, Dresden, inv. no. H 75.
31. Inv. no. IV.412. See Dufty, A.R. and Reid, W. (1968) *European Armour in the Tower of London*. London: HMSO, plate 90.
32. Imperial Armoury, Vienna, inv. no. A244. See Gamber and Thomas 1976 (cited in note 4), pp. 214–15 and pl. 108; Beaufort, C. and Pfaffenbichler, M. (2005) *Meisterwerke der Hofjagd- und Rüstkammer*. Milan: Skira, pp. 104–5.

33. Inv. no. E. 1939.95.ce. See Joubert, F. and Scott, R.L. (1924) *Catalogue of the Collection of European Arms and Armour Formed at Greenock by R.L. Scott* (privately published). Reprinted Godmanchester: Ken Trotman, pp. 82–3, A. See also Laking, G.F. (1920–22) *A Record of European Armour and Arms Through Seven Centuries*, 5 vols, vol. 3. London: G. Bell and Sons, pp. 262–3 and fig. 1044.
34. See Post 1928 (cited in note 1). This very significant example of Seusenhofer's oeuvre, preserved with its complete horse armour, was held in the former Museum für Deutsche Geschichte in Berlin (now the Deutsches Historisches Museum) until 1945; it is now in the care of the State Historical Museum, Moscow, inv. nos. 2150-1, 2209-22. The horse armour was returned to Berlin in 1959. Mann was unaware of its relocation when he referred to it in the 1962 edition of the Wallace Collection catalogue; Mann 1962 (cited in note 1), p. 27. Its post-war whereabouts were revealed in Blair, C. (1967) 'Spoils of war in the State Historical Museum, Moscow', *Connoisseur* (May): 1–3. See also Gerasimova, A.A. (2016) *Games of the Kings: European Arms and Armour of Renaissance and Mannerism in the State Historical Museum*, exh. cat. Moscow: Gosudarstvennyi istoricheskii muzei.
35. Laking 1920–22 (cited in note 33).
36. The pair in Vienna form part of a composite armour (made up of parts of several armours of Emperor Maximilian I) put together in the late 16th century; Imperial Armoury, Kunsthistorisches Museum, inv. no. A110. The pair now in the Wallace Collection (inv. nos. A285-6) come from the same (otherwise lost) armour as an armet now in the same collection (inv. no. A154).
37. See Terjanian 2011–12 (cited in note 25), pp. 346–7, image 39.
38. Ibid., pp. 358–9 and 374–5, images 62 and 91.
39. Inv. no. 190x. See Gamber 1975 (cited in note 2), p. 33; Zygulski, Z. (1982) *Stara Bron w Polskich Zbiorach*. Warsaw: Krajowa Agencja Wydawnicza, p. 138, no. 137.
40. Inv. no. 82001-2295. See Blair 1967 cited in note 34), p. 3; Gerasimova 2016 (cited in note 34).
41. Jennifer Wearden, Victoria and Albert Museum, personal communication.
42. Inv. no. 81A-1892.
43. Inv. nos. 04.3.286a and 29.150.6. For the close helmet see Laking 1920–22 (cited in note 33), pp. 255–8, fig. 1039; and La Rocca, D.J. (2017) *How to Read European Armor*. New York: Metropolitan Museum of Art, p. 130, fig. 143. For the tournament pieces, see Gamber 1975 (cited in note 2) p. 30, fig. 39; La Rocca *op. cit.*, p. 68, fig. 74.
44. Imperial Armoury, Kunsthistorisches Museum, inv. no. A295. See Gamber, O. and Thomas, B. (1954) *Die Innsbrucker Plattnerkunst*, exh. cat. Innsbruck: Tiroler Landesmuseum Ferdinandeum, p. 71 and pl. 46, cat. no. 83.
45. Inv. no. 211/W; Gamber and Thomas 1976 (cited in note 6), p. 211.
46. See note 1 above.
47. Imperial Armoury, Kunsthistorisches Museum, Vienna, inv. no. A179. Since the late 16th century this almost completely lost armour, represented purely by the surviving right pauldron, was thought to have been made for Ludwig II, King of Hungary, Croatia and Bohemia (born 1506; killed at the Battle of Mohács, 1526). This attribution was based on the documentary record of a gift armour commissioned by Emperor Maximilian I for the then Prince Ludwig in 1514/5. However, Ludwig was only 9 years-old at the time of the commission, and the surviving pauldron was made for an adult. See Gamber and Thomas 1976 (cited in note 4), p. 208.
48. Williams, A. (2003) *The Knight and the Blast Furnace*. Leiden: Brill, pp. 401–2.
49. Alan Williams, personal communication, 11 June 2021.
50. Williams 2003 (cited in note 48), p. 497.
51. Ibid., p. 499.
52. Ibid., pp. 500–501.
53. Ibid., p. 502.
54. Ibid., p. 503.
55. Ibid., pp. 498 and 504.
56. Ibid., pp. 387–8, 393–4, 403.
57. Ibid., pp. 399–400.
58. It seems that the Wallace Collection Landsknecht armour was not yet in Meyrick's possession when his collection was displayed at Upper Cadogan Place in London up to 1830, when it was relocated to Goodrich Court in Herefordshire in July of that year. There is no reference to it in Meyrick's three-volume work *A Critical Inquiry into Antient Armour* (London: Robert Jennings, 1824), nor is it found in any of the myriad sketches, drawings and watercolours executed by Eugène Delacroix and Richard Parkes Bonington during their armour study visit to Meyrick's London residence in July 1825; for the 1825 visit see Duffy, S. (2009) 'French artists and the Meyrick armoury', *The Burlington Magazine* 151(1274): 284–92. For the 1830 publication of the armour, see Skelton, J. (1830) *Engraved Illustrations of Ancient Armour from the Collection of Sir S.R. Meyrick at Goodrich Court, Herefordshire*. London: G. Schulze, plate 19. Skelton's view of the great hall at Goodrich Court, where the Landsknecht armour is shown displayed in a dedicated niche, is reproduced in Lowe, R. (2003) *Sir Samuel Meyrick and Goodrich Court*. Little Logaston: Logaston Press, p. 123, fig. 65.
59. Oxenham's, London, 29 April 1841, lot 253.

3

Henry VIII's Italian armour of 1544: a metallurgical note on alterations and improvements by the Royal Workshops at Greenwich

Simon Metcalf and Alan Williams

Introduction

The Italian armour purchased by Henry VIII in or shortly before 1544 and now in the Metropolitan Museum of Art, New York (see Figure 1), was discussed by Claude Blair and Stuart Pyhrr in 2003.[1] A.V.B. Norman also identified a pair of field spaulders and vambraces belonging to this armour (see Figure 6) still in the Royal Collection at Windsor Castle.[2] Blair and Pyhrr speculated that the pair of field spaulders and vambraces were Italian but could have been made to the king's specification because of their similarity in style to other Greenwich garnitures and differences in the etched motif to that on the rest of the armour. In addition, they identified a number of lames on the armour which they suggest might also be contemporary Greenwich alterations.

Metallographic analysis of the armour and its pair of field spaulders and vambraces by the authors has confirmed this hypothesis and suggests that the armour is of Italian make but with Greenwich alterations, and that the pair of field spaulders and vambraces were either modified in Greenwich or even made in the Royal Workshops to complement the Italian armour. This new evidence strongly supports the suggestion that the armour was altered to fit the king's requirements after it was delivered. It seems likely that for the Boulogne campaign of 1544, the armourers at Greenwich overhauled the new Italian armour for the king. They attempted to harden parts of it by heat treatment using techniques that were being developed at Greenwich to try to improve the armour's effectiveness. It is also possible that they made additional field spaulders and vambraces as well as altering and re-leathering the armour, and added linings and trimmings of the king's new red and yellow royal livery issued for the first time in that year.[3]

Samples from the Metropolitan Museum of Art, New York

Four samples of the armour were taken by the Metropolitan Museum of Art from the right arm, the bottom lame and the right gusset of the breastplate, and from the left shoulder of the backplate which has an added plate for a 'Tudor Rose' stud (see Figures 2–5). All the samples were embedded then polished to 1 μm in the usual way, and etched with nital/picral to reveal the microstructure. Their hardnesses were measured using a Vickers microhardness apparatus (100 g load) and the results are given on the Vickers Pyramid Hardness (VPH) scale, whose units are given in $kg.mm^{-2}$. For comparison, modern mild steel is around 130 VPH.

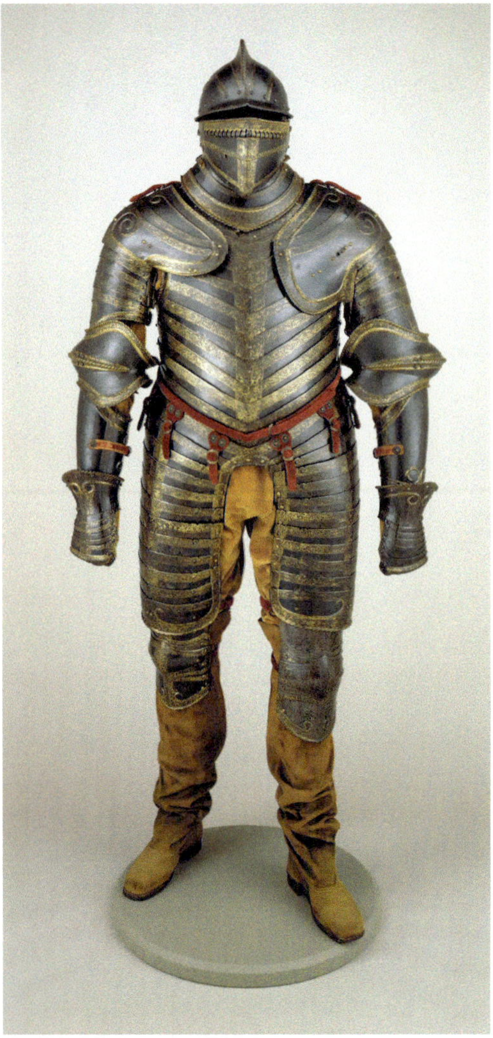

Figure 1 The Italian armour of Henry VIII c.1544, blackened surface with etched and gilt decoration. The Metropolitan Museum of Art, New York, 32.130.7. Harris Brisbane Dick Fund, 1932.

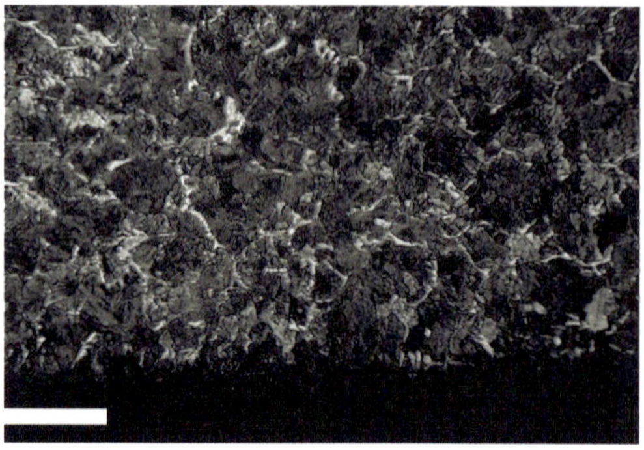

Figure 2 Sample 1 from MMA 32.130.7 from the right arm, uppermost lame, back side. Very uniform fine pearlite. Scale bar = 50 μm.

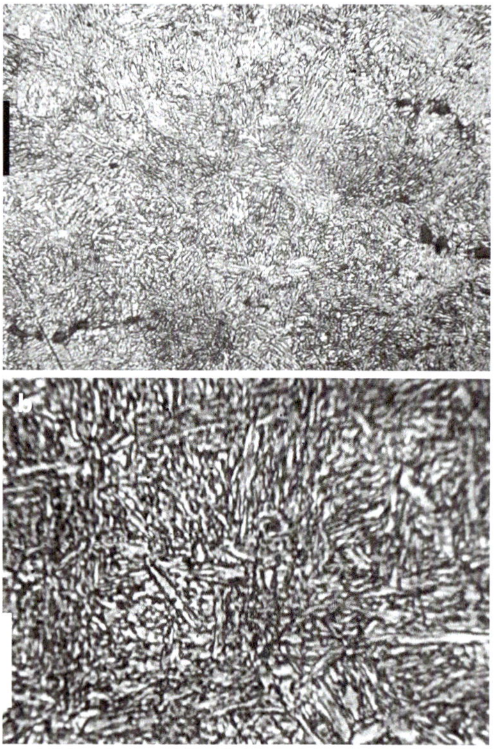

Figure 3 (a) Sample 2 from MMA 32.130.7 from the breastplate, centre bottom lame. Martensite (tempered) and slag inclusions. Scale bar = 50 μm. (b) A sample from the breastplate, centre bottom lame containing martensite and an acicular material which might be bainite. Scale bar = 10 μm.

The microstructure of sample 1 (Figure 2) consists of very uniform fine pearlite. This is a medium-carbon steel (perhaps 0.6% C) which has been air-cooled after fabrication. The microhardness range 220–261 average; = 240 VPH is consistent with the item being a product of either early 16th-century Greenwich or North Italy.

The microstructure of sample 2 (Figure 3) shows a fairly uniform acicular martensite (or perhaps bainite) which is slightly tempered. There is no visible pearlite or ferrite, and few slag inclusions. This is a medium-carbon steel that has been hardened by some form of slack quenching. The microhardness range 292–325; average = 313 VPH is not consistent with any of the products of North Italy later than

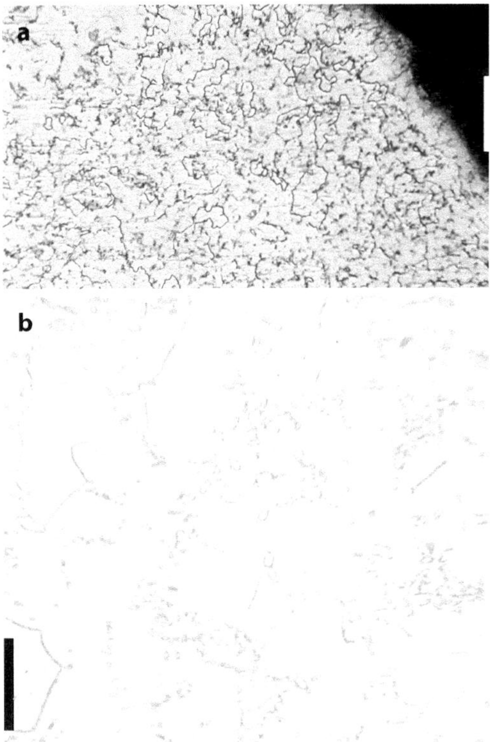

Figure 4 (a) Sample 3 from MMA 32.130.7 from the breastplate, bottom edge right gusset. Ferrite and carbides. Scale bar = 50 μm. (b) A sample from the breastplate, bottom edge proper right gusset. Scale bar = 10 μm.

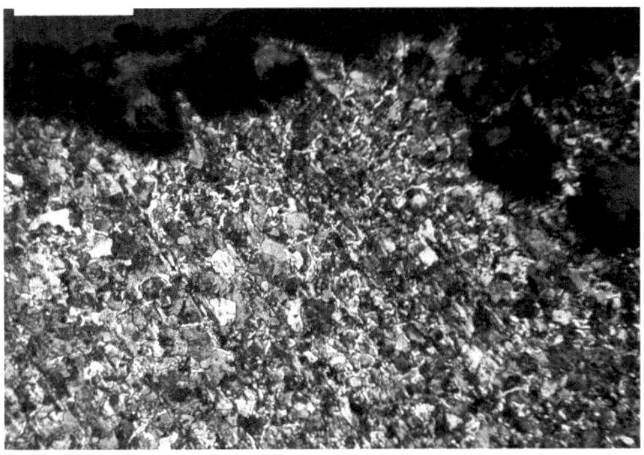

Figure 5 Sample 4 from MMA 32.130.7 from the backplate, left shoulder and added plate for the 'Tudor Rose' stud. Pearlite and ferrite. Scale bar = 50 μm.

c.1510. It is however possible that it is a product of the Greenwich workshop, since these craftsmen were experimenting with the hardening of their steel products at this time. Although unlikely, a South German origin is also possible in this case.

The microstructure of sample 3 (Figure 4) consists mostly of ferrite with some carbides (grain boundary cementite). This is a low-carbon steel (around 0.1% C) which has apparently been annealed by being kept red-hot for some hours. It is difficult to explain the use of such a procedure unless it was an attempt to rectify a failed hardening. Its microhardness range is 164–184; average = 172 VPH.

The microstructure of sample 4 (Figure 5) consists of pearlite and ferrite. This is an air-cooled, medium-carbon steel (perhaps 0.6% C). The microhardness range 256–301; average = 286 VPH is consistent with the item being a product of either early 16th-century Greenwich or North Italy. Apart from the gusset (sample 3), all the samples are made of medium-carbon steels, which means that they could have been products of either Greenwich or North Italy. However, the breastplate lame (sample 2) is definitely *not* a typical Italian product.

Samples from the Royal Collection

Two samples were taken from both of the Royal Collection field spaulders and vambraces – the first lower lame inside of the elbow of the left vambrace and on the upper cannon turning lame of the right vambrace (see Figures 6–8). Both samples were examined in the same way as samples 1–4 from the Metropolitan Museum of Art.

Microstructures of samples

The microstructure of sample 1 (Figure 7) consists largely of irresolvable carbides with some proeutectoid ferrite grains, especially where associated with a crack. There are a few ovoid slag inclusions and signs of intergranular cracking, perhaps the result of a too-drastic quenching. Its microhardness range is 581–700; average = 625 VPH. The sample appears to be a medium-carbon steel that has been quenched fully, leading to the formation of martensite (which is excessively hard) and with quenching cracks. This was probably plunged from bright-red heat into cold water or brine. It would have needed moderate

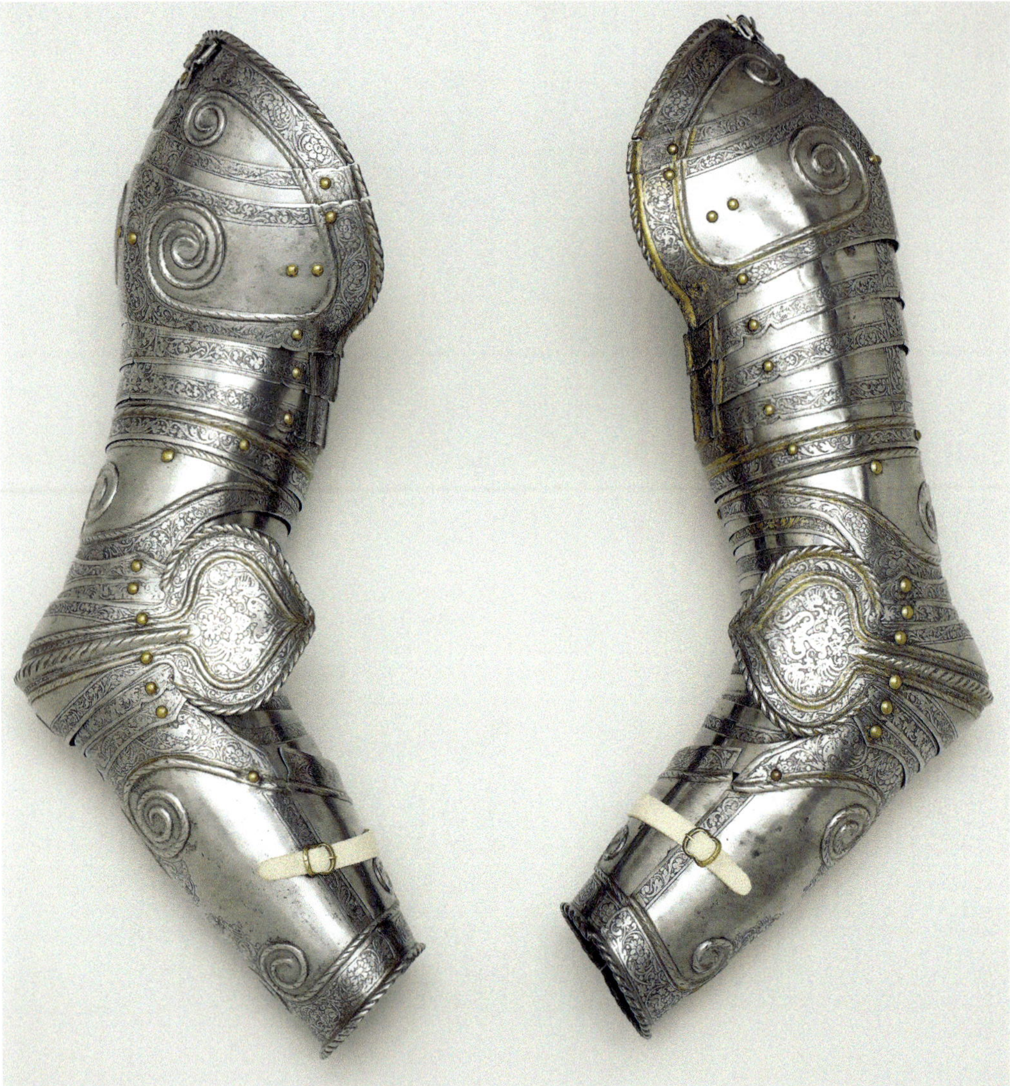

Figure 6 The pair of field spaulders and vambraces for the Italian armour made or altered in the Royal Workshops at Greenwich, originally decorated with etching, gilding and blackening. RCIN 67399, Royal Collection Trust © Her Majesty Queen Elizabeth II.

reheating to temper it. Such a second procedure was necessary but very difficult to carry out successfully and does not appear to have been practised in Greenwich until after around 1560. It also indicates that this was the last process undertaken after etching, gilding and blacking as the heating for fire gilding would have tempered the lame to some extent. In comparison, sample 2 from the other (right) vambrace has been heat treated more successfully.

The microstructure of sample 2 (Figure 8) consists of a mixture of proeutectoid ferrite grains within masses of irresolvable carbides and a few slag inclusions, with a microhardness range of 365–417; average = 388 VPH. This appears to be a medium-carbon steel which has undergone some form of heat treatment to harden it. A slightly delayed quench would seem probable as neither lamellar pearlite nor the distinctive microstructure of martensite is visible. The results indicate

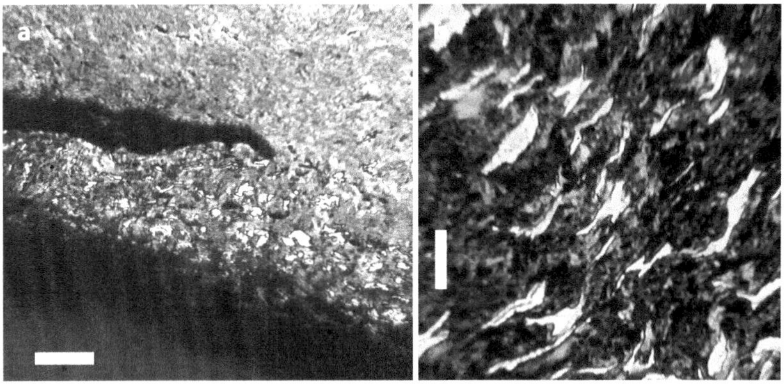

Figure 7 (a) Sample 1 from RCIN 67399 from the edge of a hole in the first lower lame of the inside of the elbow of the left vambrace. This is a hardened steel. Scale bar = 50 μm. (b) Distorted ferrite grains and granular carbides. Scale bar = 10 μm.

that both vambraces were manufactured from a medium-carbon steel and hardened by heat treatment, one more successfully than the other. Sample 1 is very hard and therefore too brittle whereas in sample 2 the quenching seems to have been delayed somewhat, resulting in a less hard but also less brittle microstructure.

Virtually all the armours produced by the Royal Workshops at Greenwich were made of medium-carbon steel (of around 0.6% C). Until the 1540s they do not appear to have been hardened successfully by any form of heat treatment. Their microstructures consist largely of pearlite and their hardnesses are in the region of 250–270 VPH. After about 1560–70 and as late as 1610, they were hardened successfully by the dual process of quenching (plunging red-hot into cold water) followed by tempering (often combined with fire gilding for decoration) to give microstructures of tempered martensite and consistent hardnesses of around 360–380 VPH. Prior to these later successes, however, between the 1540s and 60s attempts were made to harden them using some form of heat treatment with inconsistent results. It seems likely that the difficulties of full quenching (possible cracking and warping) led to various methods of slack quenching being attempted. A delayed or interrupted quench would result in a less drastic rate of cooling and therefore a different microstructure. The microstructures of a number of examples of Greenwich armour made between

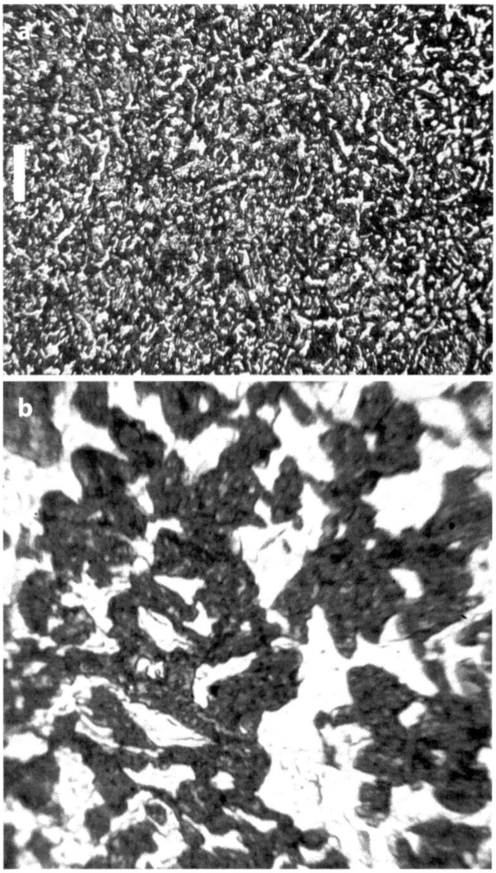

Figure 8 (a) Sample 2 from RCIN 67399 taken from the edge of a hole in the upper cannon turning lame of the right vambrace. This is a hardened steel. Scale bar = 50 μm. (b) Irresolvable carbides and ferrite. Scale bar = 10 μm.

the 1540s and 60s contain pearlite, martensite and other carbides, with hardnesses around 300 VPH. For example, the armour of Henry VIII (RCIN 72834), dating from about 1540, was manufactured from a medium-carbon steel of around 0.6% C with an average hardness between 249 and 293 VPH. It exhibits signs of varying methods used in an attempt to harden the steel by slack quenching:[4] if the quench was delayed too long, the results (pearlite) would have been indistinguishable from an air-cooled steel. It is possible, but cannot be proved, that Greenwich armours made before 1540 were treated this way – by an ineffective over-delayed quench. One lame of these vambraces was fully quenched to 700 VPH making it far too brittle. The other was hardened more successfully to 388 VPH, apparently by a delayed quench, although the carbide aggregate is irresolvable therefore the precise technique used cannot be determined.

If the parts of the armour that were heat treated using the Greenwich method are Italian, this indicates that they were hardened after etching and gilding. This is possible as there are examples of hardened and tempered Greenwich armour with etched and gilt decoration such as the armour of Sir Christopher Hatton dated 1585 (RCIN 72835). Alternatively, the heat treatment suggests that these parts are Greenwich alterations and replacements: as the exact order of process for etching, fire gilding, colouring and hardening is unclear, either proposition is possible. Sample 1 from RCIN 67399 (Figure 7) is a slightly tempered martensite of 700 VPH, therefore after quenching this lame was not reheated to temper it. The process of fire gilding needs the lame to be heated to around 250 °C, so this lame indicates that in this case, the quenching was the last process carried out after etching and gilding.

The microstructures of all the parts made of medium-carbon (perhaps 0.6% C) steel which have been heat treated closely resemble those found in examples of Greenwich armour produced in the 1540s,[5] but they *do not* resemble those found in any contemporary Italian armours.[6] Overall, the analysis strongly supports the suggestion that parts of the Metropolitan Museum of Art armour were modified in Greenwich and that the Royal Collection field spaulders and vambraces were either of Italian make but modified in Greenwich or made in Greenwich and decorated to match the rest of the armour in the Italian style. It seems that the whole armour was altered, adapted and improved for Henry VIII in the Royal Workshops at Greenwich.

Acknowledgements

We are grateful to H.M The Queen, The Royal Collection Trust and the Metropolitan Museum of Art for allowing the study of the field spaulders and vambraces and armour, and for permission to reproduce their images. We are indebted to Claude Blair, Stuart Pyhrr and Ian Eaves for their comments on the text and Edward Hunter for providing samples of the Metropolitan Museum of Art armour.

Notes and references

1. Blair, C. and Pyhrr, S. (2003) 'The Wilton "Montmorency" Armor: an Italian armor for Henry VIII', *Metropolitan Museum Journal* 38: 95–144.
2. Norman, A.V.B. and Eaves, I. (2016) *Arms & Armour in the Collection of Her Majesty the Queen. Volume 1 European Armour.* London: Royal Collection Trust, pp. 246–51 (RCIN 67399).
3. Metcalf, S., North, A.R.E. and Balfour, D. (2005) 'The conservation of a gun-shield from the arsenal of Henry VIII', in R.D. Smith (ed.), *Make All Sure: The Conservation and Restoration of Arms and Armour.* Leeds: Basiliscoe Press, pp. 76–90.
4. Blackburn, T.P.D., Edge, D. and Williams, A. (2000) 'Head protection in England before the First World War', *Neurosurgery* 47(6): 1261–85. See also Norman and Eaves 2016 (cited in note 2), p. 101 (RCIN 72834).
5. Williams, A. and de Reuck, A. (1996) *The Royal Armoury at Greenwich 1515–1649: A History of its Technology.* Leeds: Royal Armouries.
6. Williams, A. (2003) *The Knight and the Blast Furnace.* Leiden: Brill, pp. 203–329. From an examination of a total of 92 specimens of Italian armour made after 1500, only two, both of uncertain origins, had undergone any attempt at heat treatment.

4

A lance rest from a Spanish royal garniture in the Wallace Collection

Álvaro Soler del Campo

Introduction

Those responsible for the collection of arms and armour at the Wallace Collection have always provided support to the Real Armería of Madrid when requested, such as the visit of James Mann in August 1937 in order to assess the damage suffered by its collection during the first two years of the Spanish Civil War.[1] Among the items related to the Real Armería at the Wallace Collection is a small lance rest equipped with a spring and four holes for fixing a breastplate (A228). A wavy motif in the form of Chinese clouds that evokes ruffs from 16th-century male clothing (see Figures 1 and 2) is prominent in its etched and gilt decoration. On the base and side, this motif occupies the main bands complemented by other smaller bands that feature a decorative pattern in the form of bat wings alternating with small three-petalled flowers. This pattern is the key to establishing the provenance and identity of its owner.

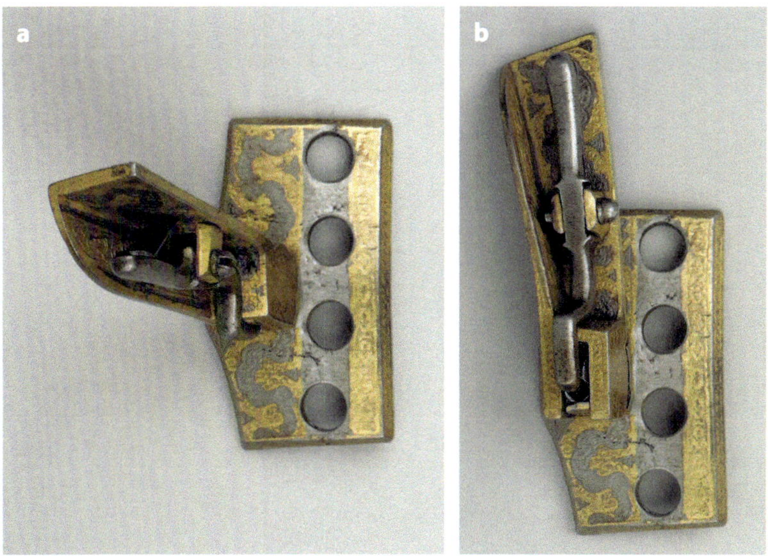

Figure 1 The lance rest, frontal view opened (a) and closed flat (b): Franz and Wolfgang Grosschedel, c.1558, Wallace Collection, A228. © Trustees of the Wallace Collection.

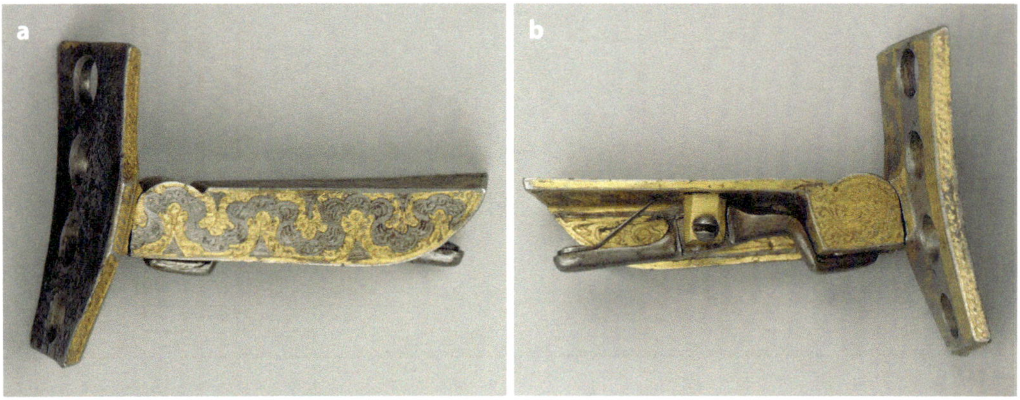

Figure 2 Lance rest, right (a) and left (b) sides: Franz and Wolfgang Grosschedel, c.1558, Wallace Collection A228. © Trustees of the Wallace Collection.

Philip II of Spain and Prince Don Carlos

In his catalogue of the collection, James Mann[2] noted that the etched decoration belonged to the 'cloud bands' garniture of King Philip II of Spain, forged by Wolfgang Grosschedel of Landshut in 1554 and preserved in the Real Armería.[3] Traditionally, the lance rest was mounted on an armour of Pompeo della Chiesa until it was recognised and removed by Vesey Norman. Years later, Norman[4] returned to this lance rest in his supplement to Mann's catalogue, reaffirming its relation with the armour of Philip II which was commissioned from Grosschedel's workshop on the occasion of his wedding to Queen Mary I of England in 1554. This attribution is not surprising, since the waves that define its main decorative pattern are very similar to those of Philip II's cloud bands armour. However, a subtle difference in the decoration of the secondary bands indicates that it could not have belonged to that armour but to a second garniture belonging to his son, Prince Don Carlos (1545–1568) – also preserved in the Real Armería (A274–A276) and dated to c.1558 – when he was 13 years old.[5]

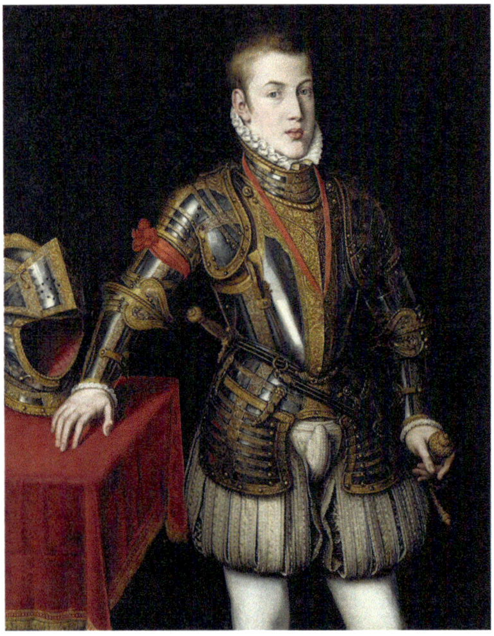

Figure 3 Joris Van der Straeten, *Portrait of the Prince Don Carlos*, c.1562, Monasterio de las Descalzas Reales, Madrid, Patrimonio Nacional, inv. no. 00612065.

For those unfamiliar with the figure of Don Carlos, he was the son of Philip II and his first wife, Princess María Manuela of Portugal (see Figure 3). Since childhood, his unstable character and ill health were evident, partially explaining his father's reservations as to his ability to govern that resulted in continual clashes between them. Despite (or because of) his problematic nature and dysfunctional relationship with his father, the prince has been the subject of attention in literature and music, including, among others, Friedrich Schiller's dramatic work *Dom Karlos, Infant von Spanien*, first performed in 1787, and Giuseppe Verdi's opera *Don Carlo*, which premiered in 1867.

The armours

Today, both the armours of Philip II and Don Carlos are considered to be the work of their favourite armourers, father and son Wolfgang and Franz Grosschedel.[6] The prince's garniture is essentially an armour for war on horseback, defined as *de seguir* according to the Spanish terminology of the time. Although incomplete, it includes light cavalry elements that can also be used on foot and which can be combined, namely: a close helmet (Figure 4), a burgonet with an associated articulated bevor (Figure 5), as well as a gorget, two sets of pauldrons and vambraces (one symmetrical and another asymmetrical), a pair of gauntlets, a breastplate, a backplate, two tassets, a full leg harness with sabatons, two half cuisses, two half greaves, a pair of toe caps and a half shaffron.

In both armours, the waves or clouds pattern predominate but in Philip II's armour it is repeated in the narrower secondary bands, a trait not found in Don Carlos' armour in which the smaller bands are covered by the aforementioned bat wings and small trifoliated flowers, a detail that differentiates his armour from that of his father. This difference allows the Wallace Collection's A228 lance rest to be linked to Don Carlos rather than Philip II. On the other hand, the design, work and quality of the waves of this lance rest correspond closely to those of the Madrid garniture.

The Real Armería only preserves one breastplate of Don Carlos' armour (Figure 6). This has two holes for fixing a lance rest, but as these are modern this indicates that it was originally a breastplate for light cavalry associated with elements that could also be used on foot. In any case, it was not the heavy cavalry breastplate which should be associated with the Wallace Collection lance rest. This breastplate is also characterised by its irregular shape: the neck droops slightly to the left to accommodate the prince's physical problems. It also displays the marks of Grosschedel and the city of Landshut close to the neck edge (Figure 7).

These marks and the garniture decoration corroborate their provenance by comparing them with the information provided by documents relating to this armour. The first is the 1594 inventory

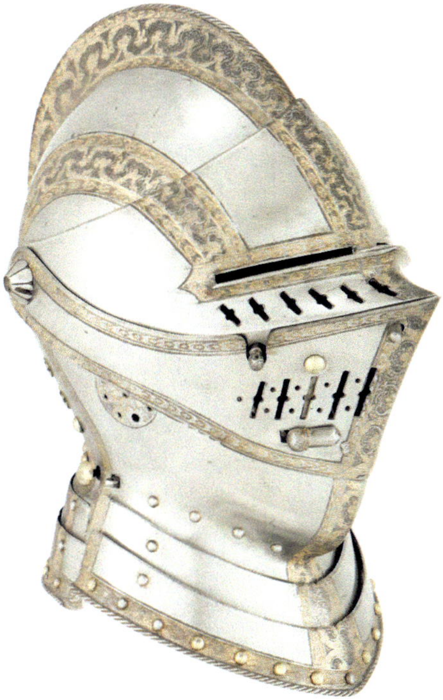

Figure 4 Close helmet from the garniture of Don Carlos: Franz and Wolfgang Grosschedel, c.1558, Real Armería, cat. A-275, Patrimonio Nacional, inv. no. 10054451.

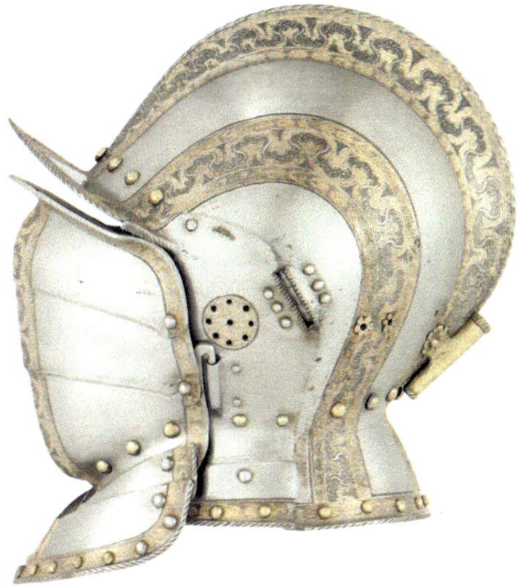

Figure 5 Burgonet with articulated bevor from the garniture of Don Carlos: Franz and Wolfgang Grosschedel, c.1558, Real Armería, cat. A-274, Patrimonio Nacional, inv. no. 10054452-10054453.

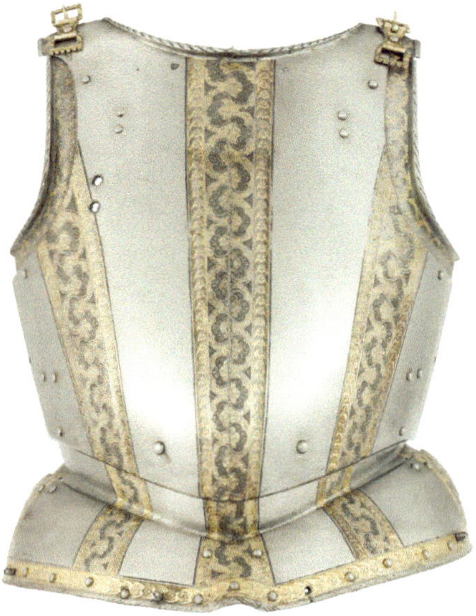

Figure 6 Light cavalry breastplate from the garniture of Don Carlos: Franz and Wolfgang Grosschedel, c.1558, Real Armería, cat. A-274, Patrimonio Nacional, inv. no. 10054435.

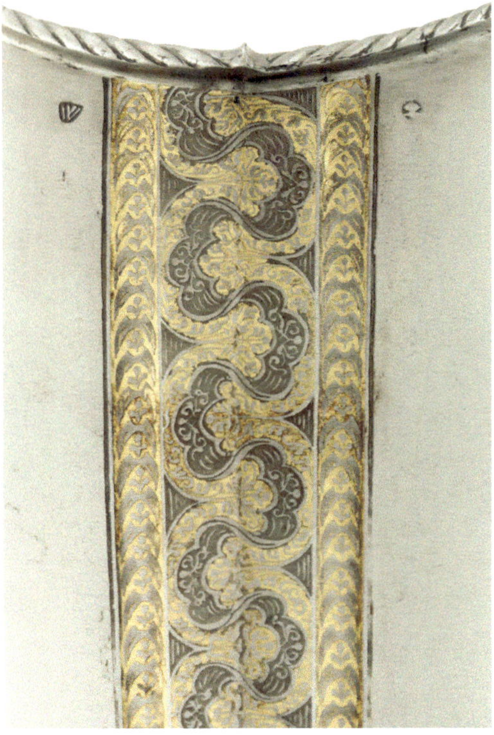

Figure 7 Maker's marks on the breastplate shown in Figure 6.

of the Real Armería in which it is described as an armour decorated with clouds (*nubes*) made in the Bavarian city of Landshut (*Lançuete*):

> An etched and gilt armour with cloud work that was made in Landshut for Prince Don Carlos with the following pieces / Breastplate and backplate and gorget, one articulate helmet [burgonet with articulate bevor] and one close helmet, two pauldrons with vambraces and its two rondels and gauntlets and two other pauldrons and two other vambraces. A couple of complete cuisses, a couple of complete greaves. Another pair of cuisses with its plates and a couple of half greaves with toe caps. Half shaffron for horse, all [these] in their red sacks. [Side note: From Prince Charles / watch out! / this harness with all their pieces is inventoried in the new inventory [1625] in cabinet number twelve where it is placed.][7]

The second reference can be found in the 1625 inventory of the collection:

> A small garniture of clouds pattern made in Germany that was in a chest of the hollow of this cabinet and were of Prince Don Carlos and could be put inside the said cabinet armed with all his war pieces.[8]

The Real Armería 1594 inventory therefore points to rondels as the unique elements that have not survived in the collection, although the rest has been preserved.[9] It also omits other parts such as tassets or sabatons, perhaps because they were fixed onto the breastplate or the greaves. The inventory does not describe the other elements that it may have had, such as a second cavalry breastplate that should relate to the Wallace Collection's lance rest, or the saddle plates that should have accompanied at least the half shaffron cited in it. Be that as it may, we should consider that it describes the condition of the garniture at the time of inventory, so we do not know the whereabouts of the pieces that may have been missing or from when they may have been lost, as is the case with this lance rest and the breastplate to which it was associated.

However, the answer might be found in two inventories compiled prior to the entry of most of the garniture elements into the Real Armería's collection: the inventory of Don Carlos' property after his death, drawn up on 28 December 1569, and the auction of his testamentary residue, an undated document that logically must have been contemporary with the former. The inventory of his property describes:

> War armour etched and gilt/ a closed helmet for war etched and gilt / one gorget and one breastplate and one backplate with a lance rest and tassets that are removed and put with their buckles / some vambraces with its pauldrons and two rondels / some greaves with its cuisses / Two gauntlets / Some sabatons / Some toe caps / Another breastplate and backplate and gorget of the same etched and gilt work that also used to serve with some pieces of those above / two counters with their straps / A horse shaffron with the royal coat of arms / Five iron plates for a saddle etched and gilt with their screws. [Side note. Watch out! / everything mentioned in this line was valued in two hundred ducats].[10]

In the auction they are described again:

> A closed helmet, gorget, breastplate and backplate. Lance rest, tassets that are removed and put with pauldrons and its vambraces, two rondels, greaves, cuisses and their gauntlets and sabatons and toe caps and another breastplate and backplate and gorget of the same etched and gilt work that also used to serve with some pieces of those above and two couters for vambraces with their straps and a shaffron with the royal coat of arms and five iron plates for an etched saddle with their screws.[11]

In neither of these documents is the identification of the armour as clear as that described in the 1594 inventory of the Real Armería, since they contain no references to either its author or its decoration, but only specify that it was an engraved and gilded war garniture. We must assume, however, that it is the same armour because it is the only one with these characteristics listed in both the inventory of his property and its auction. If we accept that it is the same armour, by contrasting these two inventories made closely following his death and that of the Real Armería from 1594 it may be noted that the latter was missing a significant number of pieces that were described earlier in the auction. These included a breastplate, a backplate, a gorget, a lance rest and the saddle plates, indicating that they never entered the Real Armería despite the short time interval between the auction and the arrival of the balance of the pieces in Madrid. We have no evidence to explain the reasons for the separation, which may have been due to a variety of circumstances. Comparison between these three texts yields another anomaly, such as the burgonet with its articulated bevor, which was listed in the 1594 inventory but not included in the earlier inventory of Don Carlos' property or its auction.

Conclusions

This lack of concordance between inventories and their comparison with the preserved pieces is a clear example of the random life of armour garnitures, even those of royal origin which were subjected to greater control in the large dynastic armouries. The processes of concentration and dispersion of their pieces cannot therefore always be attributed to having been given away in later times, perhaps stolen or subject to any other similar circumstance. This also impacts our contemporary concept of collecting: we often assume that ensembles such as a garniture came complete, when this was not necessarily the case, and depended on circumstances that were more varied than those to which we are accustomed, and a mentality which we often do not understand. Prince Don Carlos' small garniture is a clear example: it had to enter the Spanish collection without part of its pieces. On his death in 1569, the Real Armería had already been established in its Madrid hall for at least two years, where it has remained until the present day even though the original building was destroyed in 1884. Despite this, the lance rest of Don Carlos which is preserved in the Wallace Collection seems to have already been separated from the rest of the garniture in the 16th century – a span of no more than 24 years between the prince's death and the first inventory of the collection – so it is almost miraculous that it has been preserved to the present day.

Acknowledgements

I would like to thank Alan Williams for his invitation to participate in this tribute to David Edge as for me is an honour and a pleasure.

Notes and references

1. Álvarez Lopera, J. (1982) *La política de bienes culturales del gobierno republicano durante la Guerra Civil española*, vol. I. Ministerio de Cultura. Madrid.
2. Mann, J.G. (1962) *European Arms and Armour. Volume I. Armour.* London: Wallace Collection Catalogues, p. 168 (A228).
3. Crooke y Navarrot, J.B. [Conde viudo de Valencia de Don Juan] (1898) *Catálogo Histórico-Descriptivo de la Real Armería.* Madrid: Real Armería, pp. 83–6 (A.243–A.262).
4. Norman, A.V.B. (1986). *European Arms and Armour Supplement.* London: Wallace Collection Catalogues, p. 79 (A228).
5. Crooke y Navarrot 1898 (cited in note 3), pp. 89–90; Godoy, J.A. (1992) 'Piezas de justa de la armadura de nubes de Felipe II', in *Tapices y armaduras del Renacimiento: Joyas de las colecciones reales*, exh. cat. Barcelona-Madrid: Luwnberg Editores – Empresa Nacional Aldeasa, pp. 190–93.
6. Godoy, J.A. (1992) ´Piezas de justa de la armadura de nubes de Felipe II`, in *Tapices y armaduras del Renacimiento. Joyas de las colecciones reales*, exh. cat. Barcelona and Madrid: Luwnberg Editores – Empresa Nacional Aldeasa, pp. 190–93.
7. Real Armería, Madrid, *Inventario de la Real Armería de 1594*, f. 80°, Royal Palace, Madrid: 'Un Arnes dorado y Grauado de labor de nubes q se Hizo en lançuete para el príncipe don carlos q tienen las piezas siguientes / Peto y espaldar y Gola Vna celada trançada y otra borgoñona dos braçales con sus guardabrazos y con sus rrodaxas y mandiletes y otros dos guardabrazos y otros dos braçales Un par de quixotes enteros un par de greuas entereas / Otro par de quixotes con sus laonas y un par de medias Greuas con sus puntas Media testera de caballo todo en sus fundas coloradas. [Side note: del Principe don carlos / ojo / este arnes con todas sus piezas enteras esta ymbentariado en el nuevo Ybentario [1625] en el caxon numero doce donde esta puesto].' Author's translation.
8. Real Armería, Madrid, *Inventario de la Real Armería de 1625*, f. 171°, Royal Palace, Madrid: 'Un arnes pequeño De la labor De las nubes De tres ojas echo en Alemania que estaba en un cofre del gueco de este caxon y eran del príncipe don Carlos y se puso dentro del dho caxon armadas con todas sus pieças De Seguir.' Author's translation.
9. The toe caps were also missing from the Real Armería, until they were returned to it in the year 2000 by the generous donation of Peter Gwynn, who wished to give them back to their original collection.
10. Archivo General de Simancas, Valladolid – España, *Contaduría Mayor de Cuentas,* bundle 1092, fol. 65: 'Arnes de seguir grauado y dorado / Una çelada grauada y dorada de seguir / Vna gola y vn peto y espaldar con su ristre y escarçelas que se quitan y ponen con sus evillas / Vnos braçales con sus guarda braços y dos rodetes / Vnas greuas on sus quijotes / dos manoplas / vnos escarpes / vnos puntales / otro peto y espaldar y gola de la mesma lauor grauado y dorado que tamuien solía seruir con algunas pieças de arriua / dos nauajas con sus braçales con sus correas / vna testera de cauallo con el escudo de las armas reales / cinco pieças para una silla de arma açeradas grauadas y doradas con sus tornillos". Side note: "Ojo / todo lo que comprende esta rraya se taso en doscientos ducados.' Author's translation.
11. Archivo General de Simancas, Valladolid – España, *Contaduría Mayor de Cuentas,* bundle 1051, fol. 138: 'Reciuense mas en quenta al ojo Olarte vna celada gola peto y espaldar. Risttre escarcelas que se quitauan y ponían braçales con sus guardabraços dos rrodetes greuas quijotes y dos manoplas y escarpes y puntales / y otro peto y espaldar y gola de la misma labor grauado y dorado que tanbién solían seruir con algunas pieças delas de arriba y dos nauajas Para braçales con sus correas y una testera de cauallo con el escudo de las armas Reales y cinco Pieças para una silla de armas azerada y grauadas y doradas con sus tornillos.' Author's translation.

5

The remains of a rare Tibetan shaffron

Donald J. La Rocca

Introduction

Virtually everything we know today about horse armour from Tibet is derived from the examination of extant examples, which began to enter Western collections in the early 20th century but have only been studied closely during the past 25 years.[1] The existence of actual specimens of Tibetan horse armour appears to have escaped the notice of Western explorers, scholars and collectors prior to 1903 or 1904. No examples are included in the earliest and otherwise wide-ranging study of Tibetan material culture by W.W. Rockhill published in 1895, which was based on extensive fieldwork conducted in Mongolia and Tibet between 1888 and 1892.[2] Likewise, a primary survey of references to Chinese and Mongolian horse armour in Asian and European historical literature, with tangential references to armour in Tibet, published by the eminent sinologist Berthold Laufer in 1914 – informed not only by his wide reading of original source material but also by his fieldwork in the region during expeditions in 1901–1904 and 1908 – also fails to mention extant Tibetan horse armour.[3] It was during the Younghusband Expedition, a British military incursion into Tibet in 1903–1904, that actual Tibetan horse armour was encountered, photographed, and acquired (Figure 1).[4]

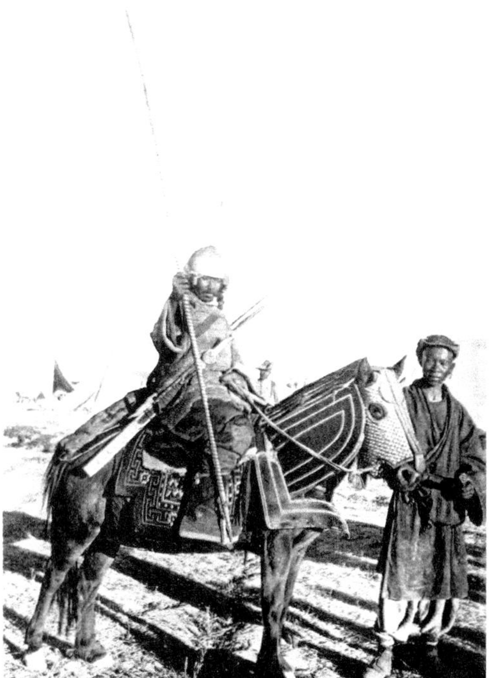

Figure 1 A staged photograph of an armoured man and horse, 1903–1904 (reproduced from L.A. Waddell, *Lhasa and its Mysteries*, New York, 1906).

Lieutenant Colonel F.M. Bailey

Apparently the first example to enter a public collection, the Royal Scottish Museum in Edinburgh, was purchased in 1909 from Lieutenant Colonel F.M. Bailey, a member of the Younghusband Expedition who remained in Tibet as a government

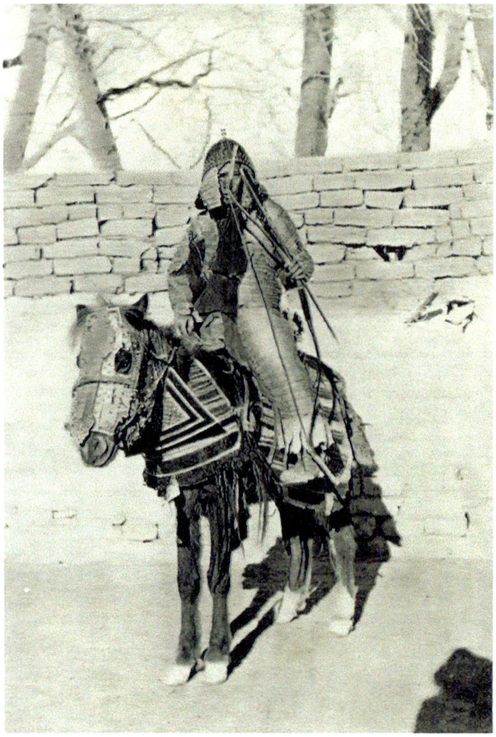

Figure 2 F.M. Bailey's photograph of the armour for the man and horse he purchased in Tibet and sold to the Royal Scottish Museum, Edinburgh, in 1909 (reproduced from W. Hutchinson (ed.), *Customs of the World*, London, 1913).

official and privately bought many Tibetan objects with the express purpose of selling them to the Edinburgh museum or other collections in the UK (Figure 2).[5] Bailey's letters from 1907 and 1908 to his parents in Edinburgh, preserved in the British Library, include several references to specific pieces, what he paid for them, how and when they would be shipped from Tibet, and for which buyers they were intended. Writing from Gyantse to his mother on 2 February 1908, Bailey comments about the armour soon to be sold to the Royal Scottish Museum:

> I have bought a suit of armour for a man and also for a horse. The horse armour is in very good condition except for the head piece which has some of the small plates missing and I can replace them. I will try to get the man's armour repaired. Waddell has a picture of a man in armour on a horse [see Figure 1] but the horse has only armour in front and mine is a complete set.[6]

A complete Tibetan horse armour was gifted by the 13th Dalai Lama in 1910 to the British diplomat Sir Charles Bell, who loaned it to the Victoria and Albert Museum, London, in 1913 and subsequently donated it in 1933.[7] This, the horse armour in Edinburgh, and two other composite examples acquired by the V&A in 1910, remain the only essentially complete Tibetan horse armours known: all other examples so far recorded comprise partial sets or individual elements, emphasising the rarity of this material both in and outside of Tibet.[8]

Tibetan horse armours

Tibetan horse armours are composed of individual pieces or panels with a textile or leather lining. On some armours these panels are covered mostly in iron lamellae like those used on Tibetan lamellar armour for men. The other principal type of panels, sometimes used in conjunction with rows of iron lamellae or pierced iron plates, are made of bands of hard leather, usually stitched together with fine copper wire and leather laces, and decorated on the exterior with a distinctive technique involving gold leaf imbedded in layers of shellac, simulating the appearance of lacquer.[9]

Based on the armours cited above, we can say in general that Tibetan horse armour comprises a shaffron covering the horse's head from the poll or forehead to the muzzle; a pair of wing or paddle-like panels to protect either side of the horse's neck; a panel at the front of the chest, which in some instances consists of a pair of panels extending over the horse's shoulders; a long rectangular panel that extends back from the saddle and sits on the top of the rump or croup; a panel across the back of the rump beneath the tail, sometimes with a hole through which the tail extends; a pair of large panels covering the sides of the rump and hips; and sometimes small panels beneath the side edges of the saddle, analogous to flanchards on a European horse armour.

Because we lacked the historical, literary, or iconographic sources necessary to provide firm

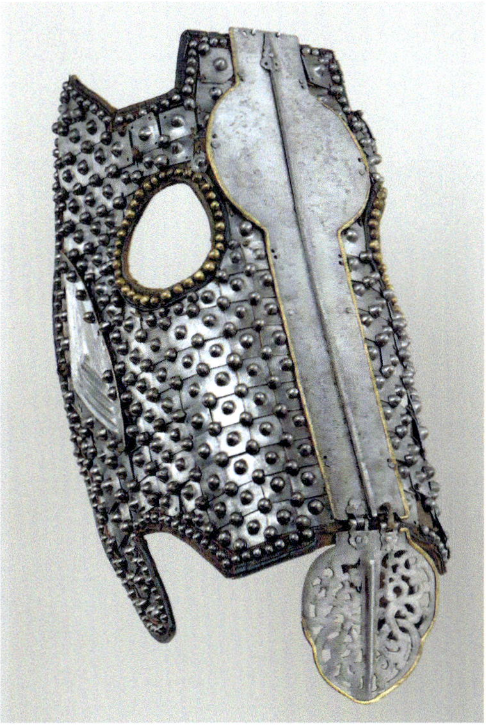

Figure 3 Shaffron, Tibetan or Mongolian, 15th–17th century, iron, leather, brass or copper alloy, h. 21 in. (53.3 cm). The Metropolitan Museum of Art, New York, Purchase, The Collection of Giovanni P. Morosini, presented by his daughter, Giulia, by exchange; Bashford Dean Memorial Collection, Funds from various donors, by exchange; and Fletcher Fund, by exchange, 1997 (1997.242d). Photo: Oi-Cheong Lee.

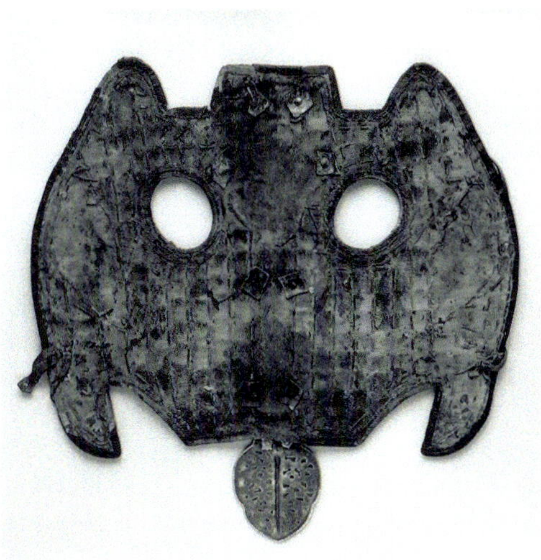

Figure 4 Interior of shaffron 1997.242d. Photo: Oi-Cheong Lee.

parameters for dating this type of armour, five carbon-14 tests were carried out on samples taken from four different pieces of Tibetan horse armour in the collection of the Metropolitan Museum of Art, New York, yielding a consistent date range from the early 15th century to the mid-17th century.[10] It is probably no coincidence that this corresponds closely to a period of frequent warfare during the last secular Tibetan monarchies before the theocracy of the Dalai Lamas, supported by Mongol allies, was consolidated in the mid-17th century.[11] Despite a statement published in 1905 by L.A. Waddell, a prominent participant in the Younghusband Expedition, claiming that some Tibetan officers rode armoured horses, all other evidence indicates that horse armour was not used in a military context during the period of the expedition.[12] In a ceremonial context, however, horse armour was included among the historical arms and armour worn in the military reviews that were held in Lhasa annually until at least the mid-20th century, as part of the month-long Tibetan New Year festival.[13] As actual fighting equipment, this material probably went out of use in Tibet during the 17th and 18th centuries, following the model of more lightly armoured cavalry used by Mongol and Manchu armies.

Tibetan shaffrons

Often, a shaffron, whether from Tibet or any other culture, is the only element of a horse armour that stands on its own as a complete or individual work; just as a helmet from a man's armour is more easily valued and understood as a dramatic and independent object than any other part of an armour. Surviving examples of Tibetan shaffrons are rare, but the majority of them resemble one basic type with, however, a few significant variations: all have a supporting structure or ground consisting of a piece of pliable leather shaped to the contours of a horse's head, with two circular eye openings.[14] On what is assumed to be the standard or most typical type (Figures 3 and 4), the exterior is covered with a series of small square iron plates with the centre of each plate raised to form a hollow boss. The plates are fastened to the ground by leather

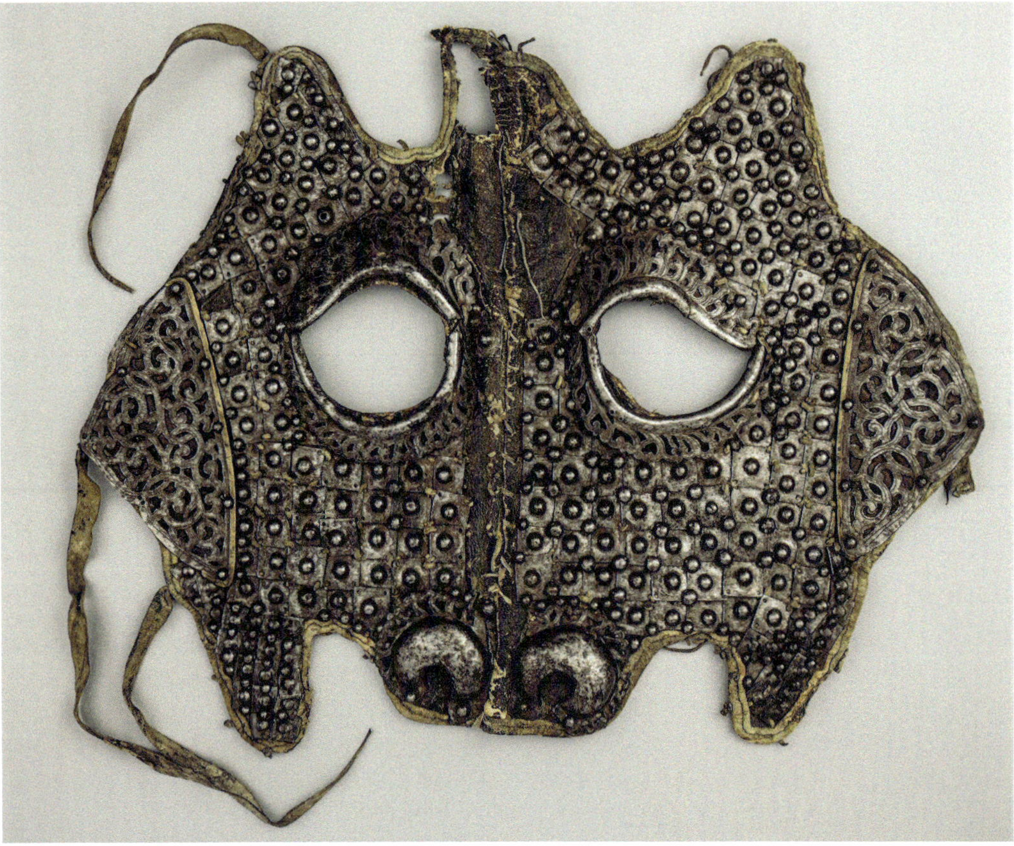

Figure 5 Shaffron, Tibetan or Mongolian, 16th–17th century, iron, leather, copper alloy, shellac, 22¾ in. × 17½ in. (57.8 × 44.5 cm): after conservation treatment in March 2021 by Sean Belair. The Metropolitan Museum of Art, New York, Purchase, Arthur Ochs Gift, by exchange, 2014 (2014.50). Photo: Stephen Bluto.

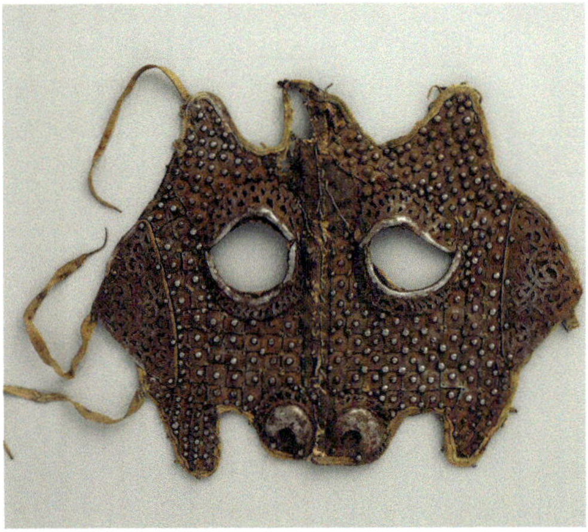

Figure 6 Shaffron 2014.50 as it appeared when acquired in 2014, prior to conservation treatment in 2021. Photo: Stephen Bluto.

laces threaded through a single hole in the corner of each plate, which loops over a transverse post inside a small hollow boss that reinforces the plates where their corners meet, giving a studded appearance to the exterior. A series of these bosses encircles the eye openings and edge the perimeter overall. A single, straight-sided, long and narrow plate covers the centre from forehead to snout, expanding to a circular section between the eye openings. There is generally a hinged and pierced plate at the base of this piece, which would dangle over the snout, with a matching plate at the top that would lay back onto the poll. The edges of the leather around the perimeter and the eye openings are trimmed with a strip of shagreen. In most examples there is a solid or pierced plate over the area of each cheek.[15]

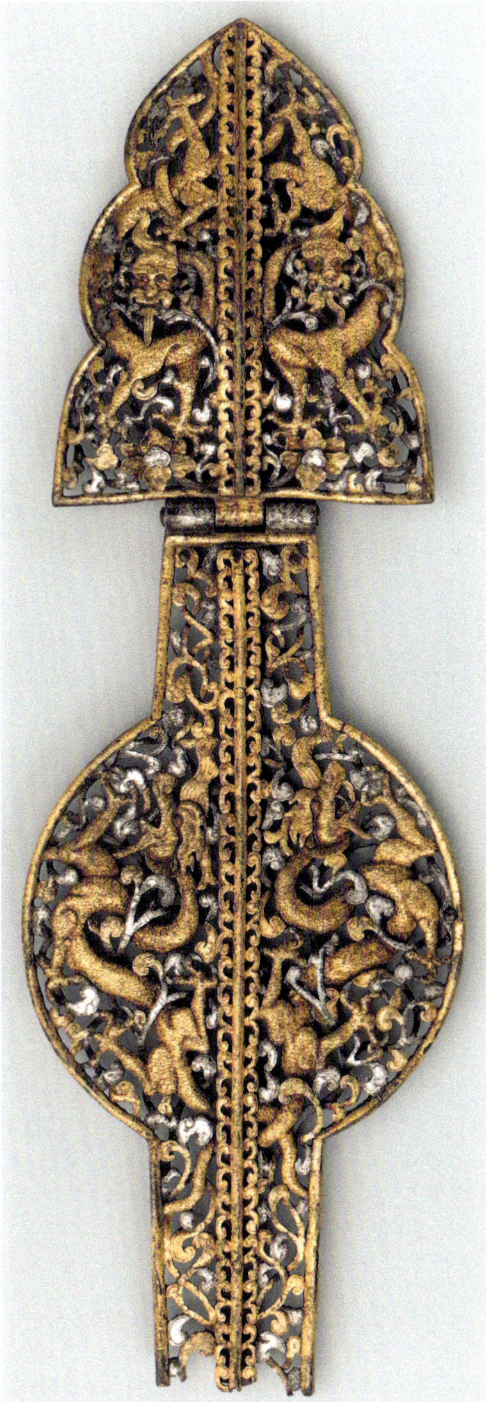

Figure 7 Centre piece from a shaffron, Tibetan, 16th–17th century, iron, gold, silver, 16¼ in. (41.3 cm). The Metropolitan Museum of Art, New York, Purchase, Arthur Ochs Sulzberger Gift, 2008 (2008.291). Photo: Juan Trujillo.

While sharing the basic features just described, the shaffron illustrated in Figures 5 and 6 is far from typical. Despite its worn and damaged condition, it includes some distinctive characteristics that are not found on any others encountered thus far, particularly the type of enframements encircling the eyes and the use of embossed plates representing nostrils.[16] To appreciate the significance of this shaffron, however, it is necessary to understand the extent of its damage, and the distortion to its appearance as a result. Most disfiguring is the loss of the shaffron's central plate and at least some of the leather underlying it, which were roughly removed at some point, cutting the originally one-piece leather ground into two halves that were subsequently rejoined by a crudely stitched vertical seam. As a consequence of losing its literal centrepiece, the shaffron was inappropriately narrowed when the two halves were sewn back together, throwing off its proportions, setting the eyes too close together, and noticeably distorting the top edge.

It is very likely that the missing centre piece resembled the type seen in Figure 3, complete with poll and snout plates that were pierced with scrollwork such as that seen on the cheek plates of the damaged shaffron and the remaining snout plate on the shaffron in Figure 3. On a rare and stunningly beautiful variation of a shaffron centre piece, a detached example in the Met is made of relatively thick iron that has been deeply pierced and chiselled, then damascened overall in gold and silver, in the style and technique normally associated with Tibetan and Sino-Tibetan saddle plates (Figure 7).[17] In another example in the Met we see the most cohesive, comprehensive, and finely rendered programme of embellishment on a complete Tibetan shaffron (Figure 8), in which the centre piece, cheek pieces, and poll plate (and presumably the missing snout plate) are entirely covered in a dense reticular pattern damascened in gold and silver.[18]

This last example also has the extraordinary feature of lively flaming eyebrows, crisply chiselled in iron, the flames damascened in gold and rising from a contoured base damascened in silver. While different in style and cruder in execution, the frames encircling the eyes of the damaged shaffron follow a similar theme,

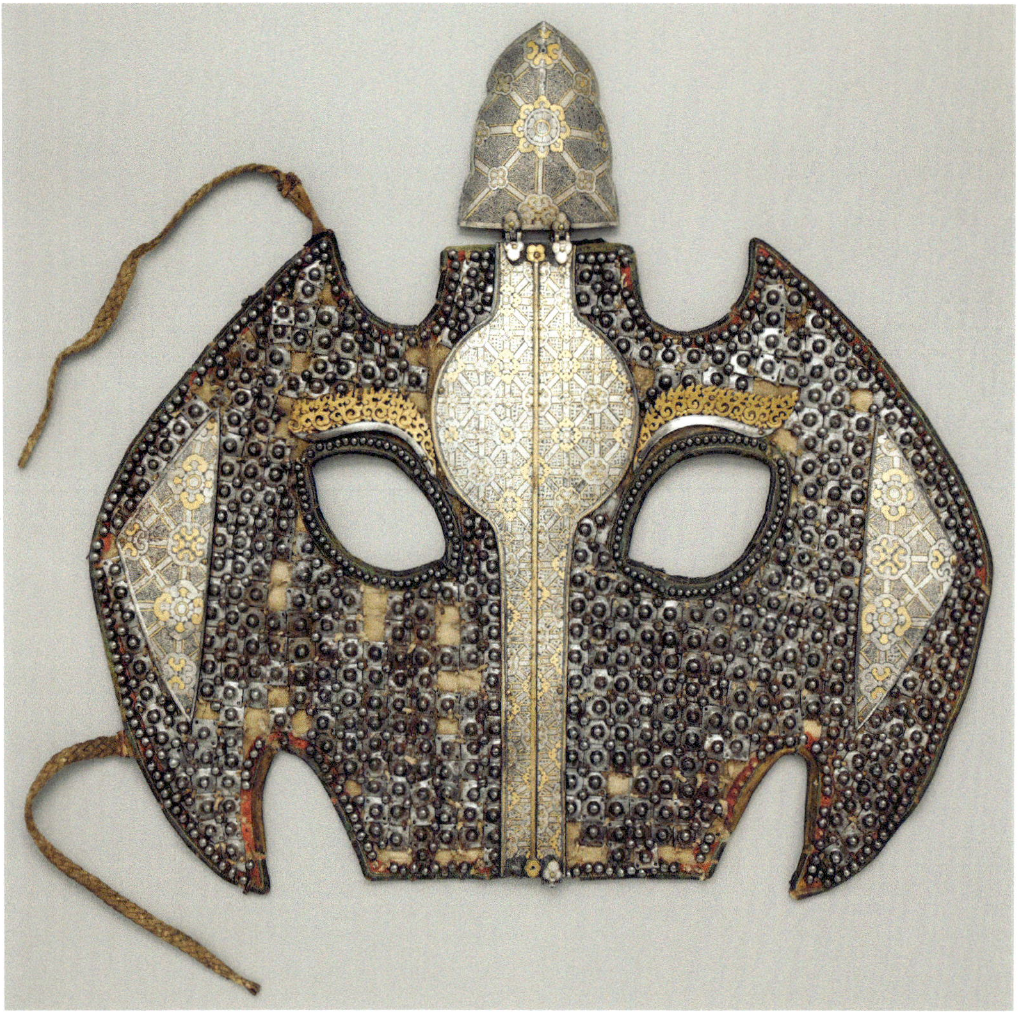

Figure 8 Shaffron, Tibetan or Mongolian, 15th–17th century, iron, leather, gold, silver, brass or copper alloy, textile. The Metropolitan Museum of Art, New York, Purchase, Arthur Ochs Sulzberger Gift, 2004 (2004.402). Photo: Oi-Cheong Lee.

making these the only two Tibetan shaffrons with eye embellishments beyond the typical border of hemispherical bosses. On the latter, however, the flames resemble the scrollwork on its adjacent cheek pieces more than the animated tongues of fire seen on the former. Nevertheless, both shaffrons have a striking parallel in a superb Tibetan war mask, with its rimmed eyes and flaming eyebrows evoking the face of a fierce guardian deity (Figure 9). The latter shaffron also has the unique feature of nostrils in the form of a pair of crescent-shaped fittings with an upper rim of pierced flames or scrollwork, as on the enframement of the eyes. Representing flaring nostrils snorting flame or steam, these fittings further transform the shaffron into the mask of a dragon or other fearsome mythical animal.

Although altered and mistreated in the past, when viewed in the context of other pertinent examples, this shaffron can be recognised as highly significant for the unusual features it exhibits and for increasing our knowledge of the nuances and variety that once existed in the decoration and construction of Himalayan horse armour.

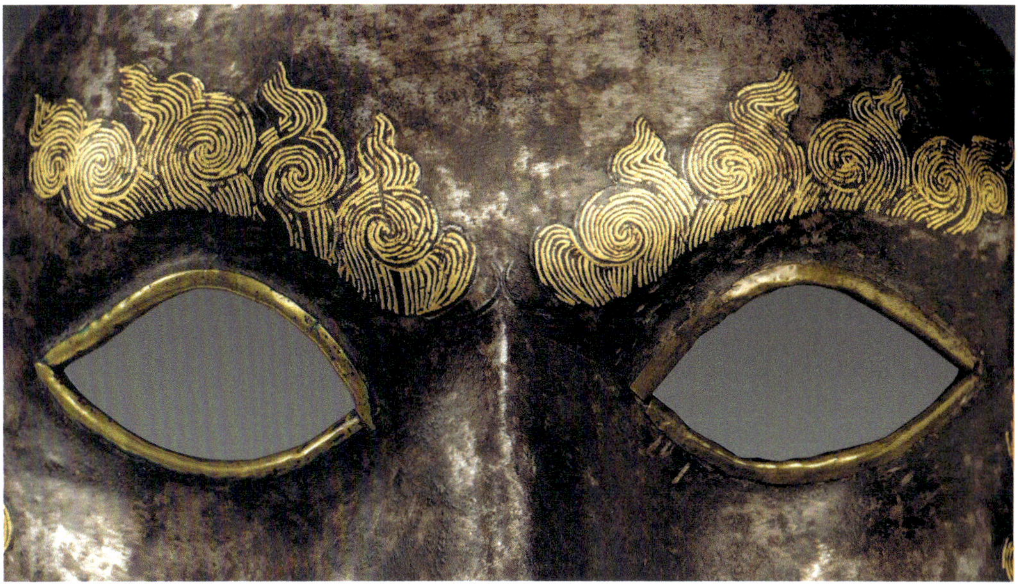

Figure 9 War mask, Tibetan, 14th–16th century, iron, gold, copper alloy, 7 11/16 in. (19.5 cm): detail. The Metropolitan Museum of Art, New York, Promised Gift of Steven Kossak, The Kronos Collections, in celebration of the Museum's 150th Anniversary. Photo: Oi-Cheong Lee.

Notes and references

1. For an overview of horse armour in Tibet see La Rocca, D.J. (2006) *Warriors of the Himalayas: Rediscovering the Arms and Armor of Tibet*. New York: Yale University Press, pp. 96–7.
2. Rockhill, W.W. (1895) 'Notes on the Ethnology of Tibet: Based on the Collections of the United States National Museum', in *Annual Report of the Board of Regents of the Smithsonian Institution ... for the Year Ending 1893: Report of the U.S. National Museum*. Rockhill includes very useful examples of Tibetan firearms, swords, and spears, but makes no mention of armour (for man or horse).
3. Laufer, B. (1914) *Chinese Clay Figures, Part I: Prolegomena on the History of Defensive Armor*. Field Museum of Natural History Publication 177, Anthropological Series, vol. 13, no. 2. Chicago. There are literary citations throughout the paper, but regarding horse armour see especially pp. 234, 280–81 and 306–15.
4. On the Younghusband Expedition, see Allen, C. (2004) *Duel in the Snows: The True Story of the Younghusband Mission to Lhasa*. London: John Murray.
5. La Rocca 2006 (cited in note 1), cat. no. 32. On the fascinating career of Bailey (1882–1967), an accomplished soldier and linguist, see McKay, A. (1997) *Tibet and the British Raj: The Frontier Cadre, 1904–1947*. Richmond, UK: Curzon Press. See also Peter Hopkirk's introduction to Bailey, F.M. (2002) *Mission to Tashkent*. Oxford: Oxford University Press.
6. British Library, MSS Eur F157/168 Letters 1908.
7. See La Rocca 2006 (cited in note 1), cat. no. 26.
8. Ibid., cat. no. 26. The other two composite examples in the V&A are inv. nos. IM.72&A–J-1910 and IM.73&A–P-1910: *op. cit.* pp. 96–7 and note 5.
9. On this technique, unique to leather armour from Tibet, see ibid., p. 96, esp. p. 116, n. 7, and passim.
10. On the carbon-14 testing see ibid., p. 96, n. 9 and the appendix compiled by Edward A. Hunter on p. 288.
11. For a summary see ibid., pp. 8–9; and for a detailed discussion on the role of Mongol forces see Venturi, F. (2020) 'Mongol and Tibetan armies on the Trans-Himalayan fronts in the second half of the 17th century, with a focus on the autobiography of the Fifth Dalai Lama', *Revue d'Etudes Tibétaines* 53: 29–55.
12. I am grateful to Alice Travers and Federica Venturi for alerting me to the fact that Waddell's statements regarding the use of historical armour by Tibetan troops at that time have been called into question by later research (personal communication, August and September 2020). For Waddell's observations on Tibetan arms and armour see La Rocca 2006 (cited in note 1), pp. 3–6.

13. For this part of the ceremonies, see Richardson, H. (1993) *Ceremonies of the Lhasa Year*. London: Serindia Publications, esp. pp. 31–7; Karsten, J. (1983) 'A note on *ya sor* and the secular festivals following the *sMon lam chen mo*', in E. Steinkellner and H. Tauscher (eds), *Contributions on Tibetan Language, History and Culture*. Vol. I of *Proceedings of the Csoma de Körös Symposium, held at Velm-Vienna, Austria, 13–19 September 1981*, pp. 117–49; and La Rocca 2006 (cited in note 1), pp. 6–7.
14. Only about a dozen examples are known to me. Beyond the six cited in La Rocca 2006 (under cat. nos. 26–29) there are: Edinburgh A.1909.407 A, part of the armour acquired from Bailey cited above; a shaffron in a private collection in the USA; a shaffron formerly in the collection of Eliane and Guy de la Boisselière, sold at Czerny's, 14 September 2018, lot 112; a shaffron in the Chojor Ancient Weapons Museum, Metok Gongkar County, Tibet; an example similar to Met shaffron 1997.242d, seen in a private collection in London in 1997; and the shaffron discussed below, Metropolitan Museum of Art acc. no. 2014.50.
15. For more details of Tibetan shaffron construction see the descriptions in La Rocca 2006 (cited in note 1), pp. 96–7, 100 and 102.
16. I am very grateful to Sean Belair, Assistant Conservator in the Department of Arms and Armor, for his skilful cleaning and conservation of the shaffron in preparation for this article.
17. La Rocca, D.J. (2014) 'Recent acquisitions of Tibetan and Mongolian arms and armor in the Metropolitan Museum of Art. Part 2', *Waffen- und Kostümkunde* 2: 187–210. pp. 199–200.
18. For a detailed description see La Rocca 2006 (cited in note 1), pp. 100–101.

6

Mannequins for armours: a conservational and aesthetic challenge

Petra Fuchs

The Imperial Armoury of the Kunsthistorisches Museum Vienna boasts one of the finest collections of armour in the world. Having agreed to loan selected armours for a number of national and international exhibitions between 2006 and 2010, we decided to improve and adapt some of the old, often unstable and inadequate mannequins used to display these armours. Our main focus was on conservation aspects such as ensuring the tension-free installation of the individual pieces which comprise the armour on the mannequin.

Complex armours function as carefully composed attire, which is why our conservators attempted to enhance the visual appearance of the wooden mannequins on which the armours are installed. The only way to do justice to the armourer's concept is to display the different pieces comprising an armour exactly how they were devised and intended to be worn. Mere adjustments, however, rarely prove satisfactory as they generally represent a compromise rather than an ideal solution, therefore in 2011 the decision was taken to collaborate with the sculptor, Bernhard Ortner, to develop new 'bespoke' display mannequins designed both to fit the size of each armour (i.e. its original owner) and accommodate conservation and aesthetic considerations. In 2016, the team of the Imperial Armoury began another collaboration with the sculptor, Gabriel Rauchegger.

Producing mannequins that also accommodated the demands of conservation proved difficult, time-consuming and expensive: many of these newly commissioned mannequins for our harnesses were financed with the help of those museums which had requested the loans. For each new mannequin we first needed to identify a human model whose size and proportions were similar to those of the armour's original (but now long-dead) owner. This model then carefully donned the individual pieces comprising the armour, which were secured with straps, before the correctly worn armour was carefully photographed from all sides. Finally, the sculptor took the model's exact measurements which, together with photographs, helped him to carve the new 'made-to-measure' limewood mannequin. Next, the finished mannequin was 'dressed' in the many pieces comprising the armour and a support construction was devised to prevent the full weight of the armour's individual (and in parts extremely heavy) pieces having to be borne solely by the original leather straps. Dressing the mannequin also had to avoid creating tensions or stresses that might have endangered any of the armour's original leather or textile elements. Each mannequin's (visible) attire and its footwear reflect contemporary fashion. The choice of colours used for the final coat of paint was informed by the original owner's heraldic colours or document his elevated rank.

The mannequins commissioned in 2011–12 include a mounted mannequin with a portrait head of Ferdinand II, Archduke of Austria (1529–1595) that was made for the so-called 'Milanese

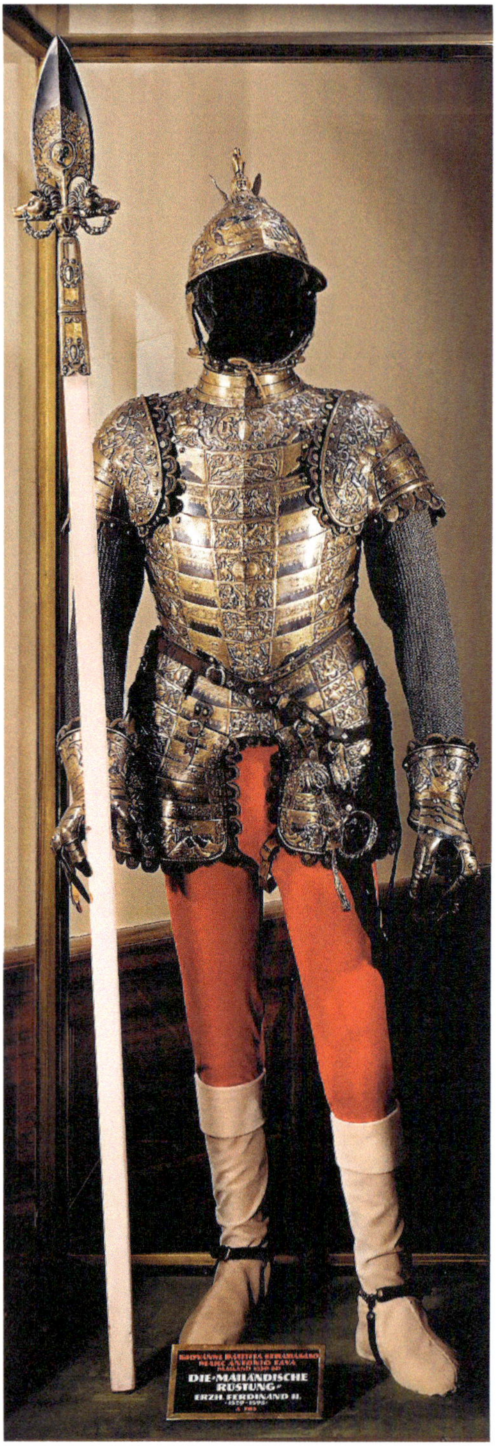

Figure 1 Armour of the 'Milanese Armour' made for Ferdinand II, Archduke of Austria, 1529–1595, produced by Giovani Battista Panzeri called Serabaglio (A 785): earlier installation on a standing mannequin.

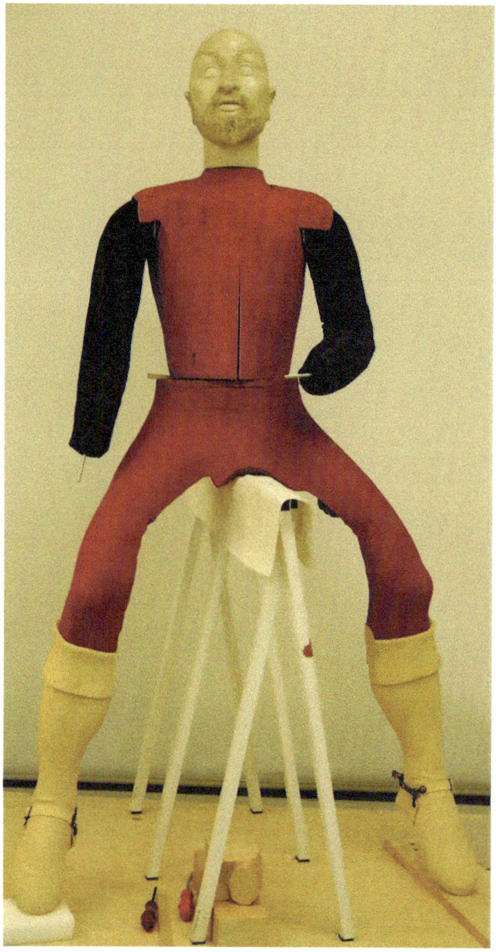

Figure 2 New mounted mannequin to display the armour of the 'Milanese Armour' (A 785) astride a charger.

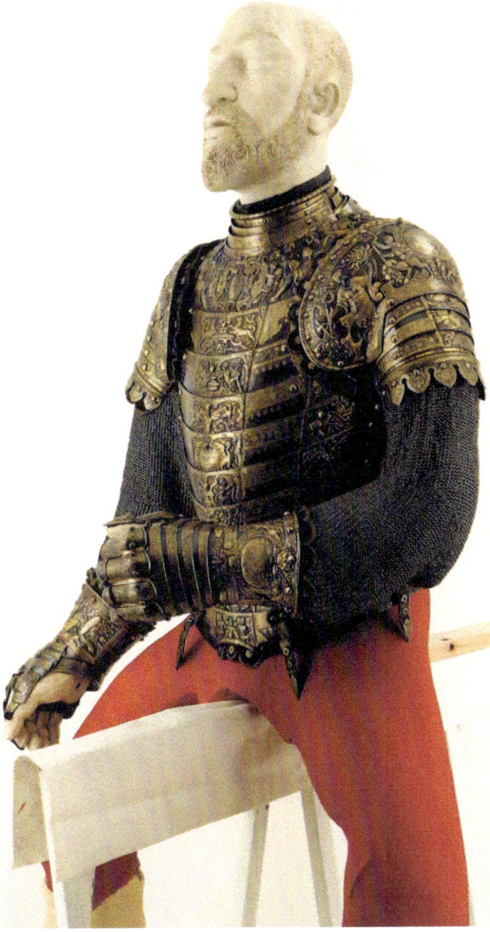

Figure 3 Pieces of the 'Milanese Armour' (A 785) installed on the torso of its new 'bespoke' mannequin.

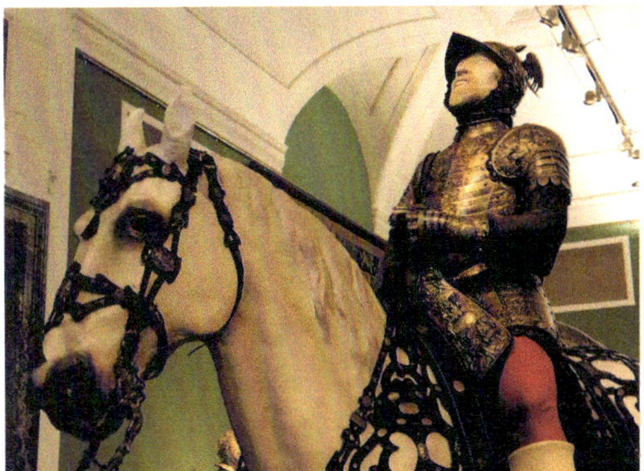

Figure 4 Armour of the 'Milanese Armour' (A 785) installed on its new mannequin on display in the galleries of the Imperial Armoury of the Kunsthistorisches Museum Vienna.

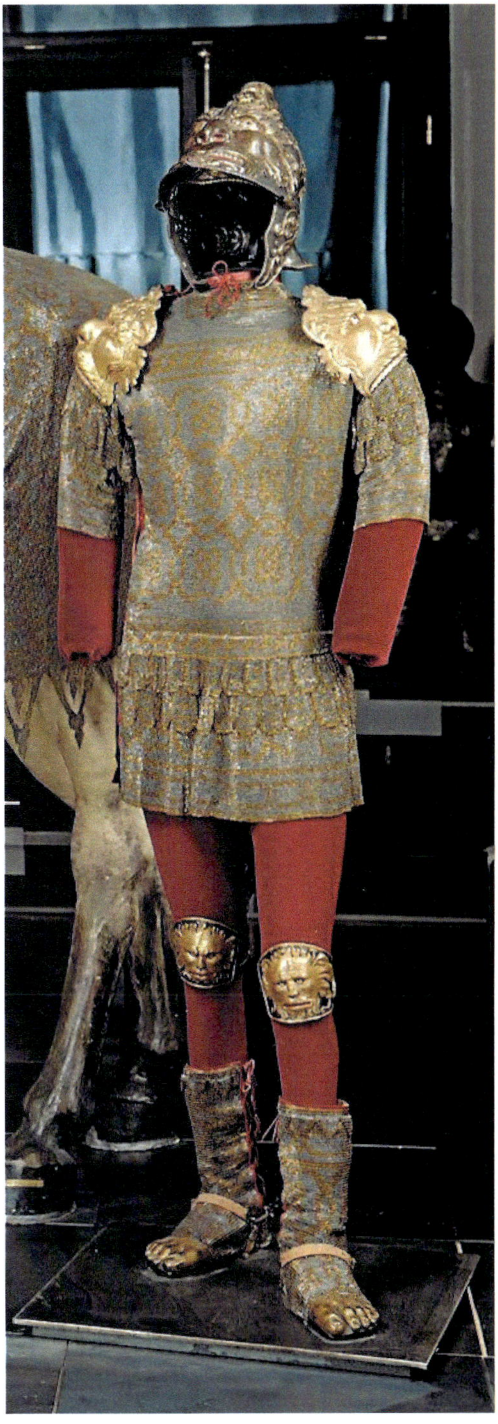

Figure 5 Armour of the 'Roman Armour' made for Ferdinand II, Archduke of Austria, 1529–1595, in the style of Giovan Paolo Negroli (A 783) installed on an old, unstable mannequin made primarily of foam. The entire weight of the chain mail is borne by the shoulders.

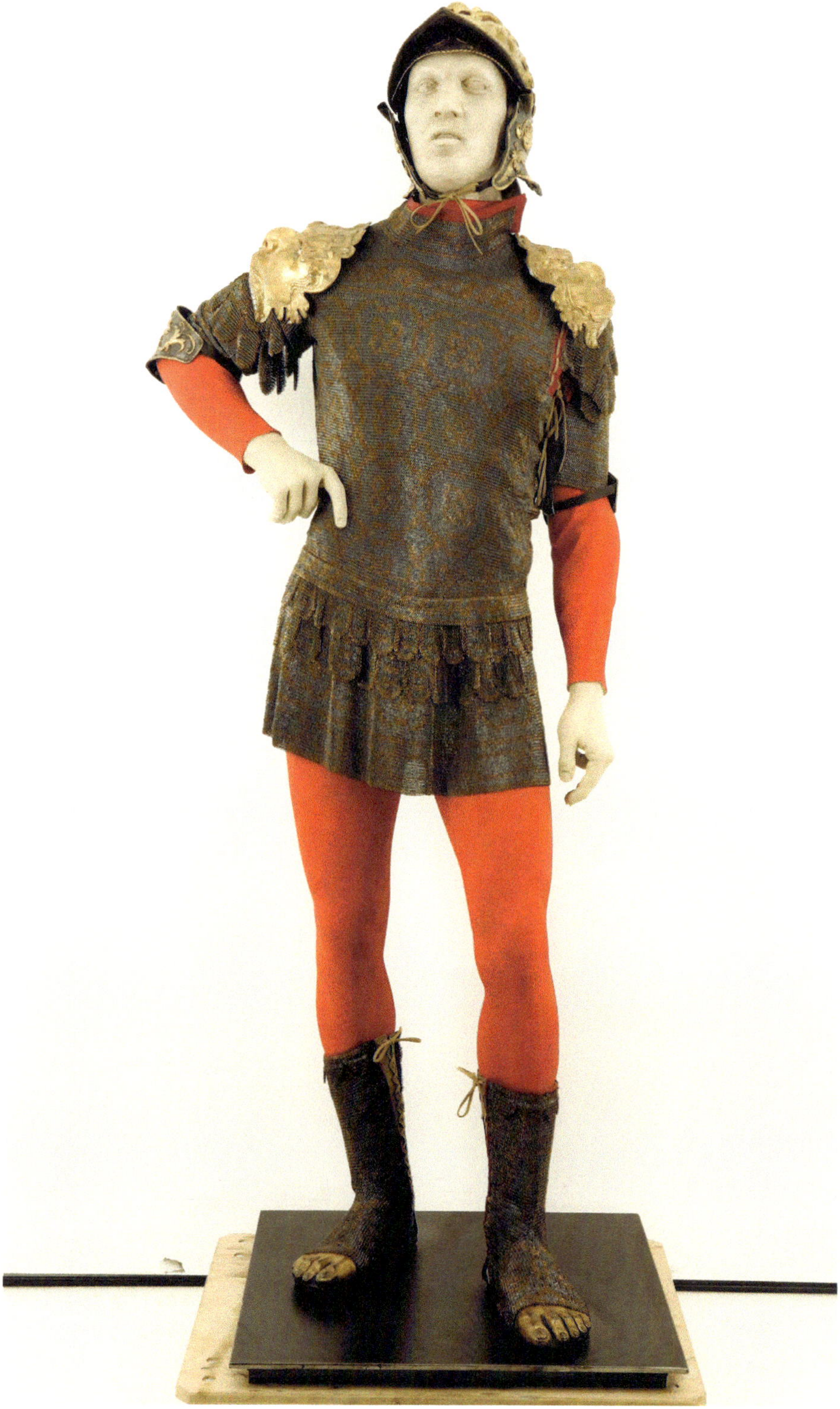

Figure 6 Armour of the 'Roman Armour' (A 783) installed on its new mannequin.

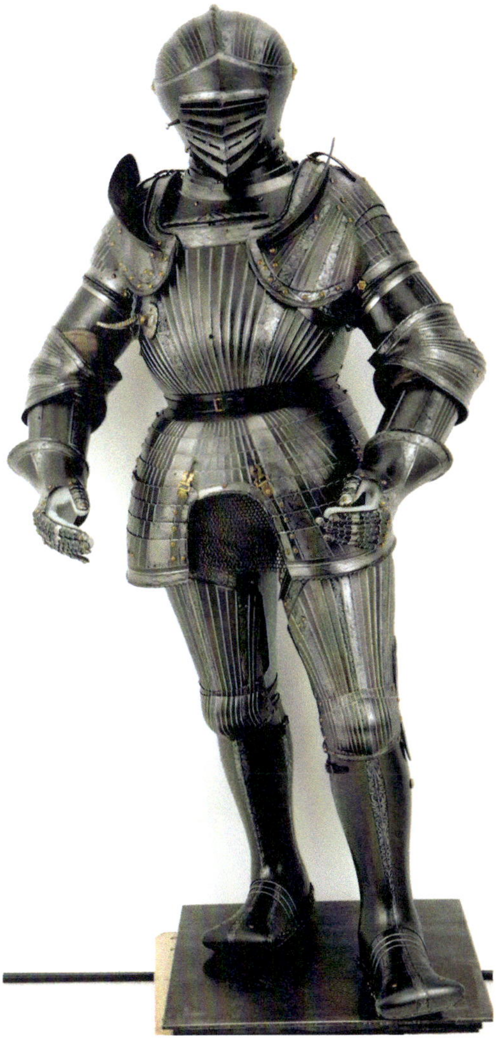

Figure 7 'Fluted Armour' of John Frederick I, Elector of Saxony-Wettin, 1503–1554 (A 347): earlier installation on a stable mannequin.

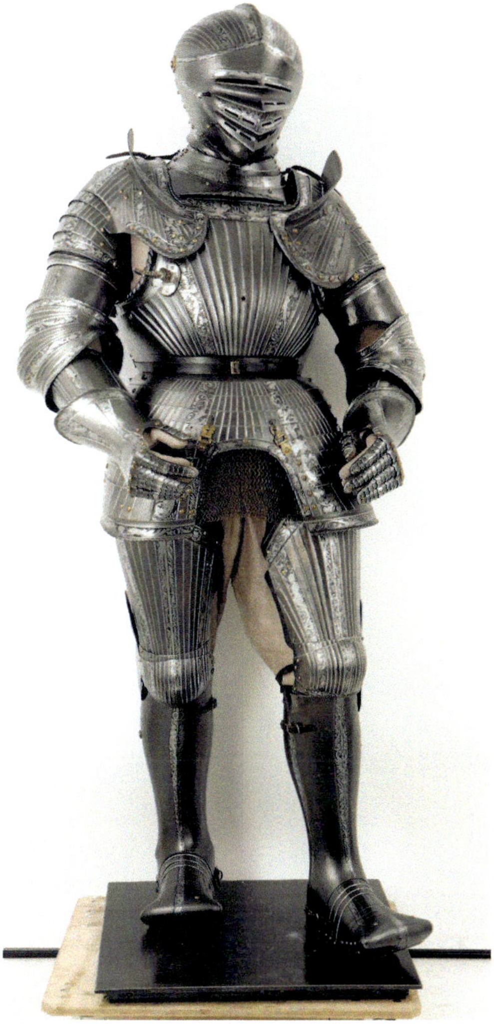

Figure 8 'Fluted Armour' (A 347) installed on its new mannequin designed to accommodate the harness's requirements.

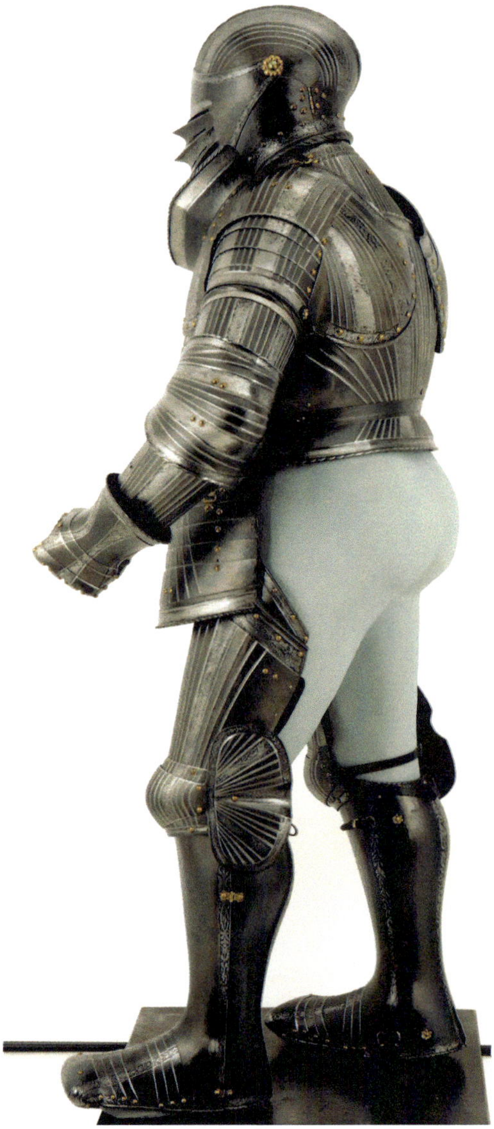

Figure 9 'Fluted Armour' (A 347): side view before the new mannequin was commissioned.

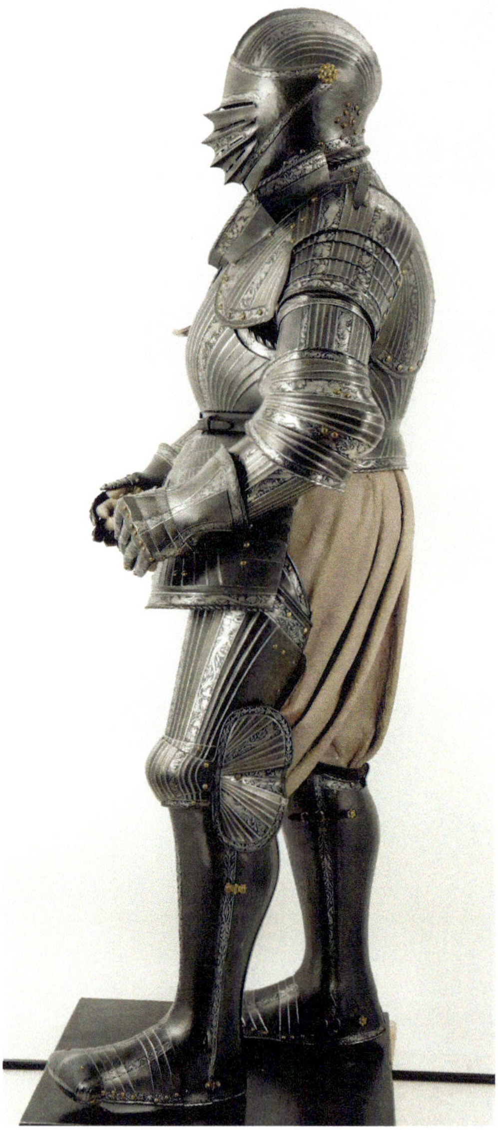

Figure 10 'Fluted Armour' (A 347): side view installed on its new mannequin.

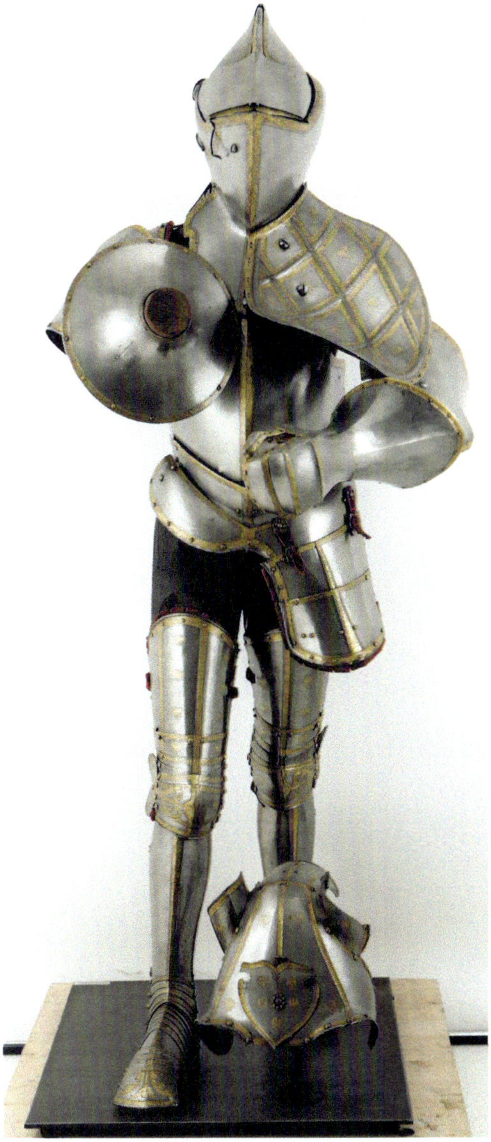

Figure 11 'Armour for the Tilt' from the 'Eagle Garniture' (A 638): old mannequin, front view.

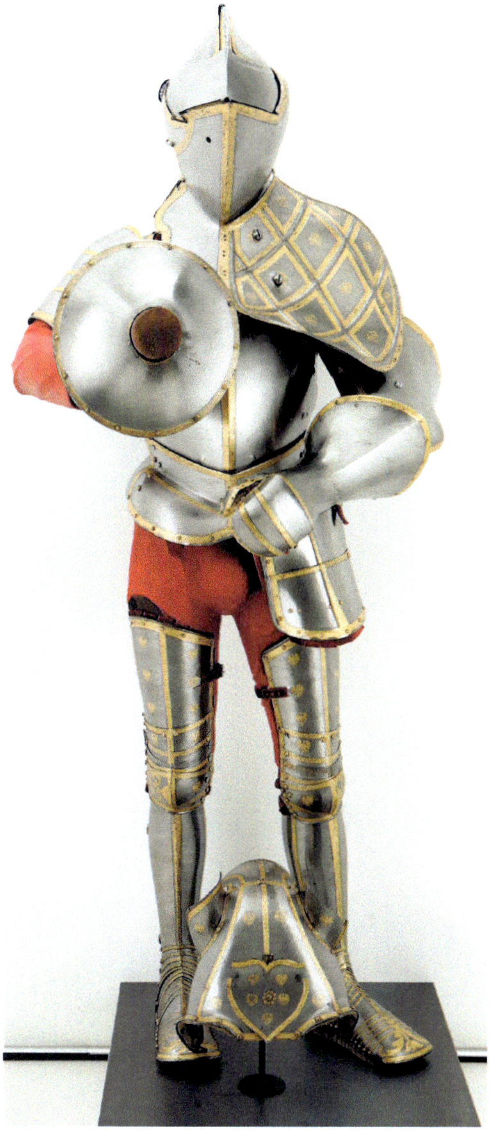

Figure 12 'Armour for the Tilt' from the 'Eagle Garniture' (A 638): installed on its new mannequin, front view.

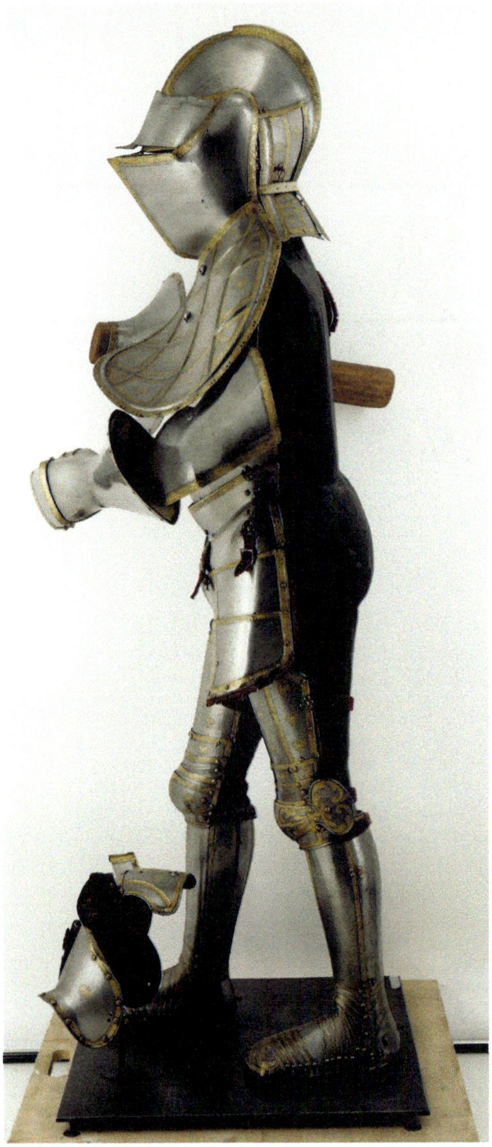

Figure 13 'Armour for the Tilt' from the 'Eagle Garniture' (A 638): old mannequin, side view.

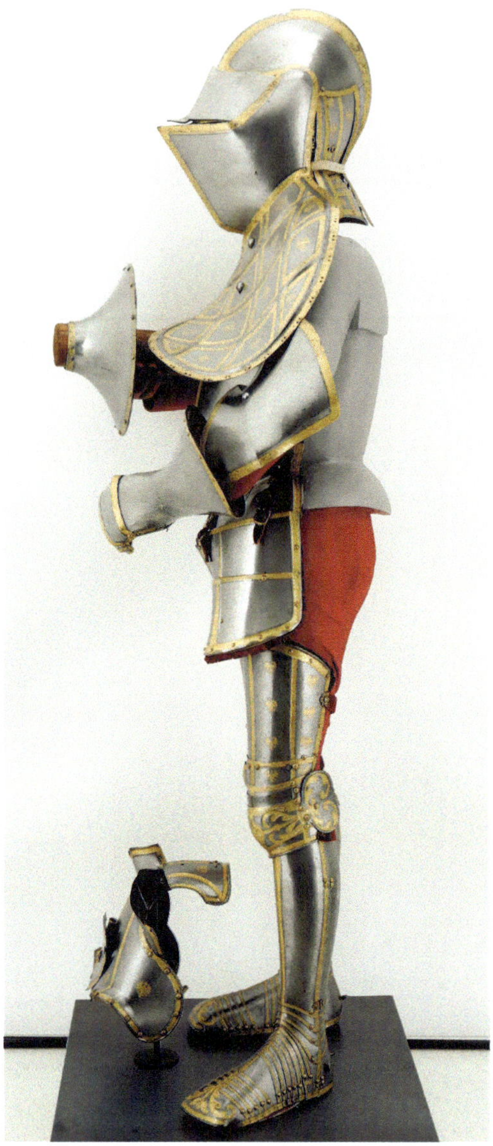

Figure 14 'Armour for the Tilt' from the 'Eagle Garniture' (A 638) installed on its new mannequin, side view.

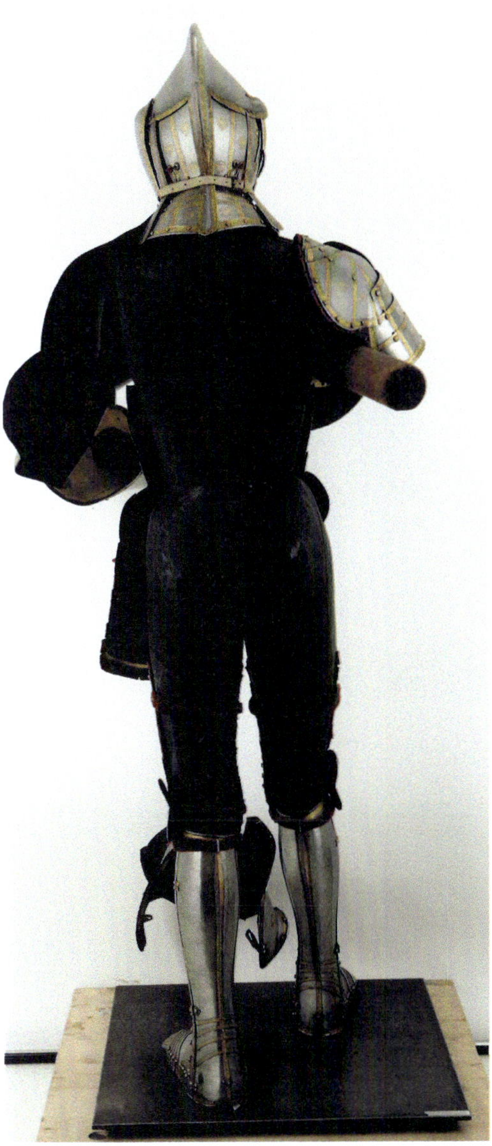

Figure 15 'Armour for the Tilt' from the 'Eagle Garniture' (A 638): old mannequin, rear view.

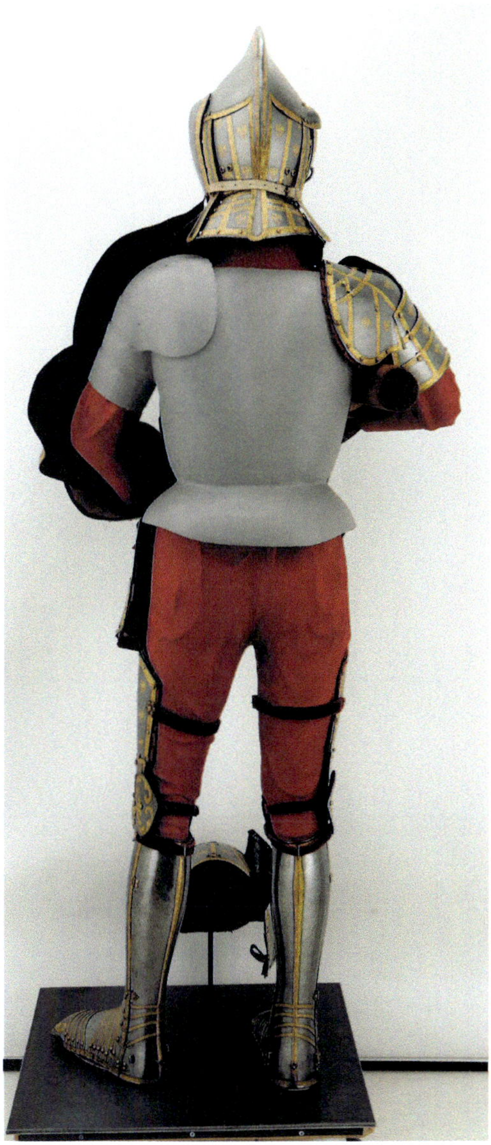

Figure 16 'Armour for the Tilt' from the 'Eagle Garniture' (A 638) installed on its new mannequin, rear view.

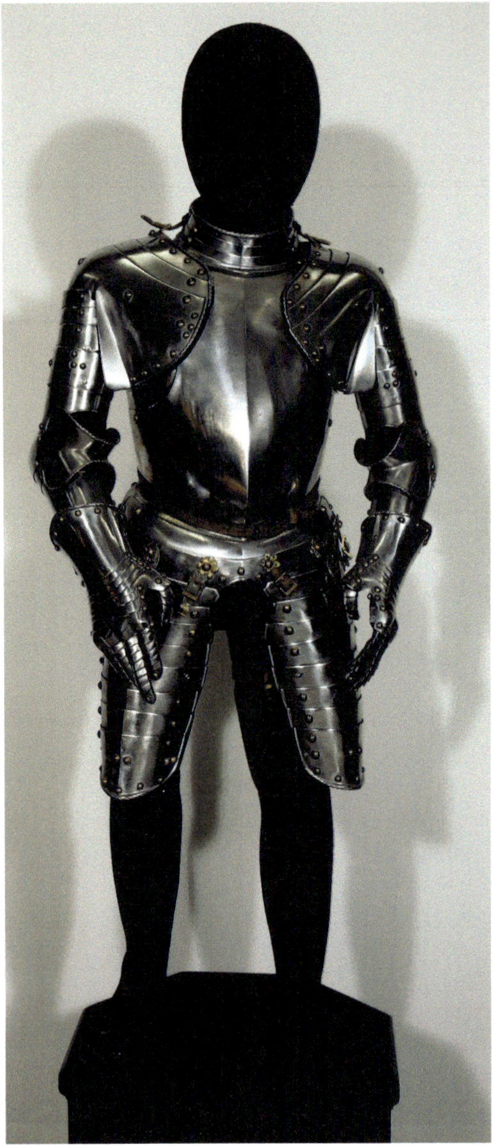

Figure 17 'Boy's Armour of Charles, Archduke of Austria (1560–1618)' (A 1186): armour pieces installed on an old mannequin made of disintegrating foam, front view.

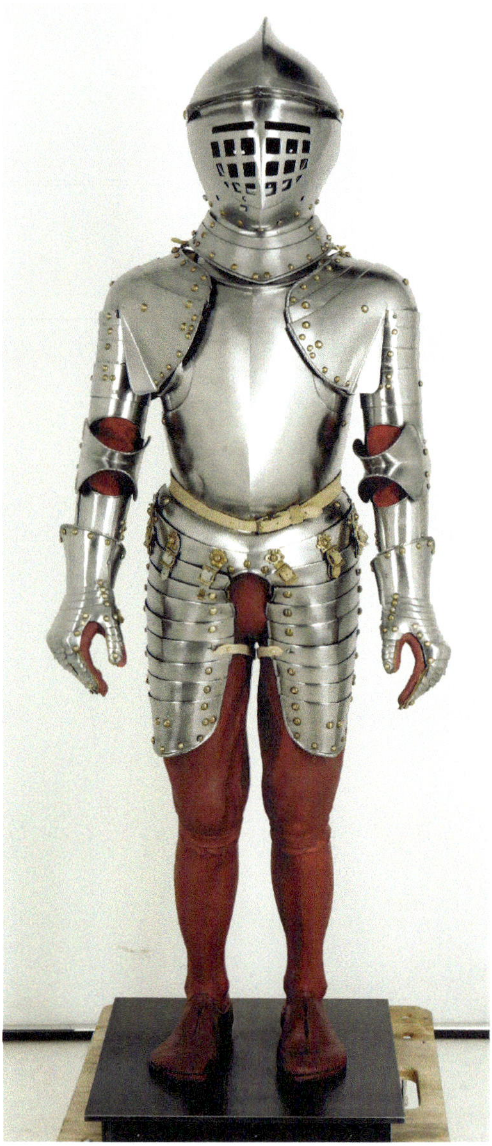

Figure 18 'Boy's Armour of Charles, Archduke of Austria (1560–1618)' (A 1186): armour pieces installed on its new mannequin, front view.

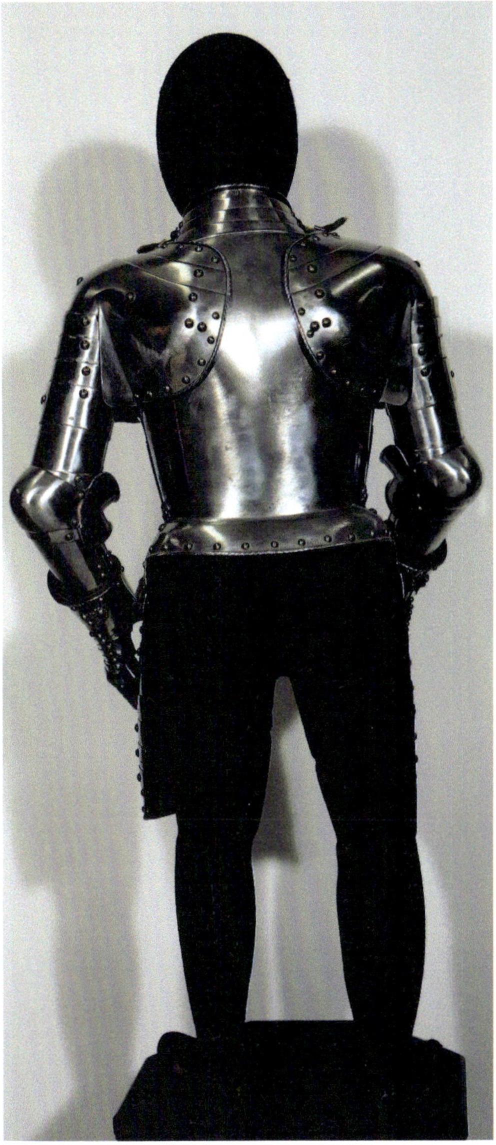

Figure 19 'Boy's Armour of Charles, Archduke of Austria (1560–1618)' (A 1186): armour pieces installed on an old mannequin made of disintegrating foam, rear view.

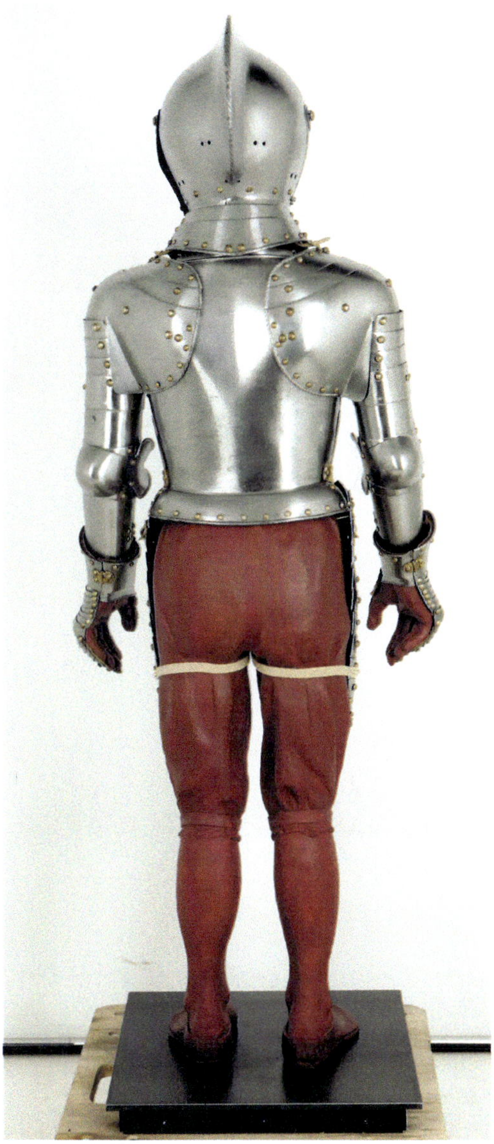

Figure 20 'Boy's Armour of Charles, Archduke of Austria (1560–1618)' (A 1186): armour pieces installed on its new mannequin, rear view.

Armour' (A 785) (Figures 1–4), and a standing youth for the 'Roman Armour' (A 783) (Figures 5 and 6) with a portrait head of Ferdinand II as a young man.

In order to facilitate handling and installation, especially for travelling/transport, this mounted mannequin comprises two halves. The lower part (fashioned after the human model) is placed astride the horse before the fully attired torso, complete with screwed-on arms and hands, is lowered onto the bottom half. Thus assembled, the mounted mannequin's weight is perfectly balanced and stable, requiring no other means to secure it.

The body shape of the standing youth made for the 'Roman Armour' (A 783) is the result of several fittings and exact measurements. Unlike the mounted dummy sporting rigid armour pieces discussed above, the body of this mannequin needs to accommodate the fluid forms of the flexible chain mail that does not determine the body's silhouette. The mannequin's dynamic construction, with one hip jutting out, ensures that the full weight of the shirt is not borne by the shoulders alone (Figures 5 and 6).

Between 2013 and 2020, we were able to commission additional mannequins for armours in the Imperial Armoury, realising long sought-after aesthetic and conservational improvements, fulfilling important conservation aims and making it easier (and safer) to lend these precious armours.

We have selected three recently commissioned new mannequins: the one created for the 'Fluted Armour' (A 347) of John Frederick I, Elector of Saxony (1503–1554) (Figures 7–10); the 'Armour for the Tilt' from the 'Eagle Garniture' (A 638) of Archduke Ferdinand II (Figures 11–16); and the 'Boy's Armour of Charles, Archduke of Austria (1560–1618)' (A 1186) (Figures 17–20). We have also commissioned new mannequins for 2021 including one for the 'Boy's Armour of Emperor Charles V (1500–1558)' (A 186), financed with the help of a crowdfunding campaign set up in collaboration with the Museum Hof van Busleyden in Malines, Brussels, which will host an exhibition in the spring of 2021 that will include this precious armour, beautifully displayed on its brand-new mannequin.

7

Six armours of the National Museum of Finland

Lasse Mattila

Introduction

The National Museum of Finland has six armours[1] that together form a group of late 16th- and early 17th-century armours, all of which relate to Finnish history and known historical figures. The latest owners, and their lives as noblemen and officers, of the 17th-century armours are known – as, to some extent, are their family and social relationships. They were all put on display as funeral achievements in the parish churches where the owners were buried during the 17th or early 18th century. Later, during the 19th century, the armours were donated to the predecessor of the current museum. Due to the renewal of the exhibitions in 2016, I planned and carried out the conservation of all the armours between 2017 and 2019, including a study of their known history, in close co-operation with the museum conservators. All their parts were thoroughly cleaned and rebuilt with new leather straps if required. New display stands completed the conservation project.

Three lives of an armour

All the armours in the National Museum of Finland seem to have been made in Northern Europe during the late 16th and early 17th century. Three of the armours are full cuirassier armours, one is a lighter half-armour and another is a less complete armour and thus difficult to define exactly, but it was apparently originally a full cuirassier armour. Such plate armours are common in the museums and armour collections of Europe,[2] however, one differs greatly from the others: although it resembles a full cuirassier armour it is made of copper alloy, therefore it was not an armour made principally for war but intended for a different purpose.

The Finnish region, part of the Swedish Empire from the Middle Ages until the beginning of the 19th century, was engaged in lengthy wars during the 17th century with Russia, Denmark, Poland and the German states. The northern countries had no significant home-based armour manufacture, so officers' armours probably originated mainly from workshops elsewhere in Europe or perhaps represent war booty. It seems that the armours of the National Museum of Finland have just such an origin.

Based on my studies of the history of the museum's armours, I suggest that as objects they had three lives: even though they differ from each other, all are important in their own ways. All lives affect the armour and its preservation to a certain degree: the first covers its use during its owner's lifetime, the second connects to the armour's role in sustaining local memory and manifestation of power in the parish church, and in its third life the armour becomes a historical object of common memory in the museum.

An armour's first life, when new or still relatively new, was a period during which it was used as protective battle gear after which it acquired the status of a prestige object both within the army and wider society. The armour was repaired and treated in several ways to maintain its useful qualities. The time span covered by this period is difficult to estimate, but it is possible to determine the date each armour was manufactured with some accuracy as well as the date of the owner's death. Assuming that owners began their military careers around the age of 20, the armour's first life lasted somewhere between 34 and 55 years.

The second life of the museum's armours was longer than the first, spanning approximately 225 years, except in one case. During this period, the armour was part of the funeral achievements of the owner's tomb or placed near it in the parish church. The armour no longer functioned as protective gear or an item of prestige, but as an object of commemoration of the owner and his family at a local level. How the armours were maintained during this period is not known, but based on my studies it seems that dust was removed to some extent and some armours may have been painted black. The second life began when the last owner was buried and ended when the congregation decided to donate the armour to the predecessor of the National Museum of Finland.

The third and final life of the museum's armours is currently ongoing. It is a life as a museum object that recalls the military history of the past, a period that began towards the end of the 19th century and has already covered about 140 years. During these years in the museum, the maintenance of the armours changed greatly. Many traces of earlier restorations and conservation can be seen but our knowledge of such earlier activities is limited due to a lack of documentation. Repairs and other measures are connected to the museum's history and its exhibitions, such as its move to a new exhibition building in 1910. The Second World War also impacted exhibition activities and required steps to be taken to protect the collections. More recently, developments in conservation science have also changed the way the armours are treated.

All three lives leave their marks on the armours' metal surfaces and their construction: close inspection reveals many signs of restoration and repairs from different periods. Numerous paint layers applied to protect the metal surface affect their general appearance. Empty rivet holes and different rivet types indicate changes of the original leather straps and rivets: before conservation, weakened leather straps and supports made of iron wire were found.

Close study of the signs of old repairs are both important and interesting: some are so good that they are difficult to discern but others are more obvious and appear to have been hurriedly carried out, perhaps between battles. This kind of repair may belong to the first life of the armour. Some repairs that seem to have been made without the skills and knowledge of an armourer may belong to the armour's second life in the parish church or even its third life in the museum. Painting the complete armour in such a way that layers of paint cover the rivets and leather straps and vambraces, and gauntlets fixed together by rivets which prevent them from moving are both indicative of later repairs. Nevertheless, every attempt to maintain the armours through the centuries has been beneficial and facilitated their preservation to the present day.

H 455 Three-quarter composite armour of the Commander of the Olavinlinna Castle Abraham Pistolekors (1635–1699)

Date:	1600–1650
Place:	North Europe
Material:	Iron or steel, leather and black paint
Weight:	22.816 kg total weight
Artist:	Unknown artist/maker
Provenance:	In the parish church of Sulkava from 1699 to 1860 and in the museum collection since 1860
Stamps:	None visible

Structure and decorations

A 17th-century composite armour with black-painted plates and iron rivets. The gauntlets may date from later centuries (Figure 1). The borders of the plates are turned but not roped. The plates are decorated with one or two incised parallel lines; the backplate has two parallel lines. A helmet, a

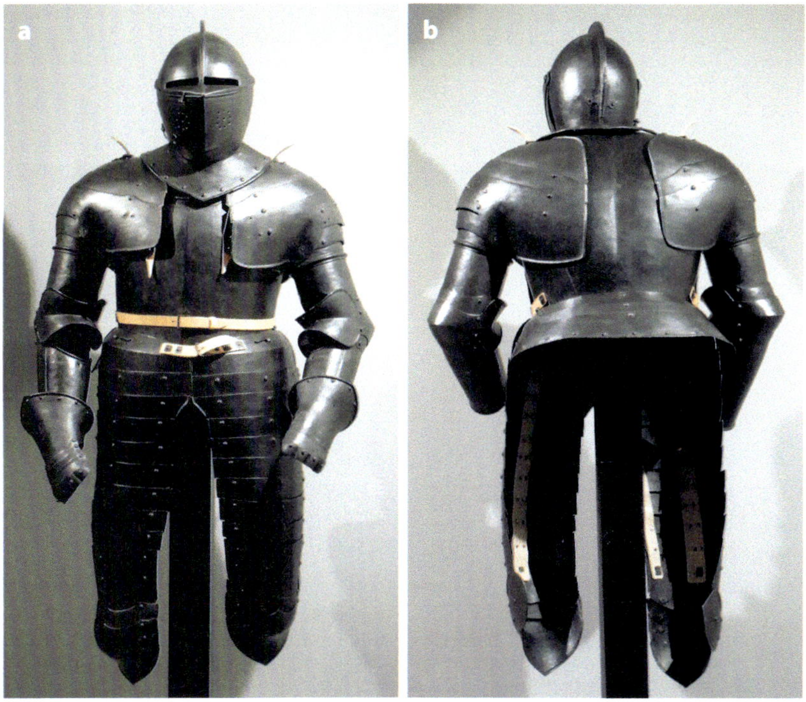

Figure 1 Abraham Pistolekors's armour (H 455) from the front (a) and back (b) after conservation.

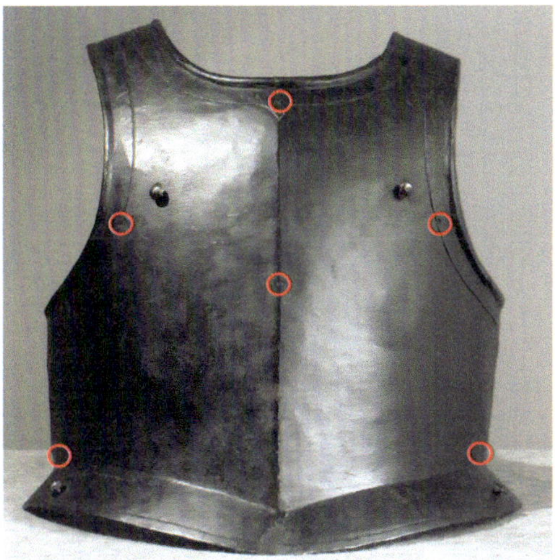

Figure 2 This armour's breastplate is marked with six single dots which lie symmetrically on the plate's surface. The dots could be measure marks made by an armourer during the forging of the breastplate.

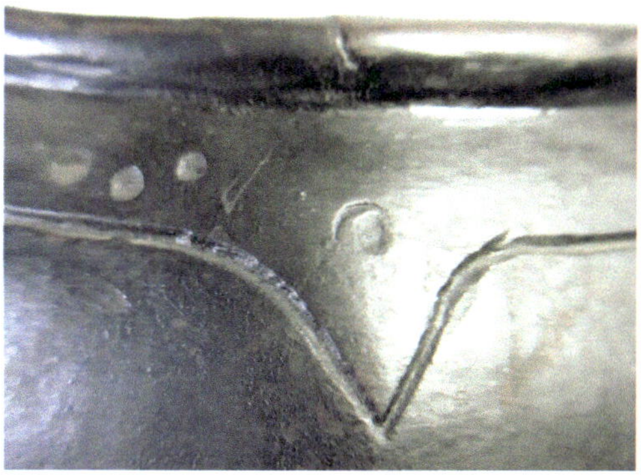

Figure 3 Group of three dots on the breastplate. They may indicate that an armour manufacturer numbered different parts of the armours during the manufacturing process.

breastplate, a culet and gauntlets have one line; the pauldrons and the vambraces have none.

Four parts of the armour are each marked with a group of small dots stamped on the outside surface of the plates (Figures 2 and 3): the backplate is marked with five dots, the right tasset with two and both the breastplate and the left tasset with three. The armour's parts that include three dots fit well together but not with the other parts. The groups of dots may have been made during the manufacturing process to indicate the numbering of different parts of the armour and their relation to each other. If this interpretation is correct, it seems that the museum's armour is a mixture of the parts from several different armours, which would explain why the right tasset is about 3 cm longer than the left. The breastplate is marked with six single dots – possibly measuring marks made by an armourer during forging – which lie symmetrically on the plate's surface.

H 1621 Three-quarter armour of Cavalry Captain Mauritz Kristersson Horn (1593?–1671)

Date: 1600–1650
Place: North Europe
Material: Iron or steel, leather and black paint
Weight: 22.924 kg total weight without a helmet; 26.422 kg with a great helmet H738
Artist: Unknown artist/maker
Provenance: In the parish churches of Uskela from 1671 to 1874 and in the museum collection since 1874. The close helmet is from the Vihti parish church and has been part of the museum collection since 1858
Stamps: Both pauldrons are stamped with an unidentified 8 ×10 mm stamp (Figure 5)

Structure and decorations

A 17th-century composite armour with plates and iron rivets painted in black. The borders of the plates, which are turned but not roped, are decorated with one incised line. In addition, there are parallel and diagonal incised lines over the breastplate and the backplate (Figure 4). Both pauldrons have two decorative rosettes, not set

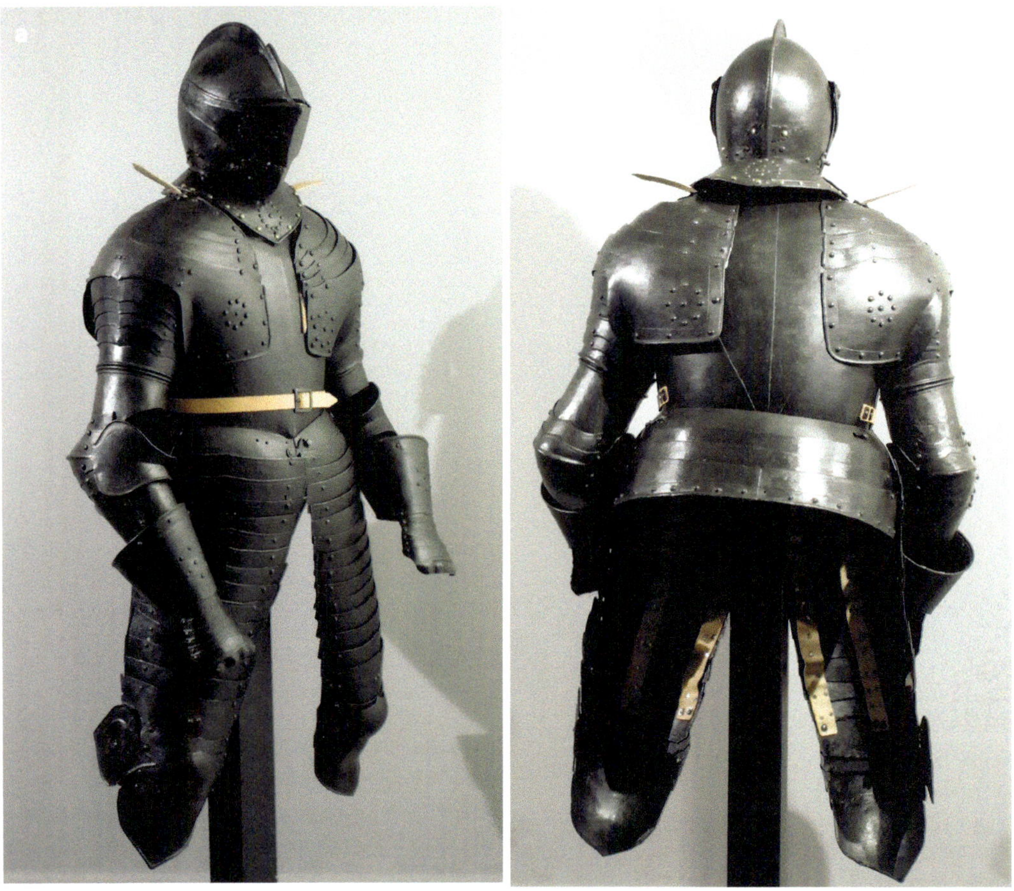

Figure 4 Mauritz Kristersson Horn's armour (H 1621) from the front (a) and back (b) after conservation.

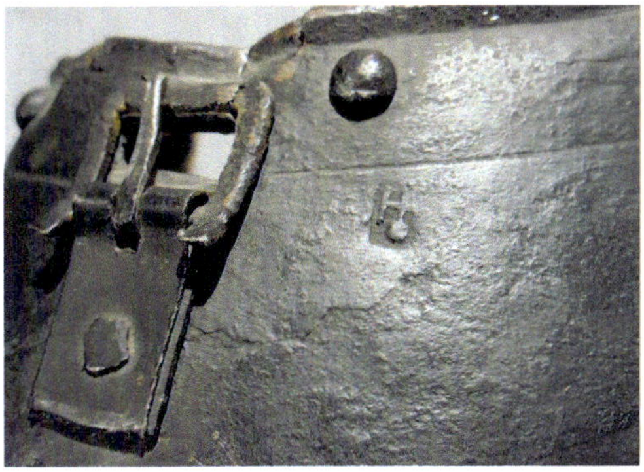

Figure 5 Both pauldrons are stamped with unidentified 8 × 10 mm stamp.

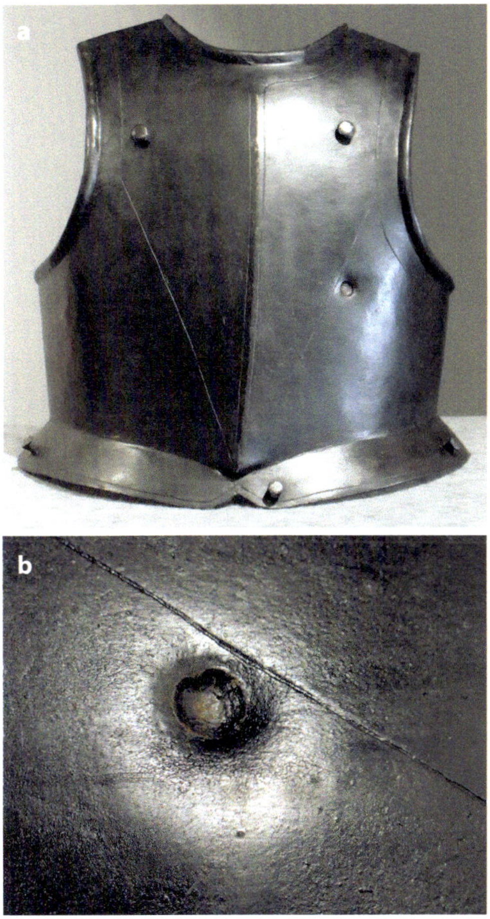

Figure 6 (a) The heavy breastplate with (b) a circular 'bullet dent' on the left-hand side.

symmetrically, comprising nine iron rivets with soldered copper alloy cups. A belt and a belt buckle are restorations carried out during conservation in 2019. The heavy (5.7 kg) breastplate has a circular 'bullet dent' on the left-hand side,[3] which can also be seen on the inside of the breastplate (Figure 6).

The two-piece gorget does not fit with the breastplate or the backplate. In addition, the decorations of the plates of the gorget differ compared to other parts of the armour. The borders of the plates are turned and have a very fine roped decoration. This is the only armour in the museum that also has a gorget. According to the museum's archives, the gorget has the same origin and inventory number as the other parts of the armour. It is possible that the armour's owner used the gorget separately and it was added to the armour later.

The composite helmet HM 738 is not an original part of the armour nor is its first owner known. The helmet is constructed of a well-made two-piece skull and five other plates. The lowest part of the visor is also well constructed and fits well to the skull, but the upper part and gorget lames, which also have some empty rivet holes, do not fit properly. In general, the visor and gorget seem to be from another helmet. Furthermore, the visor's middle plate appears to be a modern restoration. The plate is made of very thin iron plate and does not fit accurately to the other parts of the helmet.

H 1289:1 Three-quarter armour of Lieutenant Colonel Carl Bäck (1609–1663)

Date:	1600–1650
Place:	North Europe
Material:	Iron or steel, copper alloy, leather and black paint
Weight:	27.965 kg total weight
Artist:	Unknown artist/maker
Provenance:	In the parish church of Urjala from 1663 to 1873 and in the museum collection since 1873
Stamps:	None visible

Structure and decorations

An armour from the 17th century with black-painted plates (Figure 7), this is the heaviest of the museum's armours. The borders of the plates are

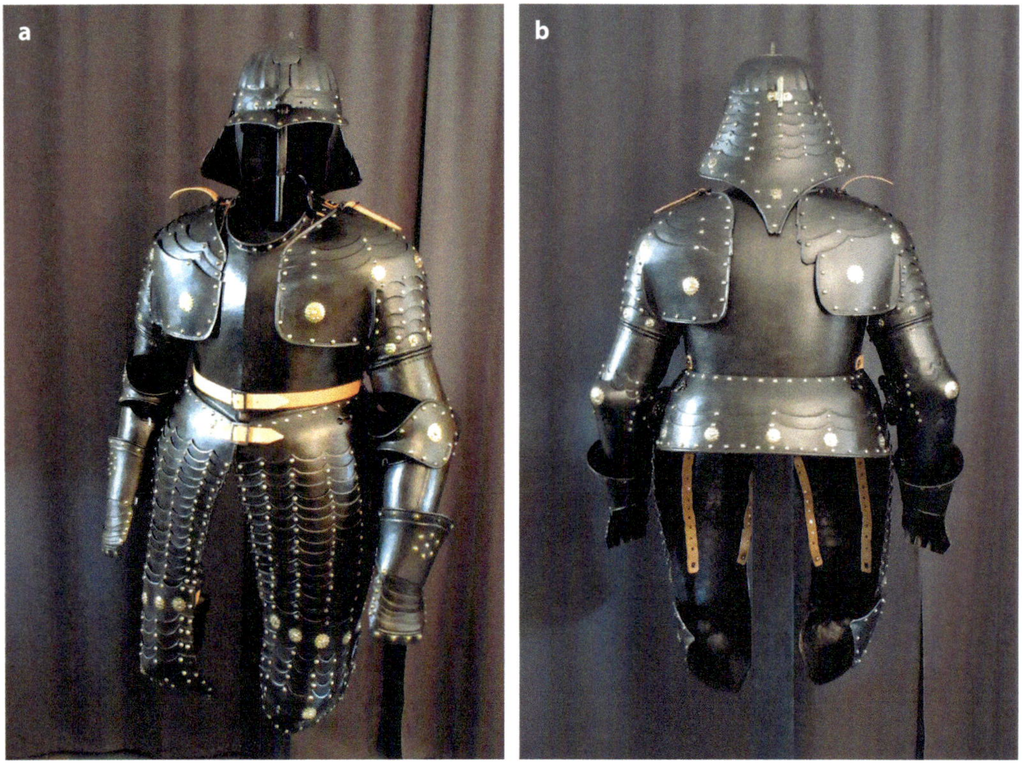

Figure 7 Carl Bäck's armour (H 1289:1) from the front (a) and back (b) after conservation.

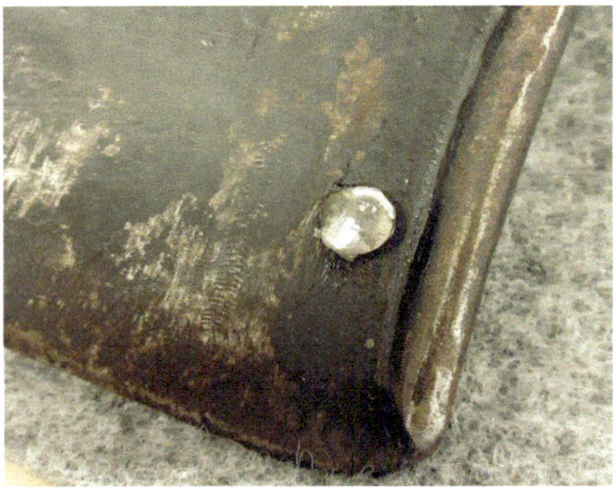

Figure 8 An engraved wrigglework decoration runs along on the borders of the breastplate, mostly covered by a layer of black paint.

turned but not roped. As a decoration there seems to be a double engraved wrigglework line running parallel on the borders of the plates except the helmet and gauntlets (Figure 8). Both wrigglework lines are 2.5 mm wide and lie 13 mm apart. As this decoration is under the black paint it is difficult to study thoroughly – it is possible that in some parts of the armour there is only a single wrigglework line. The borders of the gauntlets are decorated with one straight incised line. The only part of the armour without incised decorations is the Zischägge helmet.

Most of the iron rivets have soldered copper alloy caps. The armour also has 46 decorative copper alloy rosettes in the form of a star or sun. The rosettes are attached to the armour by iron rivets with soldered copper alloy caps. The shiny polished rosettes and rivet caps stand out on the black surface. As the small rosettes of the lower parts of the pauldrons are not as well made as the others, it is possible that they are not original but later additions.

The armour has a very heavy (7.5 kg) breastplate. The thickest measurement near the edge was 7 mm but it is thicker in the middle: the breastplate was X-rayed by the museum's conservator and seems to be a solid piece of iron without any layers or structures. The black paint has worn away from some plates and there appear to be signs of high-quality blueing: a peacock blue, probably blued when new. In general, it is likely that the armour has not always been painted black: the paint completely obliterates the wrigglework decoration although originally it was undoubtedly meant to be seen. Furthermore, the latest layer of paint covers the rivets and leather straps and some rosettes. It is possible that the latest paint layer was added during the armour's second life in the parish church or its third life in the museum.

The main plate of the right-hand side pauldron and one of the lames of the left-hand side pauldron are not original but were replaced or restored later. Both cheek pieces of the helmet appear to be replacements and the right-hand side gauntlet is possibly a restoration. Some of the replacements could belong to the end of the armour's first life when the owner Lieutenant Colonel Carl Bäck was still alive. During his long career as an officer, he was involved in battles that might have required the protection of armour. In a cavalry regiment he held the ranks of lieutenant in 1637, a captain in 1638, a major in 1643 and a lieutenant colonel in 1647. He fought in the Thirty Years' War (1618–1648) and in the war against Poland (1655–1659) during which he was seriously wounded in 1658, by a bullet to his arm and possibly a lance thrust to his chest from which he never recovered. He died of his wounds in 1663.[4]

It is possible that the main plate of the right-hand side pauldron was repaired because the wearer was wounded in battle: this plate differs in size and is not a perfect fit to the pauldron. In addition, the rivets are not iron but copper alloy. The repair may have been carried out at the time of the funeral when the armour was placed in the parish church.

H 1072 Half-armour of Cavalry Captain Gustaf Falkenfelt (?–1704)

Date:	1600–1650
Place:	North Europe
Material:	Iron or steel, copper alloy, leather and black paint
Weight:	20.45 kg total weight
Artist:	Unknown artist/maker

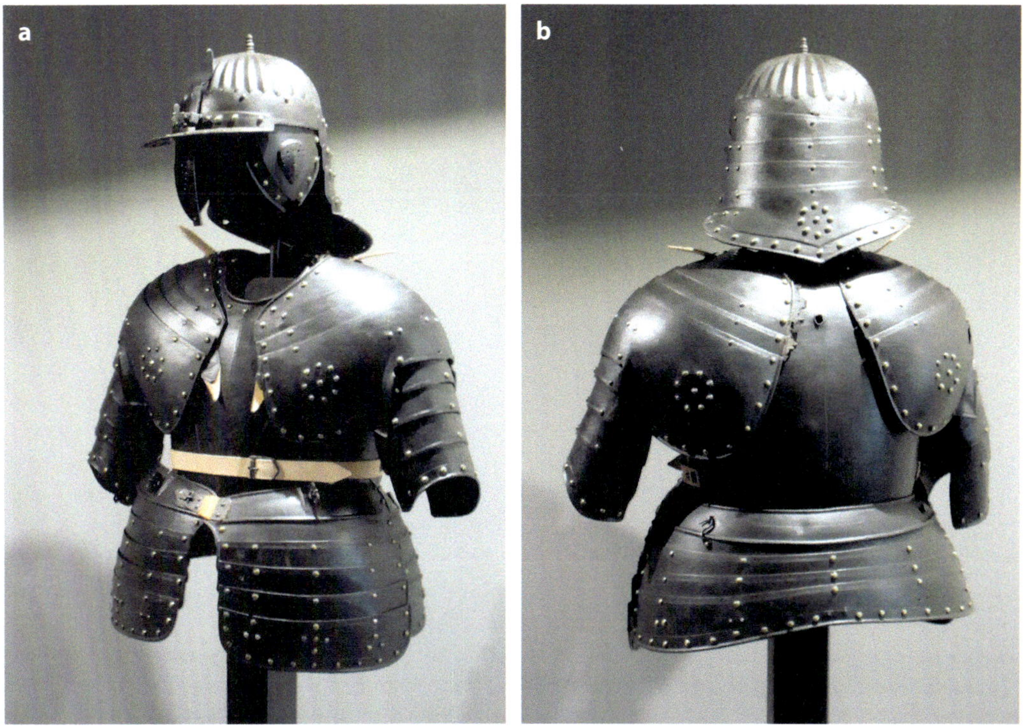

Figure 9 Gustav Falkenfelt's armour (H 1072) from the front (a) and back (b) after conservation.

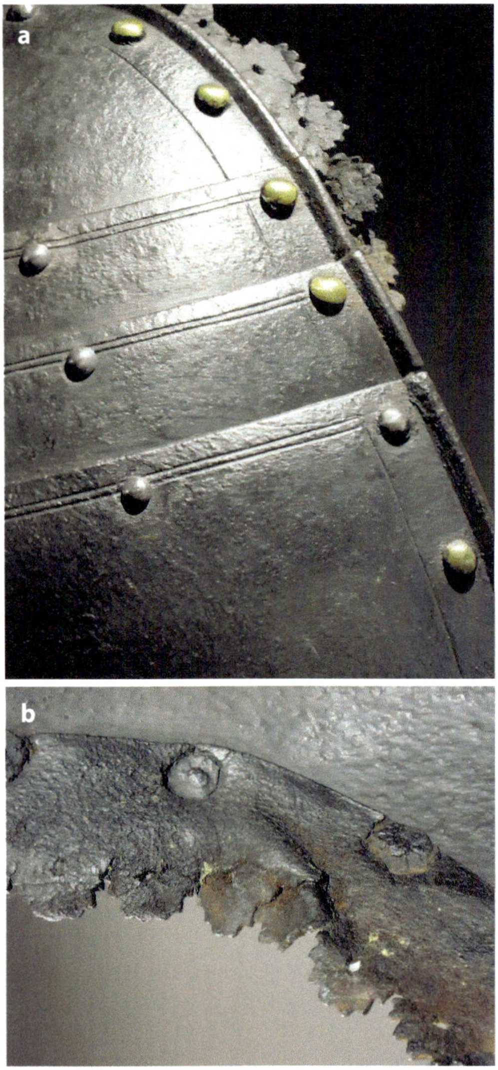

Figure 10 Fragments of a leather piccadill decoration inside the pauldron: (a) front and (b) back.

Provenance:	In the parish church of Perniö from 1705 to the late 19th century and in the museum collection since the late 19th century (the exact year is unknown)
Stamps:	None visible

Structure and decorations

An armour with plates painted in black and iron rivets with soldered copper alloy caps (Figure 9). The borders of the plates, which are turned but not roped, are decorated with two incised lines. There are some fragments of a leather piccadill as a decoration inside the pauldrons (Figure 10).[5] The width of the preserved piccadill varies from 35 to 40 mm. Two missing parts – the Zischägge helmet's nasal bar and the breastplate's left-hand side winged nut – were reproduced and added during conservation in 2018. The winged nut was made to accord with the nut of the right-hand side of the breastplate while the Zischägge helmet of Carl Bäck's armour was used as the template for the nasal bar (H 1289:1).

H 1686 Three-quarter armour of Colonel Axel Kurck (1555–1630)

Date:	1550–1575
Place:	North Europe
Material:	Iron or steel, leather and black paint
Weight:	14.970 kg total weight
Artist:	Unknown artist/maker
Provenance:	In the parish church of Ulvila from 1630 to 1875 and in the museum collection since 1875
Stamps:	None visible

Structure and decorations

An armour painted in black (Figure 11). The borders of the plates are turned and roped with a file. There are three incised parallel lines along the borders. Both the breastplate and the backplate have incised, curved and straight line decorations. This is the earliest and perhaps the finest in quality of the museum's armours. Unfortunately, only the breastplate, backplate, culet, tassets and polyns are preserved. The armourer was very skilful: the armour was perfectly formed with a hammer. In addition, it is very simple and practical even though it was carefully manufactured. The outer surfaces of the plates were rendered smooth after hammering and almost without finishing by grinding. Furthermore, all the plates were hammered into three-dimensional shapes, not just bent. The turned borders of the breastplate's openings at the neck and arms are very good examples of the quality: in the middle they rise up almost 90 degrees. They are 15 mm wide and turned to a round tube. Before conservation, the armour had no belt or belt buckle but the missing parts were reproduced in 2018

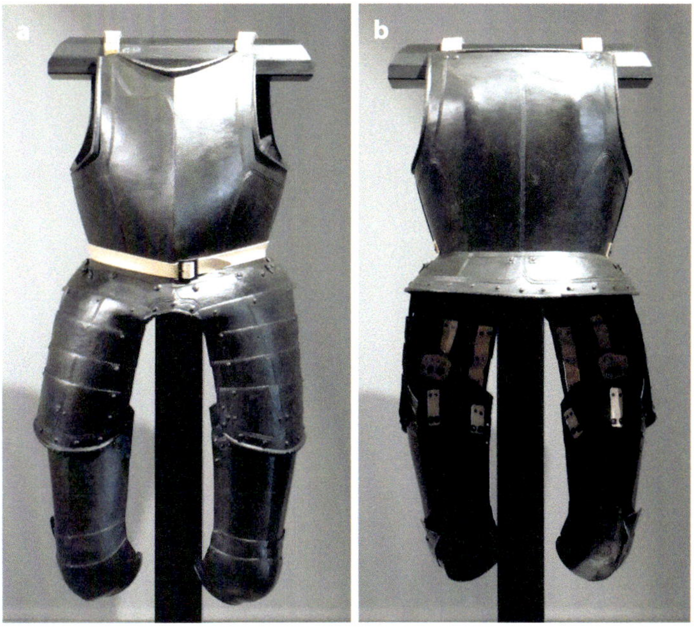

Figure 11 Axel Kurck's armour (H 1686) from the front (a) and back (b) after conservation.

using examples from Carl Bäck's armour (H 1289:1) – simple and functional, they were regarded as good models for reproduction.

H 1544 Funeral armour of Admiral Lord Marshal Henrik Claesson Fleming (1584–1650)

Date:	1620–1635
Place:	North Europe
Material:	Forged copper alloy plates, black paint, leather and iron rivets with shouldered copper alloy caps
Weight:	23.921 kg total weight
Artist:	Unknown artist/maker
Provenance:	In the parish church of Mynämäki from 1635 to 1874 and in the museum collection since 1874
Stamps:	None visible

Structure and decorations

An armour with copper alloy (= brass) plates and iron rivets with copper alloy caps (Figure 12). The borders of plates are turned but not roped. There are two incised parallel lines both along the borders of the plates and in the middle of the breastplate and backplate. The surfaces inside the armour are painted black. After conservation, the colour of the newly polished copper alloy plates varied from shades of bright gold to darker yellow and brownish tones. The armour may have been intended to imitate European royal gilded steel armours as it looks very much like them when glimpsed from a distance.[6]

During conservation the 23 pieces of copper alloy plates were studied by the museum's conservators. X-ray fluorescence (XRF) elemental analysis was performed with an Oxford X-MET 7500 handheld measuring device. The chemical concentration of the samples varied: Cu 70–72%, Zu 27–29%, Pd 0.6–0.9%, Sn 0.1–0.2% and some parts had less than 0.1% Ag, Al, Sd, Ni and Co. The hardness of a 70/30 brass is in the region of 165 VPH, comparable to 17th-century armours made of iron or low-carbon steel.

In general, it appears that the armour was not made principally for battle, but perhaps for display. Although the quality of the work is not as high as that generally found in royal armours,[7] it is perfectly

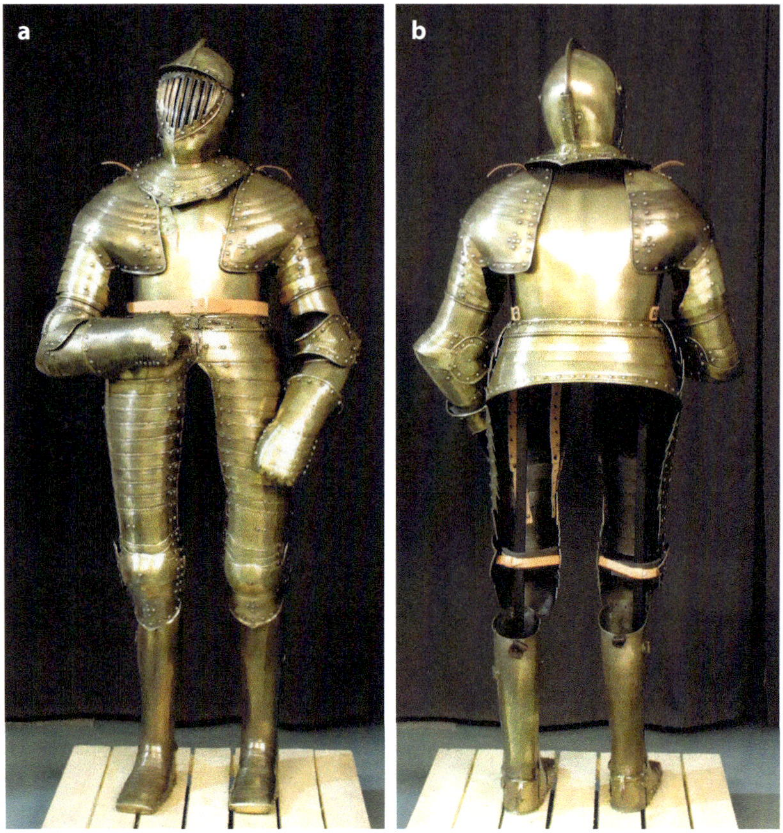

Figure 12 Henrik Fleming's armour (H 1544) from the front (a) and back (b) after conservation.

wearable. The origin of the armour, or whether it was ever worn, is not yet known: had Fleming worn it he would perhaps have elevated himself to an almost regal position in the eyes of society. The armour was part of the funeral achievements of Fleming's funeral in 1651 at Mynämäki. First placed under the floor of the parish church in the crypt, it was later erected in the church near the tomb and then afterwards in the nave. It remained in the parish church for some 223 years until it was donated by the congregation to the National Museum of Finland in 1874.

From the middle of the 16th century ceremonial funerals for kings and nobles from Central Europe were carried out in the area of Swedish control,[8] during which a parade was held for the deceased: a great funeral convoy travelled with the deceased's coffin, his horse and his armour to his tomb. The armour was carried by someone close to the deceased, who rode his horse and carried his sword. This rider was called a *kyritzryttare* (a cuirass rider) in the Swedish area and a *Freudenküriss* (an armour of joy) in the German-speaking area. After the ceremony, the armour was left in the church at the deceased's tomb. This type of ceremony began in the Swedish area with the funeral of King Gustav Vasa in 1560 and continued until the funeral of King Charles X Gustav in 1660. In this context, it is tempting to try to interpret the purpose of Henrik Fleming's copper alloy armour. As a man of note, he probably also had a funeral convoy and his funeral armour may have resembled a gilded king's armour.

The Fleming armour is among the small number of 17th-century armours made of plates of copper alloy in the northern countries, one example being a gilded copper alloy funeral armour of King Charles X Gustav of Sweden (1622–1660) in the collection of the Royal Armoury (Livrustkammaren) in Stockholm.[9]

Another is a gilded copper alloy armour of King Christian V of Denmark (1646–1699), made first for carousel riding and later used for the king's funeral. Today the armour survives only as a helmet and breastplate in the Royal Collection in Denmark.[10] In addition, a silvered copper armour of King Charles IX of Sweden (1550–1611) is displayed in the cathedral of Strängnäs.[11]

Fleming's armour is the only one from the National Museum of Finland about which some information on its restoration history is available. The museum's archives record that the armour was restored, possibly in 1874, at a cost of 150 Finnish marks. If that year is correct, the restoration was carried out during the year that the congregation bestowed the armour on the National Museum's predecessor. Apparently, a death mask made of plaster of Paris was added to the armour in that restoration, but it was removed in 1964.

Conclusions

From the point of view of preservation, the comprehensive conservation of the National Museum of Finland's armours for the new exhibition was essential. The leather straps were dry and breaking and many had become disconnected from the rivets. Most of the armour was held together with iron wires and weak leather straps. Many of the older repairs would not have been able to withstand the armour's use as protective gear – some repairs would even have prevented the owner from putting it on or walking when wearing it. In general, it was relatively easy to see an order of repairs as well as the different layers of paint: these have helped to date repairs relative to each other when the paint was spread over rivets and leather straps. I suggest that many of the repairs were connected to the armours' second or third lives.

In some cases, it was possible to discern the different phases of repair by studying the types of leather straps and iron wire, the width and thickness of materials being indicative of a certain phase. For instance, on close inspection five different types of leather straps were found on Henrik Fleming's armour (H 1544) which appeared to have been applied in four different phases. Some of the straps showed signs of stitches and one leather strap was apparently a binding, suggesting that some of the straps had been recycled from objects other than armours.

In addition, some rivets and plates differed in age. Exact dating of this kind of repair is extremely difficult but they appear to have been done before the armour was displayed in the parish church, indicating that they could date to the armour's first life: it does not seem very probable that the armour plates would have been broken during its second or third life.

Both conservation and restoration was carried out for the new exhibition with the aim of restoring the armours as closely as possible to their 17th-century condition while keeping the treatments to the minimum.[12] Some work was required to enable the armours to be put on display therefore almost all the repairs from their second and third lives – mainly the supporting iron wires and leather straps with their rivets – were removed and replaced. All the original rivets and remains of leather straps were preserved. Other than being cleaned and protected, plates and other metal parts of the armours were left untouched. All new constructions of leather straps, rivets and washers were marked by stamping the year with 1 mm high numbers and my triangle seal containing the letters LMM. In addition, I made a complete study and conservation report of every armour to be stored in the museum's archives in order to aid future conservation and study of the armours.

Many questions still remain to be answered concerning the museum's armours, but research into historical sources such as letters and diaries could provide information as to where the owner acquired the armour and details of the kind of memorial service held. The owner's will may further elaborate the funeral preparations and the property that he held. Furthermore, thorough study together with metallurgical research could shed light on the quality of the armours.[13]

Currently, how common the practice of placing armour near or over the tomb of deceased officers was during the 17th century in the Finnish region is not known, although it would appear that this happened relatively infrequently. However, it is very likely that more armours have been placed in parish churches of which we unaware. For instance, parts of different armours not donated to museums may remain in storage in these churches.[14] The congregation archives may reveal more information concerning funeral achievements, a subject that needs further research.

Another source of information worth further consideration is the armours owned by the Finnish nobility, whose families may be able to shed some light on their history including whether they had a role in the funeral convoy. In 1555, Olaus Magnus described how noble Nordic families brought back funeral achievements such as arms and armour from the church immediately following the funeral.[15] Aristocratic families in the Finnish region may also have acted likewise.

Apart from the collection of the National Museum of Finland there are six other armours in Finnish museums and churches of which I am aware: two in Turku Cathedral as funeral achievements and one in the museum of Turku Castle. They all date from the 17th century, have connections to Finnish history and their last owners are known. Three other armours in the museum are old armours collected in the 20th century. The six armours in the National Museum represent the best surviving collection of late 16th- and early 17th-century armours in Finland, covering most of the armour types in use in the Finnish region at that time, and representative of the kinds of armour officers in the region owned and used during the 17th century.

Notes and references

1. Detailed conservation reports on all these armours are deposited in the archives of the National Museum of Finland.
2. Norman, A.V.B. and Eaves, I. (2016) *Arms & Armour in the Collection of Her Majesty the Queen. Volume 1 European Armour.* London: Royal Collection Trust, pp. 328–43; Oakeshott, E. (1980) *European Weapons and Armour from the Renaissance to the Industrial Revolution.* Woodbridge: Boydell Press (reissued 2000), pp. 206–7, 209; Puype, J.P. and Wiekart, A.A. (1998) *Van Maurits naar Munster, Tactier en triomf het Staatse leger.* Delft: Leger Museum, pp. 166–217; Richardson, T. and Rimer, G. (2012) *Littlecote: The English Civil War Armoury.* Leeds: Royal Armouries, p. 89; Capwell, T. (2007) *The Real Fighting Stuff.* Glasgow: Glasgow City Council, pp. 68–71; Blackmore, D. (1990) *Arms & Armour of the English Civil Wars.* London: Royal Armouries, pp. 11–13; La Rocca, D.J. (2017) *How to Read European Armor.* New York: Metropolitan Museum of Art, pp. 24–5; Quaas, G. (1992) *Eisenkleider, Plattnerarbeiten aus drei Jahrhunderten aus der Sammlung des Deutsches Historisches Museum.* Berlin.
3. Williams, A., Edge, D. and Atkins, T. (2006) 'Bullet dents – "Proof marks" or battle damage?', *Gladius* 26(1): 175–209. Available at: http://gladius.revistas.csic.es/index.php/gladius/article/view/8/8 (accessed 6 June 2021).
4. Ramsay, J. (1909–1916) TAB. III. Bäck i Finland. Frälsesläkter i Finland intill stora ofreden. http://runeberg.org.frfril/0066.html, p. 61. Adelsvapen-Wiki: Bäck i Finland nr 193.
5. Oakeshott 1980 (cited in note 2), pp. 206–7.
6. Capwell, T. 2011. *Masterpieces of European Arms and Armour in the Wallace Collection.* London: Wallace Collection, pp. 126–9 (close helmet of Emperor Ferdinand I, The Wallace Collection, A188); Norman and Eaves 2016 (cited in note 2), pp. 174–7 (armour of Henry, Prince of Wales, RCIN 72831); Richardson, T., Woosnam-Savage, R.C. and Rimer, G. (2011) *Masterpieces of Medieval and Renaissance Arms and Armour from the National Collection*, vol. 2. Milan: Hans Prunner edition, pp. 236–41, 296–7 (Charles I gilded steel armour No 11.91, The Royal Armouries, Leeds).
7. See, for example, the armour for a Prince of Wales c.1625 (Norman and Eaves 2016 [cited in note 2], p. 205) which had a hardness of 164–197 VPH. For the properties of brass, see Rollason, E.C. (1977) *Metallurgy for Engineers.* London: Edward Arnold, pp. 304–6.
8. Rangström, L. (2015) *Dödens Teater. Kungliga svenska begravningar genom fem århundraden.* Stockholm: Livrustkammaren, pp. 39–41.
9. Ibid., p. 41.
10. Kongernes Samling Rosenborg.
11. Rangström 2015 (cited in note 8), p. 37.
12. Smith, R.D. (ed.) (2005) *Make All Sure: The Conservation and Restoration of Arms and Armour.* Leeds: Basiliscoe Press, pp. 51–5.
13. Williams, A. and de Reuck, A. (1996) *The Royal Armoury at Greenwich 1515–1649: A History of its Technology.* Leeds: Royal Armouries, pp. 18–25.
14. According to Markus Hiekkanen, personal communication, 25 September 2019.
15. Magnus, O. (1555) *Historia om de nordiska folken.* Möklinta: Gidlunds Förlag (reprinted 2010), p. 762.

8

Pierced metalwork on Iranian vambraces and their cleaning

Francesca Levey

Introduction

Fourteen pieces of Iranian armour in the Wallace Collection, London, are embellished with pierced steelwork. This often has a complex layered construction and a decorative red coating painted on the metal underlay (Figure 1) which has subsequently been obscured by an old (petroleum-based) conservation coating. The usual method used for the removal of the conservation coating involved immersing the object in a solvent bath. This study aimed to develop a new method to remove the degraded conservation coating and associated dust and corrosion products, while minimising the use of hydrocarbon solvents, to reveal the decorative red underlay (Figure 2).

This new method was tested on model substrate tokens and examined with the aid of a graticule to assess its efficacy. Three pierced Iranian vambraces (OA1514, OA1468 and OA1507) from the Wallace Collection were then selected as case studies (Figures 3 and 4). The new method of treatment resulted in a significant reduction of solvent use as well as an improvement to the removal of the degraded coatings with their dust and corrosion

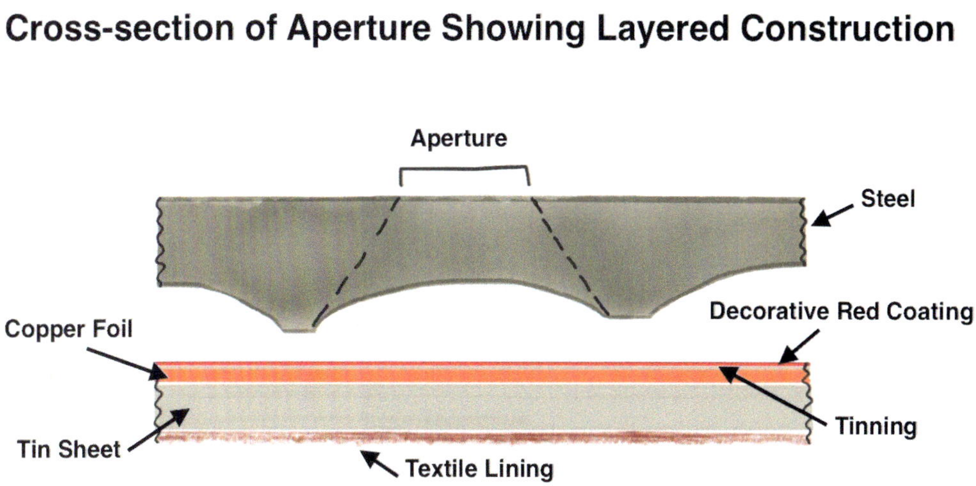

Figure 1 Cross-section of the aperture showing the layered construction and decorative coating of the pierced steelwork.

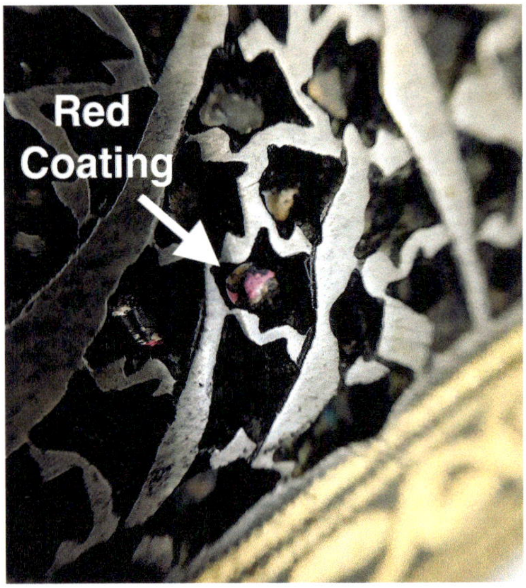

Figure 2 Detail of a fragment of red decorative coating. © Trustees of the Wallace Collection.

products. The decorative red coating that had been obscured was revealed on these underlays and analysed using Fourier transform infrared spectroscopy (FTIR). This new method has further highlighted several areas for future study.

For the past decade, the Wallace Collection has been undertaking a major cataloguing project of around 1000 items of arms and armour in the Ottoman, Middle Eastern and Asian (OMEA) collection. These objects have been photographed in colour for the first time for this publication, making the aesthetic considerations of their conservation treatments particularly important.

Corrosion protection

Historically, the arms and armour in the Wallace Collection were protected against corrosion by a coating of an in-house blend of Vaseline petroleum jelly and 3-in-1 mineral oil. This was used from at least the 1950s, if not earlier, until it was replaced by microcrystalline wax and environmental controls in the late 1980s. When this coating was first applied, it sat on the surface of the object as an oily film. The coating was effective but required regular maintenance as atmospheric dust settled into it – not only was this unsightly, but as dust is hygroscopic it also put the object at greater risk of corrosion from moisture and pollutants. The resident armourer maintained and reapplied these coatings annually. A soft cotton cloth was used to wipe off the soiled layer of coating (with accumulated dust) before a fresh layer was applied *in situ*.[1] This was problematic for objects with pierced metalwork as the apertures became filled with a mixture of coating and dirt, masking the decorative underlay and encouraging corrosion. A further problem was that organic fats, oils and waxes oxidise over time to form carboxylic acids ('fatty acids'), which interact with copper to form copper soaps:[2] these are often produced on historic objects by the breakdown products of organic coatings with copper or copper alloys in an oxidising atmosphere.[3] The deep green of the degraded coating on vambrace OA1514 (Figure 5) led to the suggestion that copper corrosion products, in the form of copper soaps, were present in the coating.[4]

The source of the fatty acids is not known but given its composition, the oxidation of organic oils in this coating was considered to be their source. In addition to the detrimental effects on the objects, the degraded coating led to the misidentification of materials and techniques, which has wider implications for visitors, curators and the conservators tasked with their care and study. Figure 6 demonstrates the likely mechanism for deterioration of the old coating into the apertures of pierced work.

Removal of this old coating often necessitated immersion of the entire object in a bath of hydrocarbon solvent. Visual examination of pierced work objects in the Wallace Collection in 2014 showed that although this immersion cleaning was an effective method for removing the old coating on various object types, it was unsuccessful in eliminating all the coating from inside the apertures of pierced metalwork. Immersion of an entire object not only used significant amounts of solvent, but also made control difficult when treating complex objects of mixed media. An alternative method of cleaning was therefore developed using syringes to deliver measured amounts of solvent into the apertures of pierced work.

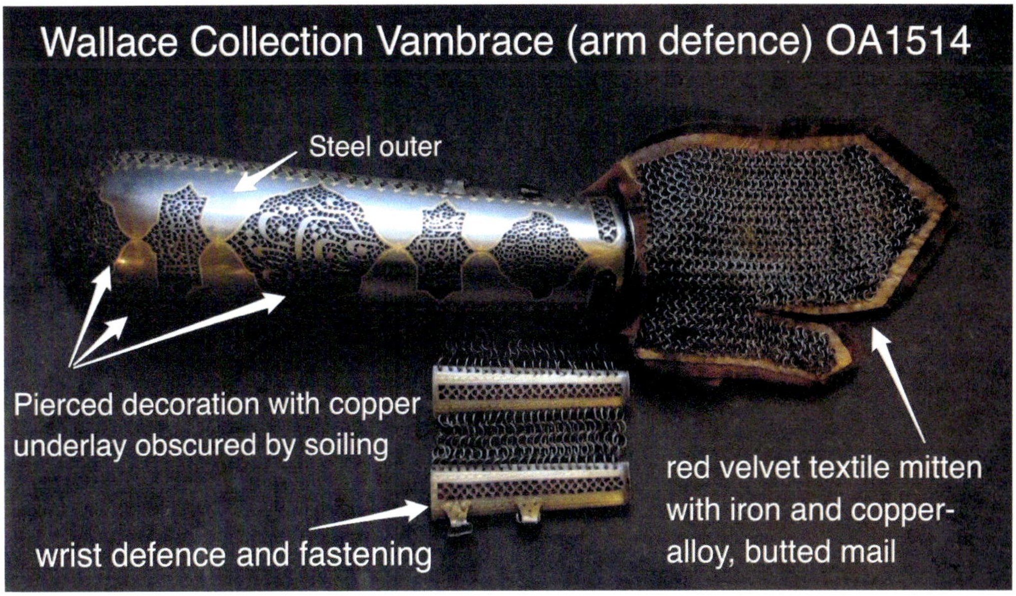

Figure 3 Annotated photograph of the exterior of OA1514, one of the three vambraces selected for study, showing pierced areas and fittings. © Trustees of the Wallace Collection.

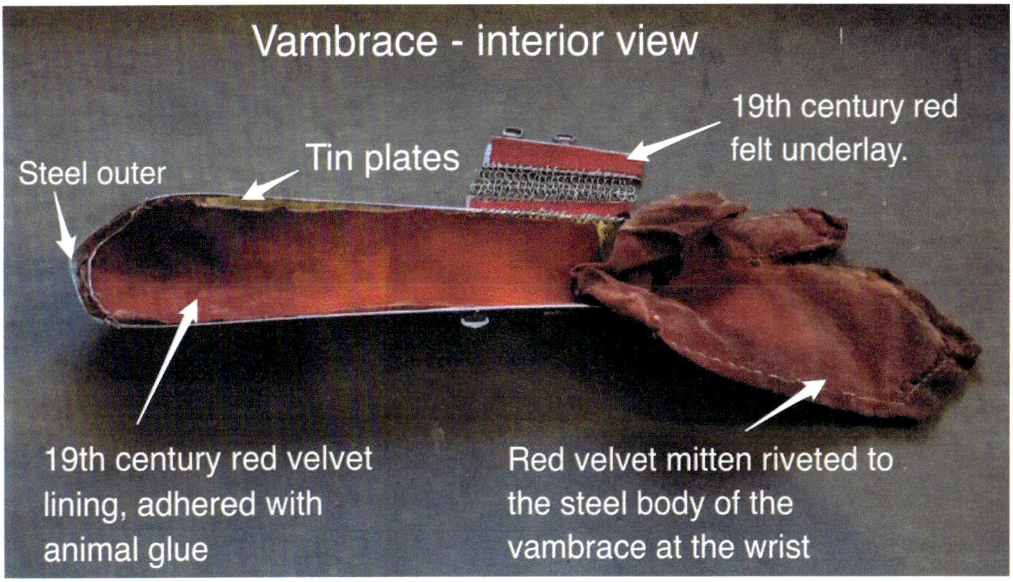

Figure 4 Annotated photograph of the interior of OA1514 showing the textile lining and mitten. © Trustees of the Wallace Collection.

In 2017, an investigation began into the removal of these degraded coatings from the abovementioned three 18th–19th-century Iranian vambraces. All three are decorated with 'pierced work' whereby apertures were cut out of steel plate to create a delicate fretwork. Decorative coloured materials were placed under this pierced steelwork. Visual examination suggested that the decorative underlay was a tinned copper foil, covered with a ruby red translucent

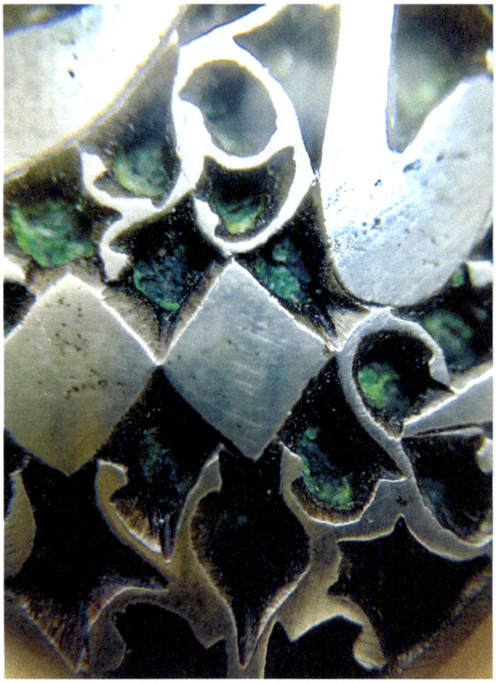

Figure 5 Detail of OA1514 showing apertures congealed with degraded green coating. © Trustees of the Wallace Collection.

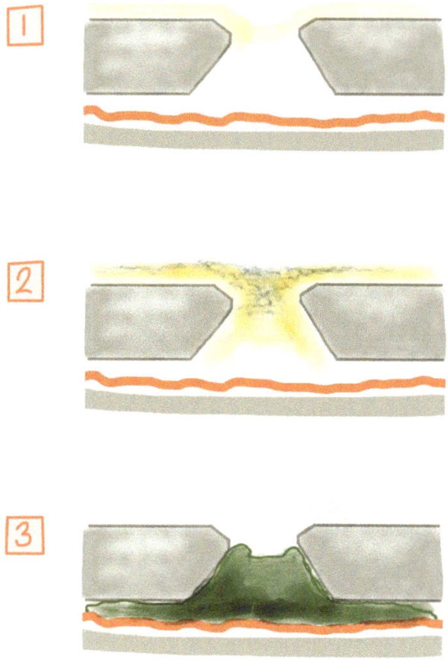

Figure 6 Cross-section showing the travel and degradation of the coating in one of the three aperture types. (1) The transparent coating sits on the surface of the object and collects around the apertures. Obstruction of detail is minimal. (2) Dust sticks to the coating. Temperature fluctuations due to lack of environmental controls mean that the coating becomes more fluid and creeps further into the aperture. (3) The dust-saturated coating now reaches the copper foil. The copper catalyses the oxidation of the coating and accelerates copper corrosion. The coating, now green and impregnated with dust, obscures the foil underlay.

coating, possibly shellac. It is not uncommon to find copper underlay on pierced work covered in a red translucent coating, however, in many cases, the underlay is so dirty that its presence cannot be identified at all, let alone any potential decorative coating. The copper foil was held in place by sheets of tin that were riveted to the steel and then covered by a textile lining (Figures 7–9). The aim was to ascertain whether the new cleaning method would remove the degraded coating and reveal the coloured underlay without damaging it. However, as the composition and stability of the decorative red coating on the surface of the foil underlay was unknown, care had to be taken to ensure it would be disturbed as little as possible.

Discussion of underlays on Iranian pierced work suggests that some pierced work may have been designed to be seen against a golden ground.[5] Revealing the original colours of these underlays to understand the intended appearance of these vambraces was a focus of this project. Research on examples such as the Benin Bronzes in the Pitt Rivers Museum, Oxford, has highlighted the issue of the degradation of protective coatings to the detriment of the object. Emphasis was placed on the necessity to remove such coatings that might be masking unchecked corrosion.[6]

Previous conservation treatments

Since the start of the cataloguing project in 2005, the treatment for removing such degraded coatings had been to immerse the object in a bath of Stoddard solvent[7] and agitate the surface with a soft bristle brush. Frequently there was so much accumulated coating that multiple solvent baths

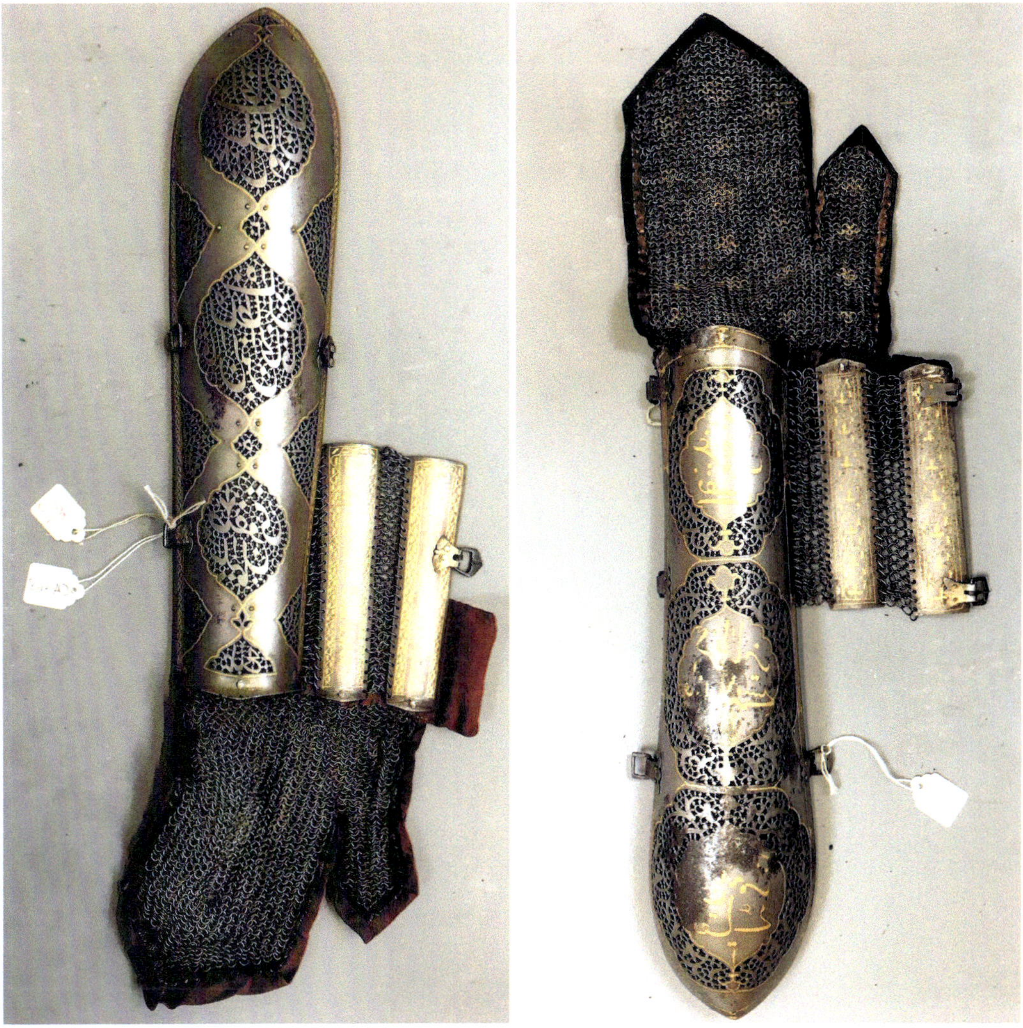

Figure 7 The construction of OA1468 (*left*) and OA1507 (*right*) matches that of OA1514. © Trustees of the Wallace Collection.

were required to clean a single object; 59 helmets in the OMEA Armour collection were treated this way. Although this treatment was effective for areas that might be reachable with a bristle brush, it was not suitable for those objects where the soiling lay behind pierced steel or for surfaces with friable decorative layers. It also generated a large quantity of solvent waste.

The desire to reduce this waste led to the trial of other techniques. Cutulle and Kim tested the efficacy of dry ice blasting on the removal of this coating from 38 helmets in the Wallace Collection.[8] Although they found that dry ice treatment did reduce the amount of solvent needed, solvent cleaning was still required to remove the coating. Poultices of cotton wool and white spirit applied to the pierced areas were also tested. Although poulticing treatment revealed some of the decorative underlay on vambrace OA1468,[9] as with immersion, poulticing could not extract the larger particles of dust and corrosion products, and it is doubtful how much of the poultice was able to reach the copper foil behind the steel given the 2–3 mm distance between them.

A direct injection of a solvent through gaps in the pierced fretwork by means of a syringe

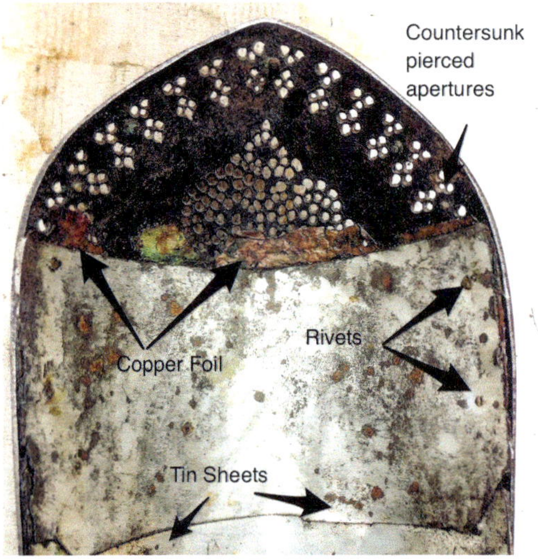

Figure 8 Interior of OA1514 with the textile lining removed, showing the layered construction. © Trustees of the Wallace Collection.

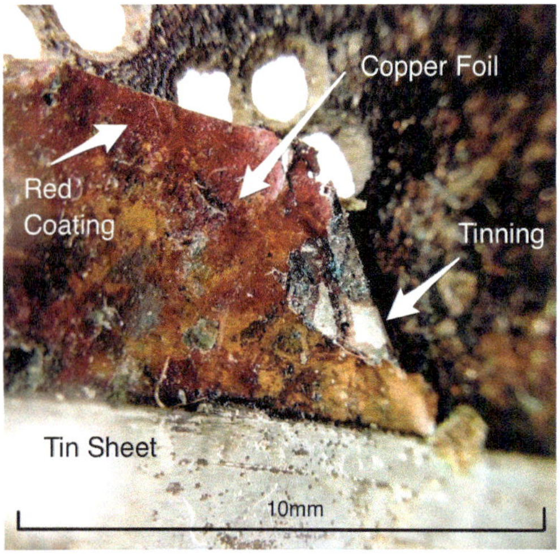

Figure 9 Detail of OA1514 interior showing the copper foil, tinning and red coating. © Trustees of the Wallace Collection.

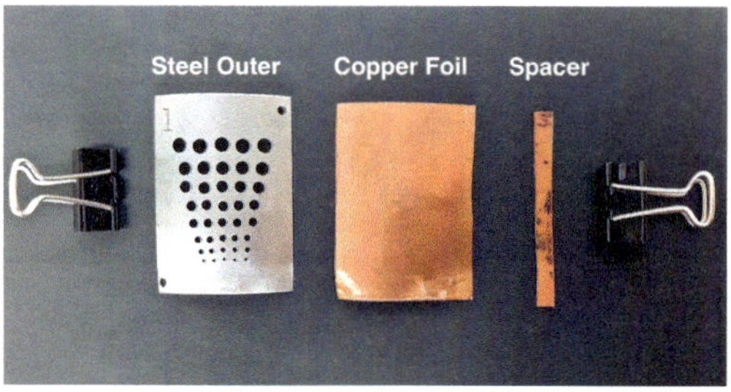

Figure 10 Deconstructed test token showing the components.

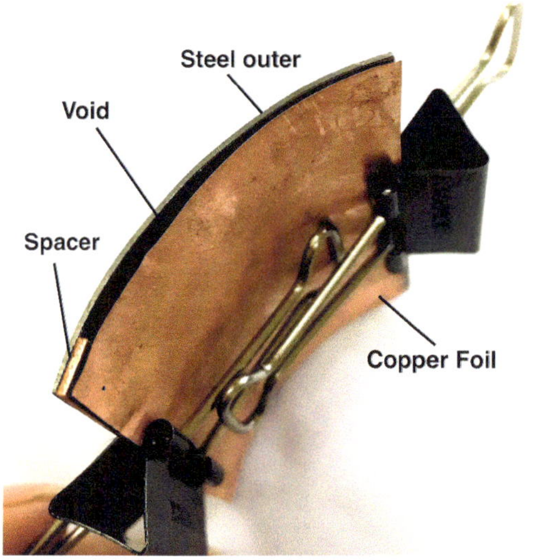

Figure 11 Top view of the test token showing the copper spacer.

was then proposed to enable the solvent to flow through the object across the underlay, creating a gentle flushing action. The solvent would then dissolve the degraded coating on the interior surfaces, freeing the accumulated dust, corrosion products, etc. and washing them out of the object. Werner had performed extensive tests and recommended a solution of 90% low odour varsol[10] and 10% ethanol, which proved extremely effective at removing copper soaps in another case study.[11] Unfortunately, this blend of solvents was inappropriate for use in this instance as it was considered that the ethanol might cause the decorative red coating (assumed to be shellac) on the underlay to swell or even dissolve.

Testing

Test substrates or 'tokens' constructed from 1 mm thick mild steel to model the vambraces were developed for cleaning trials. A series of measurements was taken of the original apertures and a range of aperture sizes was calculated based on measurements taken from the vambraces for use on the test tokens, which were then drilled

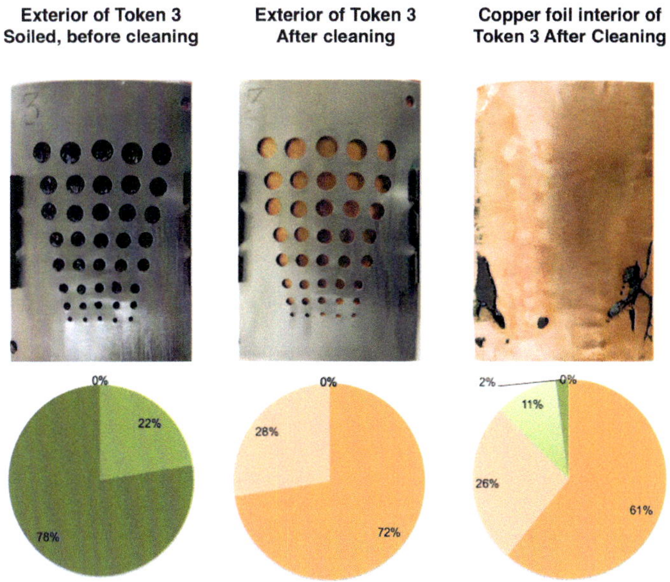

Figure 12 Results from cleaning token 3.

into the steel. The reverse of each aperture was countersunk with a hand drill: countersunk areas, observed on the case study objects, catch soiling in the tooled ridges left by drilling. They can also act as funnels, helping to direct the solvent out of the object during flushing.

Copper was rolled down to 0.2 mm thickness for the underlay and a 0.9 mm copper spacer was employed to create a space between the copper foil and the steel carcass as found on the original objects. This helped to imitate the tight layering of materials where rivets held them together and the loose layering in areas without rivets. The pieces were held in place with bulldog clips to allow the tokens to be taken apart and the cleanliness of the interior to be examined after testing (Figures 10 and 11). Three tokens were used for ease of repeating the tests.

Method

A pigment was added to 70 g of replica 'degraded coating' to aid the visibility of the replica coating on the test substrates. This coating was applied to all the apertures with a hog bristle brush, then wiped off with a soft cloth. Stoddard solvent (another form of white spirit) was considered to be the best solvent to remove the coating as, unlike acetone and IMS, a hydrocarbon solvent can dissolve oily layers without posing a significant risk to the decorative red coating on the underlay, then assumed to be shellac, and more susceptible to polar solvents. A 10 ml glass syringe was used to inject the solvent with a curve-nosed nozzle that was adapted to fit the glass syringe. The token was held securely above a drip tray by a clamp stand. The solvent was reused throughout the cleaning procedure to minimise solvent waste.

Results of preliminary cleaning tests

All the soiled tokens were examined after their solvent injections to assess the efficacy of cleaning. Five categories of soiling were defined for scoring, from 1–5 ('clean'–'dirty'), each aperture on the tokens being assigned a value. These scores are presented in pie charts to show the percentage area of soiling present on the sample pieces before and after cleaning (Figure 12).

Tokens were photographed and scored while 'assembled', meaning that only the areas of foil visible through the apertures were scored. The

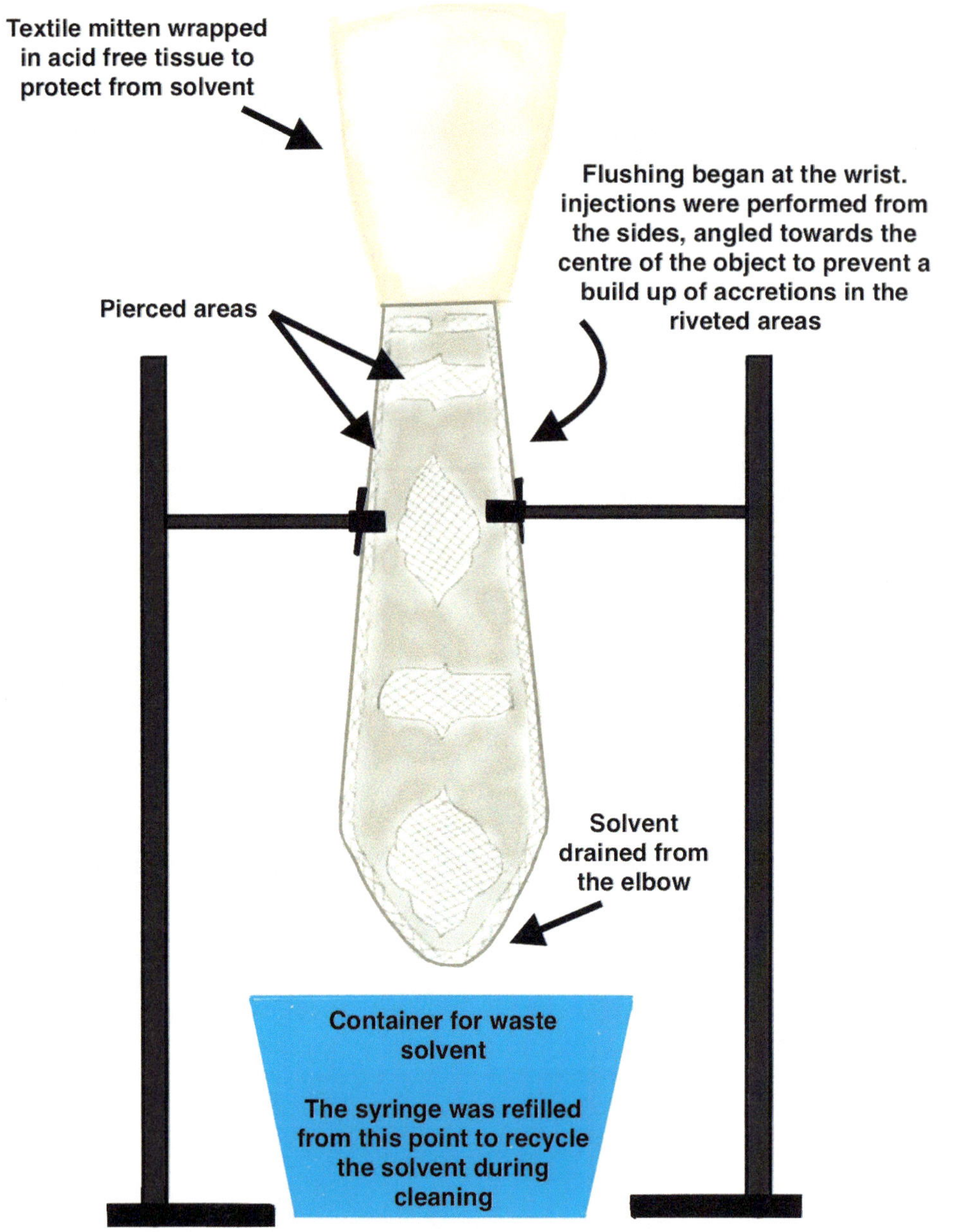

Figure 13 Diagram of the vertical cleaning setup for case study objects.

tokens were then dismantled to reveal their interiors. A graticule was imposed onto these sheets of foil and the grid squares scored according to the same criteria. These data were presented alongside data from the apertures to ascertain internal cleanliness after treatment.

The test tokens proved invaluable for predicting the potential success of this treatment, but as the vambraces are unlikely ever to be dismantled it must be accepted that the true extent of their internal cleanliness is never likely to be known. It was found that the orientation of the object when injecting solvent was critical: if the tokens were injected 'face up', this carried the soiling to the simulated rivet points at the sides where it could become trapped. The convex vambraces would therefore have to be carefully oriented to avoid flushing the soiling further into the interior.

Case studies

Vambrace OA1514 was used first as a case study and from the success of its initial treatment, the Wallace Collection offered vambraces OA1468 and OA1507 (both constructed identically to OA1514), for further testing. All three vambraces have ferrous bodies (apparently crucible steel) ornamented with pierced cartouches decorated with foliate designs or inscriptions. The pierced areas are underlaid with copper (or copper alloy) foil, which appears to have been tinned and covered with a translucent ruby red coating. The silvery layer on the copper foil was examined by X-ray fluorescence (XRF) to establish whether it was tin, but the results were inconclusive due to the presence of tin sheet elsewhere on the vambrace causing interference. The absence of noticeable black tarnish was held to rule out the presence of silver.

This 'tinned' and coated foil was held in place by large sheets of (1 mm thick) tin, which were riveted to the steel outer plate. The interior was then lined with velvet and fastened to the tin with animal glue. Textile mittens were riveted to the main plate at the wrist with copper alloy rivets and covered in mail made of butted ferrous and copper alloy links. Wrist straps of steel plates were also joined with butted ferrous links. The tinned and coated foil, designed to be seen from behind the pierced decoration, was exceptionally thin and fragile, and crinkled like a confectionery wrapper when probed with a swab. The degraded coating was so thick that the foil was barely visible, and because it was sitting on the red decorative layer, it was impossible to assess if there were any losses to the latter. The textile linings had already been removed for previous conservation treatments but detaching the tin plates by removing the rivets was considered too invasive (Figures 13 and 14).

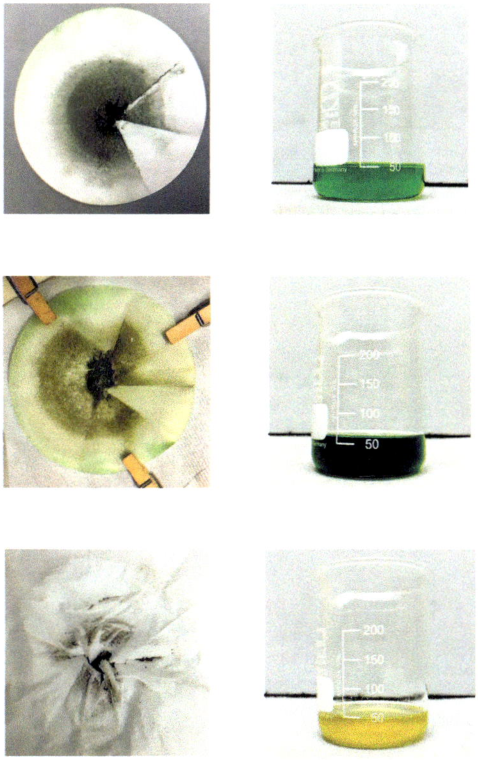

Figure 14 Debris and waste solvent from cleaning case study objects: (*top row*) OA1514, filter paper and waste solvent; (*middle row*) OA1468, filter paper and waste solvent; (*bottom row*) OA1507, filter paper and waste solvent.

Method

The setup for cleaning was arranged in a fume cupboard (see Figure 13) with a Plastazote pad supporting the textile mitten, wrapped in acid-free tissue. During the process, the waste solvent was observed for changes in colour and the presence of debris (see Figure 14). Small fragments were rinsed from the substrate: visual analysis under magnification showed this to be a mixture of dust, fibres (thought to be cotton wool from previous treatments) and small flakes of the decorative coating. Microscopy revealed that the reverse of the flakes were covered in copper corrosion products (Figure 15), suggesting that these were responsible for detaching the flakes from the foil, rather than pressure from the solvent flushing.

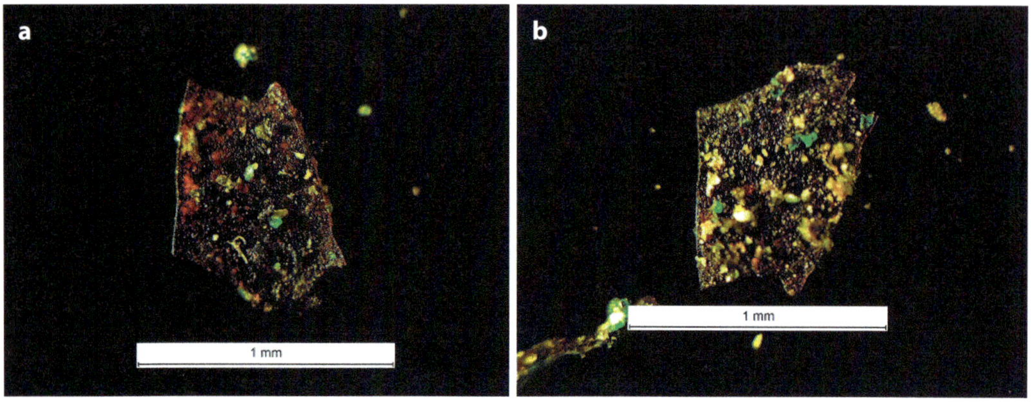

Figure 15 Microscopic image of a flake of red coating: (a) front and (b) back.

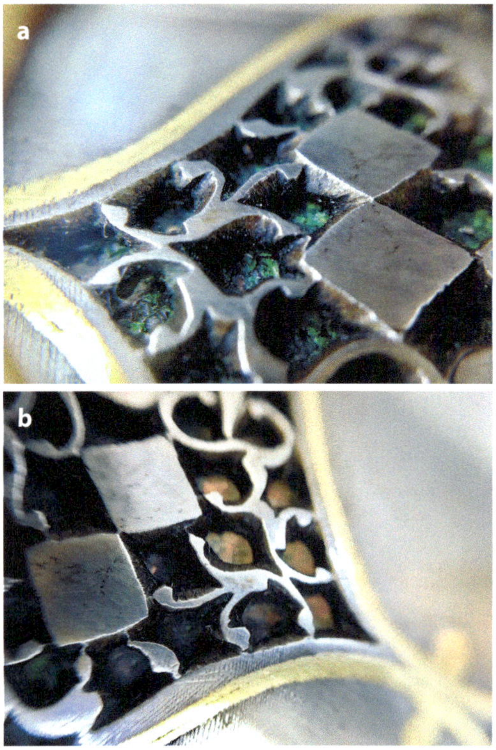

Figure 16 Detail of OA1514 before (a) and after (b) cleaning of petroleum coating. Note the increased visibility of the copper foil: this fragment has no tinning or red coating. © Trustees of the Wallace Collection.

Many of the accretions removed were over 1 mm in length. In several incidences, dust mats (2 mm wide × 4 mm long) were rinsed out of the larger apertures. Initially, there was a concern that flushing would result in a build-up of soiling in one part of the object but the only problem experienced was trying to flush apertures close to rivets: at these edges, the syringe tip was angled into the sides of the object to flush the soiling out of these areas. The vambraces were flushed systematically from the centre to the edges and top to bottom until no more debris appeared. At this point, 50 ml of clean solvent was used to rinse any contaminated solvent and dissolved coating from the object.

Results of cleaning case studies

The procedure for OA1514 took five hours during which solvent was injected into the object around 900 times: the 400 ml of solvent used was recycled throughout the cleaning process, resulting in 50 ml of waste. The treatment of OA1468 took four hours and approximately 720 injections: 300 ml of solvent was used resulting in 50 ml of waste solvent. On the other hand, OA1507 was only treated for one hour because after 180 injections the treatment had produced no change: 100 ml of solvent had been used, resulting in 50 ml of waste. Minimal debris was seen and the yellow colour of the waste solvent suggested that Fe^{3+} ions were present, signifying ferrous corrosion, rather than copper. The waste solvents from OA1514 and OA1468 were a deep green-blue colour, typical of copper corrosion products (see Figure 14). This would seem to indicate that

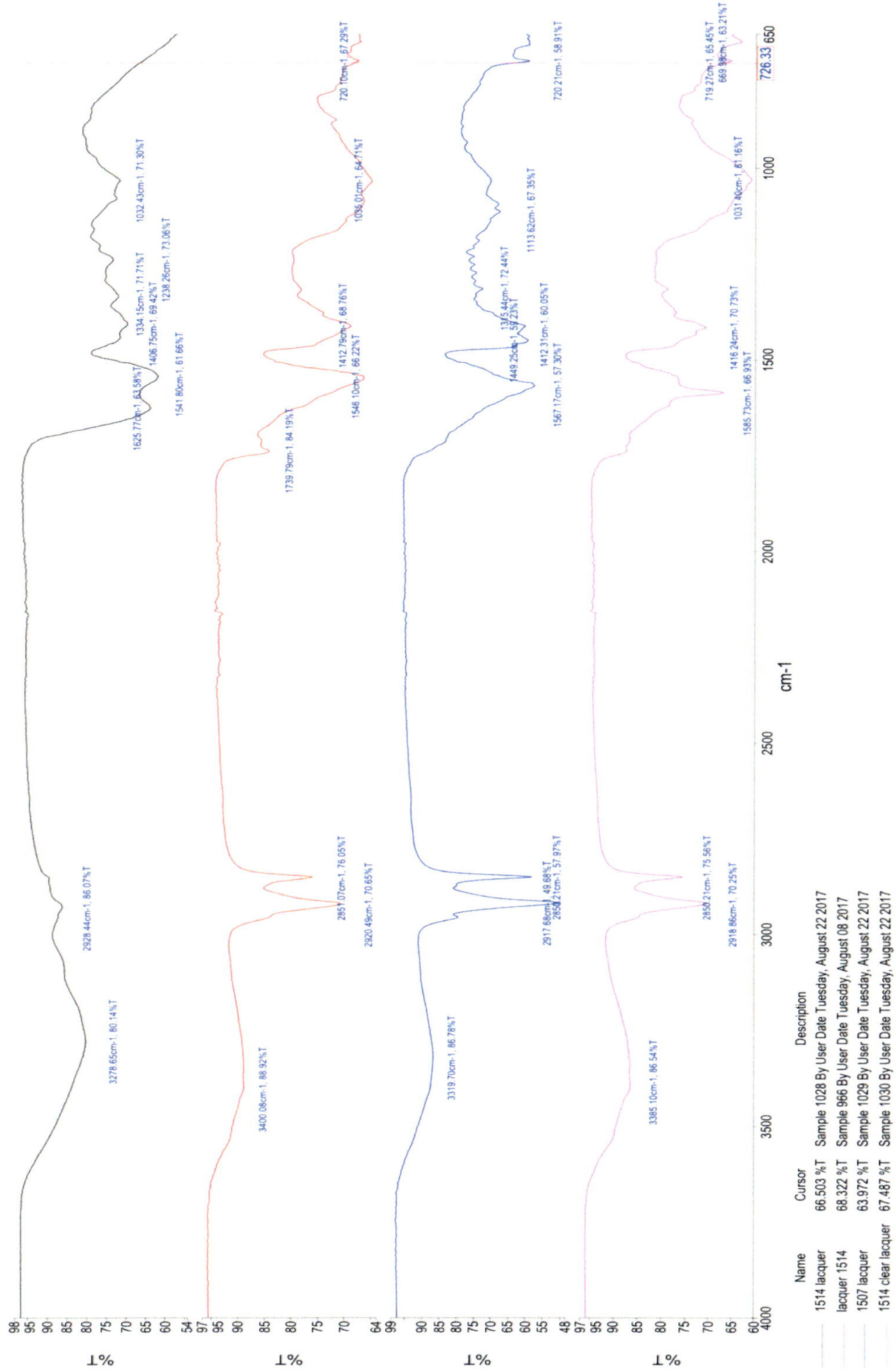

Figure 17 Four spectra from flakes of red coating, stacked for comparison.

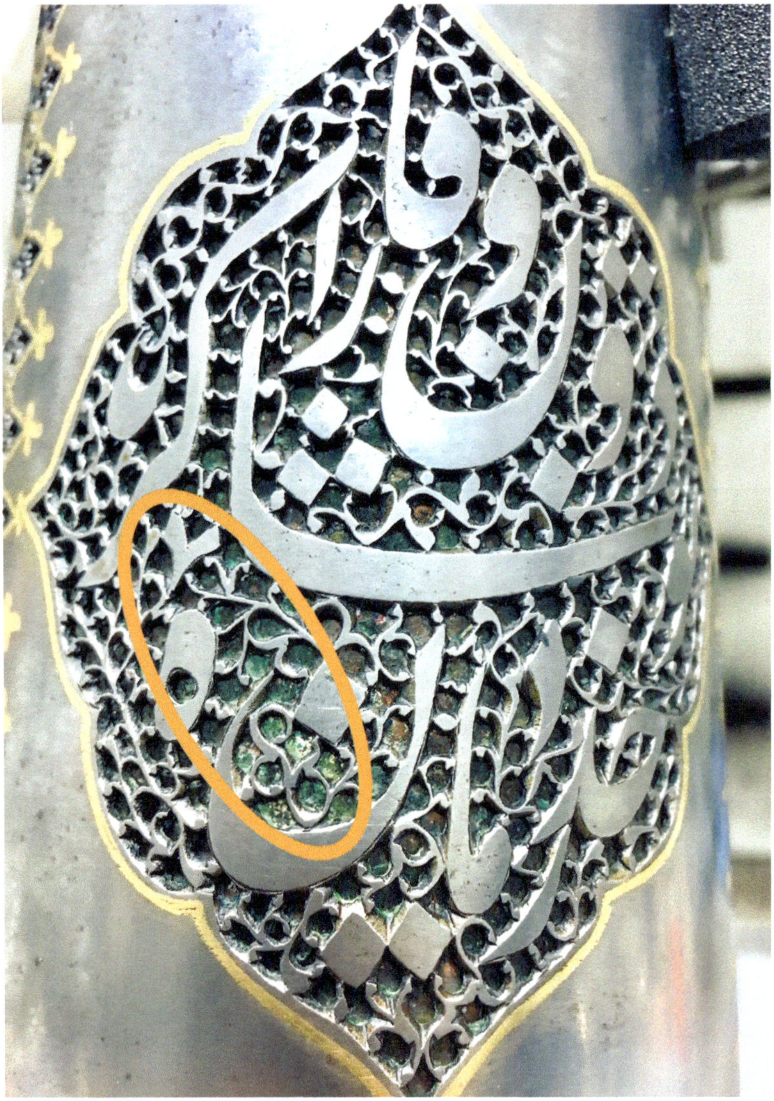

Figure 18 Detail of the central cartouche from OA1514 with the highlighted area showing particularly severe copper corrosion. © Trustees of the Wallace Collection.

the arrangement of the metals within OA1507 or exposure to the coating differs from that in the other two vambraces so that the coating has formed different products.

Visual examination

A significant amount of debris was removed from the objects. This was evident to the naked eye, but documenting this through photography was difficult due to the construction of these objects. The decorative foil of the vambraces sits as much as 3 mm behind the steel fretwork and this made meaningful graticule scoring impossible. Images taken with a macro lens were more illustrative and the removal of the petroleum-based coating is evident in Figure 16. However, a significant amount of corrosion products remained.

Debris

After treatment, the waste solvent was filtered, and the fragments washed from the objects collected for study. The debris from OA1468 was the most voluminous and contained the most copper corrosion products. Detritus from OA1507 was minimal and the media present was brown rather than green, supporting a suggestion of iron corrosion products. The filter paper from OA1514 contained the most fragments of the decorative red coating.

Decorative red coating

Microscopy revealed that flakes of the decorative red coating on the foil underlay had been pushed from the surface by the copper corrosion products that were forming underneath the decorative surface (see Figure 15). Although the loss of original material is to be avoided, the only way of consolidating the coloured coating would have been to dismantle the whole object at the risk of creating even more damage. After consideration, a small loss was considered acceptable, and the flakes from the object were viewed as a unique opportunity to study and analyse the red layer in isolation. The debris from each object was examined under the Nikon AZ100 microscope and analysed using FTIR. Preliminary FTIR results have shown that the coating is collagen-based, rather than shellac as previously thought. Although the decision to use Stoddard solvent was based on the hypothesis that the coloured coating was shellac-based, fortunately, collagen coatings are also insoluble in this solvent.

Figure 17 shows a comparison of the absorption spectra of four different samples of the decorative coating. The uppermost spectrum has strong bands at 1540 and 1620 cm^{-1}, which are the amide I and II bands of a protein. There are also characteristic amide bands at 3080 and 3280 cm^{-1}.[12] The other three spectra follow this basic pattern but become more difficult to analyse due to the presence of bands in the region of 1585 cm^{-1}, suggesting carboxylates (from copper soaps) or contaminated hydrocarbon oil.[13]

Although this analysis shows the decorative coating to be collagen-based, it is not possible to discern from these data from where this collagen is derived: it could be from an egg medium or animal glue. Robinet and Corbeil have warned of the limitations of using FTIR on degraded historic samples, and further research should be conducted bearing these in mind.[14] However, the knowledge that the coating is collagen-based can be of immediate practical use in the conservation of objects with similar coatings.

Corrosion products

There was a concern that the degraded conservation coating could be masking unchecked corrosion. Once the coating was removed, it became evident that the copper corrosion on the decorative underlays was indeed more extensive than previously realised, particularly on OA1514 (Figure 18). The remaining decorative surfaces can now be assessed and further treatments to deal with the corrosion of the copper considered. In the case of OA1507, the collagen coating appeared to have interacted with the brownish soiling that had been lying on top of it, forming a crust. This left areas of red, only just visible, under a layer that did not appear to have been removed by the solvent. This suggests that it may be the result of iron corrosion products reacting with organic material to cross-link and form an insoluble resin.

Conclusions

This project aimed to examine whether an injected solvent delivery system might be able to remove a degraded coating from Iranian pierced work. This was tested successfully on the model substrates, which were assessed using a graticule. However, on dismantling these tokens, it was clear that clean apertures did not guarantee a fully clean interior. This was particularly evident on token 3: its apertures scored 100% cleanliness, but the interior showed that 13% soiling remained. The treatment was slightly modified with more systematic and directed injecting for the case study vambraces, focusing mainly on the sides of the objects. This system enabled the conservator to access parts of the object inaccessible to mechanical cleaning and allowed a more controlled cleaning than immersion in solvent.

The objects' pierced apertures were congested with a degraded coating that obscured a decorative underlay. Debris flushed from the objects, the colour of the waste solvent and visual analysis demonstrated that much of the degraded coating had been removed thereby revealing the decorative underlay and aggressive copper corrosion. It was hoped that the removal of the degraded coating would also improve the visibility of the coloured coating on the copper foil: the visibility of the underlay was increased, but not as much as was expected given the success of an earlier (2014) test treatment. The cleaning procedure also uncovered corrosion products on the surface of the underlay. Instead of revealing the surface of the foil, it was realised that much of the original coloured layer had been lost due to a previous treatment, showing instead the partially tinned surface of the copper. This surface, although visible, retained a green tinge from contact with the copper corrosion products. However, now that the underlay and the extent of the corrosion is visible, conservators have more information with which to make informed decisions about further treatment.

Reducing the amount of waste solvent produced in the cleaning of these objects was also a conscious aim of the project. Cutulle and Kim[15] needed 22 litres of solvent to clean 38 helmets in the Wallace Collection, amounting to around 0.57 litres of white spirit per helmet. The direct injection of the solvent, in comparison, produced 150 ml of waste solvent per object and used only 800 ml to clean three objects. This averages 266 ml per object, which is a reduction per object of about half the volume of solvent.

Acknowledgements

I would like to thank the Wallace Collection and David Edge for permitting and facilitating the study of the vambraces and allowing the reproduction of these images. Thanks are also extended to the Royal Collection Trust, Dr Alan Williams, Graeme McArthur and Simon Metcalf for their support and guidance.

Notes and references

1. Lemmer, G.M. (1972) 'The cleaning and protective coating of ferrous metals', *Bulletin of the American Group. International Institute for Conservation of Historic and Artistic Works* 12(2): 106.
2. Brühl, B. (2012) 'Copper soaps on ethnographic and decorative art objects', in *The Decorative: Conservation and the Applied Arts*, 2012 IIC Congress, Vienna. Available at: https://www.iiconservation.org/congress/sites/iiconservation.org.congress/files/IIC%202012%20Vienna%20Congress/student-posters/student-poster-511.pdf (accessed 9 June 2021). Neatsfoot oil, Ballistol, lanolin, beeswax, stearin, olive oil were all tested on commercial brass, bronze and copper for copper soap production.
3. Dalewicz-Kitto, S.J. (2016) 'Stopping corrosion around the world: an investigation into protective surface coatings on arms and armour', *Journal of the Institute of Conservation* 39(1): 49.
4. This was confirmed by FTIR analysis undertaken on several samples by Letitia Steer in other areas of the OMEA collection (unpublished research).
5. Allan, J.W. and Gilmour, B.J.J. (2000) *Persian Steel: The Tanavoli Collection*, Oxford Studies in Islamic Art 15. Oxford: Oxford University Press, p. 297; Alexander, D. and Khalili, N.D. (1992) *The Arts of War: Arms and Armour of the 7th to 19th Centuries*, Nour Foundation in association with Azimuth Editions and Oxford University Press. See catalogue entries 91 and 92 for examples of pierced work with gold foil underlay.
6. Weston, J. and Johnson, M. (1975) 'Benin bronze preservation', *African Arts* 8(4): 50.
7. Stoddard solvent is another name for white spirit, a desulfurised hydrocarbon solvent.
8. Cutulle, C. and Kim, S. (2015) 'Dry ice blasting in the conservation of metals: a technical assessment as a conservation technique and practical application in the removal of surface coatings', *Objects Specialty Group Postprints* 22: 77–100, esp. p. 89.
9. McArthur, G., 'OA1468 Conservation Report', The Wallace Collection, London, 8 August 2012.
10. Another solvent similar to white spirit.
11. For a full table of their solvent tests and results see: Werner, U., Selwyn, L.S., Stone, T., Ross McKinnon, W., Mackay A. and Grant, T. (2012) 'The removal of metal soaps from brass beads on a leather belt', *Studies in Conservation* 57(1): 20–23, esp. p. 14.
12. Norman Billingham, personal communication, 24 August 2017.
13. Ibid., 18 August 2017.
14. Robinet, L. and Corbeil, M-C (2003) 'The characterization of metal soaps', *Studies in Conservation* 48(1): 23–40.
15. Cutulle and Kim 2015 (cited in note 8).

9

Viking-era swords investigated by medical X-radiography and CT scanning: a new pommel classification and the assessment of blade geometry

Robert A. Hill

Introduction

The use of X-rays in archaeology is well established as part of the investigation of metallic artefacts.[1] X-radiography can reveal the presence and nature of an artefact even when hidden in a concretion of soil and rust, and it has also been used to investigate the structure of swords, including the blade, as in the 1989 study of pattern-welded swords by Lang and Ager.[2] Computerised tomography (CT) scanning is a routine medical technique and, like X-radiography, can also be used in archaeology. A CT scanner also employs X-rays but in this case the sources (X-ray tubes) and detectors are mounted inside a circular housing within the scanner, and the object (or patient) to be scanned lies on a table that slides gradually into the scanner.

A series of X-radiographs is taken from around 360 degrees in the vertical plane as the tubes and detectors rotate around the object to obtain a succession of cross-sectional images which a computer then processes to produce the final results. Although CT scanning is used principally to produce cross-sectional images, advances in medical technology and image processing mean that any sequence of planes and 3D images can be produced. Image processing offers other advantages such as the virtual slicing of an object or the subtraction of areas of the same radiological density. For example, it is not only possible to identify metal artefacts incarcerated in a soil block,[3] but also to produce 3D images of them that can be rotated and viewed from any angle, in other words 'virtual excavation'.

The most well-known archaeological use of medical CT scanning is for the analysis of Egyptian mummies for which it is particularly suitable because the radiological density of a mummy is almost identical to the intended use of the scanner. The radiological density of metal artefacts is much greater than that of animal or human tissue. As measured by attenuation of the X-rays, the radiological density of hard cortical bone is in the region of 3000 Hounsfield units (HU) whereas that of iron is around 24,000 HU. The energy of the X-rays produced by a medical scanner is in the region of 140 keV but in contrast a typical industrial scanner, such as those used to image car engine blocks for cracks, employs X-rays with a higher energy, in excess of 300–400 keV. Obviously the lower energies of medical CT scanners place some limitation on what might be achieved but provided the metal object is not too thick – and more importantly is composed of materials of different radiological densities such as wooden grips and even cracks or hollow cavities – then much useful information can be obtained. Medical CT scanners also

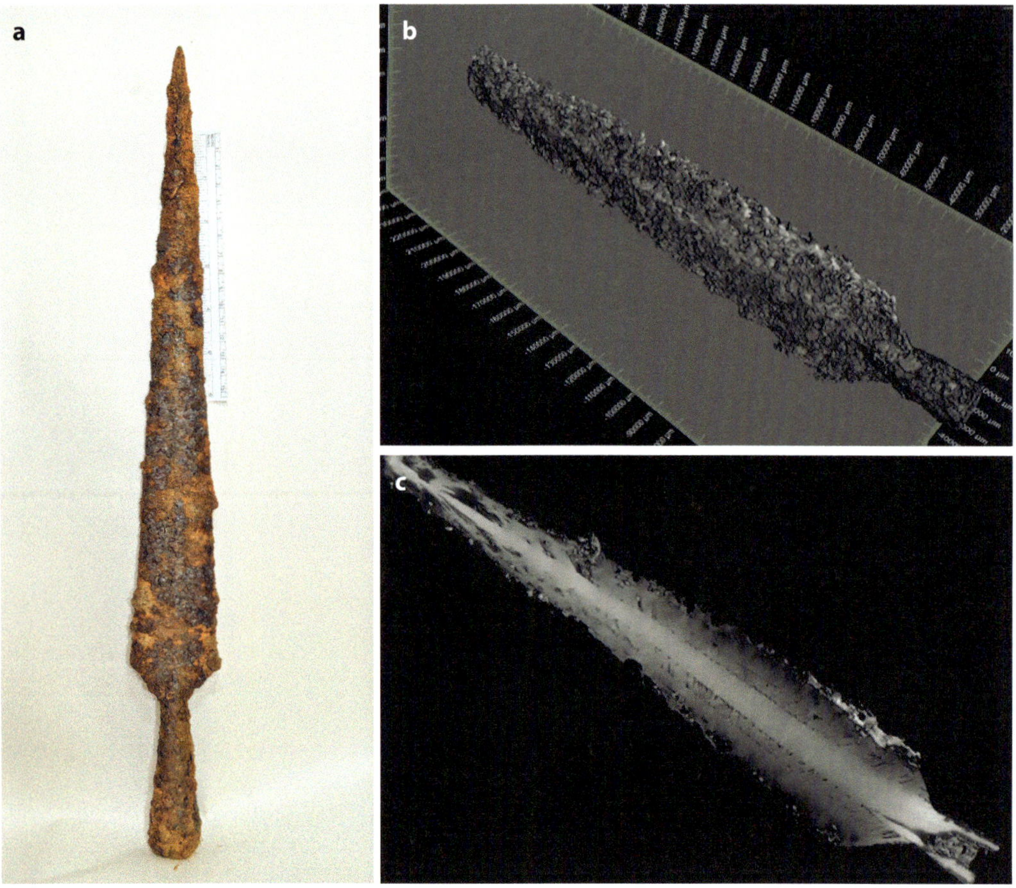

Figure 1 (a) A Viking period spear head; (b) micro CT 3D image; (c) micro CT section showing a panel of pattern welding.

have the advantage of being widely available and generally have sophisticated image processing packages.

More recently, micro CT scanners have been introduced into medical practice which have greatly enhanced resolution compared to conventional medical CT scanners. The most sophisticated of these scanners can work at a resolution of 1 μm (equivalent to 0.001 mm) with suitable specimens. This compares to the resolution of a medical scanner which is about 0.5 mm. However, from the point of view of practicality they are only suitable for small specimens – smaller than a sword – and use high energy X-rays, necessitating enhanced shielding; in medical practice they are restricted to post-mortem tissue examinations. A spear head is small enough to fit into a micro CT scanner and an example of the quality of such a scan is shown in Figure 1. In the future this will undoubtedly be a very useful technique in archaeology but at present there is very limited availability.

CT scanning and conventional X-radiography are complementary techniques and where possible both should be used in the investigation of swords. It is also important to emphasise at the outset that it is highly desirable to X-radiograph and scan excavated swords before attempting cleaning that might destroy evidence of any inscriptions preserved in the rust not visible to the naked eye but observable on X-radiographs. In general, conventional X-radiography does not have the same penetration as a CT scan so is less useful for the assessment of thick pommels. It can, however, give good definition of blades and may be preferable for the identification and

Table 1 CT scan compared with X-radiography.

Modality	Information	Technique/comments
CT topogram	Blade shape including presence of fuller	Long X-ray cassettes now not widely available. Topogram up to 200 cm
CT topogram	Measurements of length and width of blade and fuller and taper on blade	Accurate measurement using software. Blade angle useful in comparing types of blade
CT scan	Pommel construction	Not visible on X-ray. Can see tang inside pommel and any bolts. 3D multiplanar rendering
CT scan	Blade construction and pattern welding	Difficult but in some cases can identify layers such as pattern welding around central iron core. 3D volumetric rendering with section
CT scan	Separate visualisation of markings on each side of blade	X-rays superimpose markings. CT can separate if markings in good condition but minimal advantage over visible unless blade incarcerated. 2D or 3D volumetric rendering
CT scan	Measurement of radiological attenuation to deduce nature of metal, e.g. iron or bronze	Theoretically possible at any point but badly affected by radiological artefact and not reliable
CT scan	Virtual excavation	Good for swords stuck in scabbards. Windowing 3D volumetric rendered scan. 3D mulitplanar rendering
CT scan	Repairs especially hidden under grip etc.	May identify rehilting and composites
Digital X-ray	Hidden or obscure inscriptions	Can be seen by CT but not so easily
Digital X-ray	Nature of blade	Blades made from bloomery iron have heterogeneous streaky appearance. Identification of repairs especially resin or similar

assessment of inscriptions. But if there is an inscription on both sides of the blade, letters or symbols will be superimposed, making identification more difficult: with image manipulation, CT scanning can separate out both sides of the blade.

In this study of Viking and early medieval swords, a Siemens Somatom Definition AS+ CT scanner was used. This machine produces an initial topogram of the object and then self-calibrates the settings appropriately. However, in most cases these settings were adjusted manually for each object to maximise image quality. For metal artefacts, the X-ray energy was usually set at around 120–140 keV with a tube current in the range 36–74 mA. The slice interval was set at 0.6 mm. The images were processed with the Osirex 32 and subsequently the 64 bit post-processing package. The post-processing packages were used to reduce the radiological artefact and maximise image quality, carry out cross-sectional analysis, measurements and 3D reconstructions. Objects were also X-radiographed in most cases using a digital X-radiographic apparatus with adjustment of the keV and mA to produce the best images. As the images are digital, subsequent manipulations such as magnification or inversion of the image are possible. With CT scanning there may be a particular problem with radiological artefacts degrading the image and care has to be taken not to misinterpret these scan images.

A preliminary study was carried out to assess the feasibility of using the CT scanner in the examination of metallic artefacts and to determine what useful information could be obtained.[4] As part of this study, custom settings were adjusted to obtain the best image quality and three different post-processing packages were compared for ease of use. The better preserved the sword, the better the images: initially it had been hoped that CT scanning would be useful in identifying worn makers' marks and separating out layers of pattern welding but this proved difficult. A CT scan cannot reverse the effects of corrosion and once a mark is obliterated the scan cannot bring it back. Phase contrast neutron imaging has recently been shown to be of value in revealing makers' marks obscured by over-polishing,[5] however, this preliminary study did identify some useful areas for further research and helped to clarify the possible roles of X-radiography and CT scanning in the examination of swords (Table 1). In particular, the CT topogram was helpful for assessing blade morphology including the taper, the extent of the fuller and for accurate measurements of length and width. The CT scans also facilitated assessment of the construction of the pommel and upper guard, which led directly to the current study.

Pommel classification by CT and problems of rehilting

Viking-era swords have been classified on the basis of the appearance of the pommel by Petersen et al.[6] and on the shape of the blade by Geibig.[7] These classifications are essentially descriptive and do not take construction into account, although Geibig did define three types of pommel construction in 1991, which he termed *Konstruckiontyp I - III.* Wilson[8] also described three types of pommel construction in late Anglo-Saxon swords which he termed A, B and C.

CT scanning permits a non-destructive assessment of the method of pommel construction and hence enables a new classification based on construction. It has generally been believed that the manufacture of sword blades and the fitting of their hilts were separate activities in this era. There is particular interest in 'Vlfberht' marked blades and it was thought interesting to determine whether different 'Vlfberht' blades were all hilted in the same way and if certain pommel designs shared the same type of construction. With the limited technology available in the 9th–11th centuries, there could only have been a finite number of ways to attach a pommel to a tang, but by adding a knowledge of construction to design it might be possible to identify regional or individual variations in manufacture. Blades may have been rehilted or their pommels repaired, both events that can be identified by CT scanning. In addition, the congruency of the fit of the tang to the pommel slot may be assessed.

It is inevitable in a study such as this that some swords may fall under suspicion of being modern copies, concerns that can be addressed by both CT scanning and X-radiography. The only definite way of identifying modern copies or restorations is by metallurgical analysis but this is both destructive and time-consuming. Radiological investigation can help to determine which swords may need metallurgical analysis.

A total of 44 Viking era and 6 early medieval swords from private collections were CT scanned to determine their pommel construction; most were also X-radiographed. The results of the pommel study are listed in Table 2 and all these pommels appear in the appendix (a catalogue of pommels accompanied by a CT image). The terms 'coronal', 'sagittal' and 'transverse' are used to denote the plane of the image. Most of the CT images are coronal, i.e. 'face on', analogous to a human form. The cross-hilt or guard is like the outstretched arms and the flat surface of the pommel is the face of the body. Sagittal images are taken from the side and transverse images are cross-sectional. There are two types of pommel construction in this heterogeneous group:

- *Type 1* in which the pommel is made of one component.
- *Type 2* in which there are two separate components – a pommel and an upper guard.

Type 1 is the simplest form of construction and may be found in some swords of the Viking era and even pre-Viking swords. But later in the Middle Ages, the simpler Type 1 construction becomes almost universal – the more complex Type 2 pommels seem to be confined to the Viking era and may depend upon the individuality of swords of this period, their fashion, or perhaps their prestige as a weapon. Many Type 2 pommels and upper guards of this era are beautifully decorated and this seems to have parallels with Anglo-Saxon England, although many Type 1 pommels of the Viking era also feature decoration. In the medieval era, the style of decoration becomes less elaborate, with some

Table 2 List of pommels.

Type	Description	Number of this type
One piece neither ferrous nor bronze		
1	One example – ivory but possible reconstruction. One piece pommel/upper guard.	1
One piece ferrous		
1Fa	Tang in – pre-Viking	1
1Fb	Tang out peened flush or capped - simple, widespread, medieval type	14
1Fc	False two-piece pommel	8
One piece bronze		
1Ba	Tang in	1
1Bb	Tang out	1
1Bc	False two-piece pommel, ? Curonian type	1
Two piece ferrous		
2F	No rivets – tapering tang peened over pommel. 1 of known Baltic origin. Others *	4
2Fa	Straight rivet – common Viking type	10
2Fb	Angled rivet	1
2Fc	Bent rivet *	4
2Fd	U rivet – ornate pommel possibly hollow	2
2Fe	D rivet/staple – attached to solid pommel	1
Two piece bronze (Categories may need extension)		
2B	No rivets, tapering tang, Curonian	1

* Denotes sword where some doubt as to provenance or appearance of the blade on the X-ray or CT scan

rare, possibly royal or noble, exceptions, and this, together with the adoption of the simpler Type 1 construction, might indicate a trend towards the mass production of swords rather than individual commissions.

Both Type 1 and Type 2 pommels may be ferrous or bronze as denoted here by the suffix 'F' or 'B'. In the rare instance the pommel is neither ferrous nor bronze (e.g. for sword Viking 6), no suffix is added. Type 1 pommels may be further subdivided by their method of construction into three types (Figure 2):

- *Type 1a* Tang wholly inside the pommel – this corresponds to Geibig I.
- *Type 1b* Tang protrudes out of the pommel and is peened or capped – this corresponds to Geibig III. If peened it may be flush to the top of the pommel.
- *Type 1c* The pommel is made to resemble a two-piece construction and may even include false rivets. In this study, only a few of this type were found, and in those cases the tang did protrude from the top of the pommel. This group would require further subdivision should any pommels be discovered in the future with tangs located within them.

Type 2 pommels may be similarly divided according to the method of construction but the technique used to attach the upper guard to the pommel must also be taken into account:

- *Type 2* No attachment between the upper guard and the pommel. The tang is trapezoid preventing the upper guard from sliding down and secured by pommel with the tang peened over.

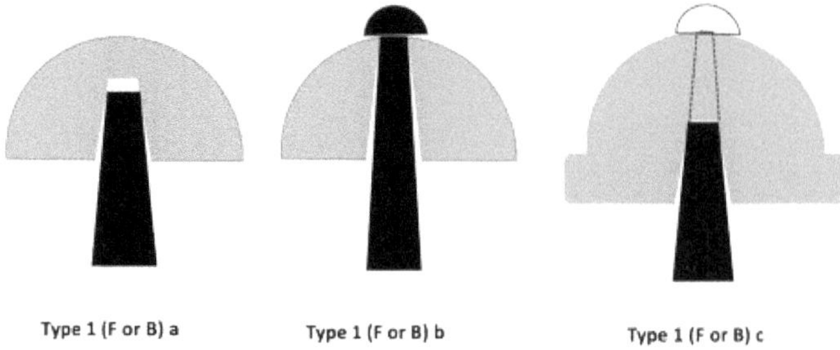

Figure 2 Type 1 pommels.

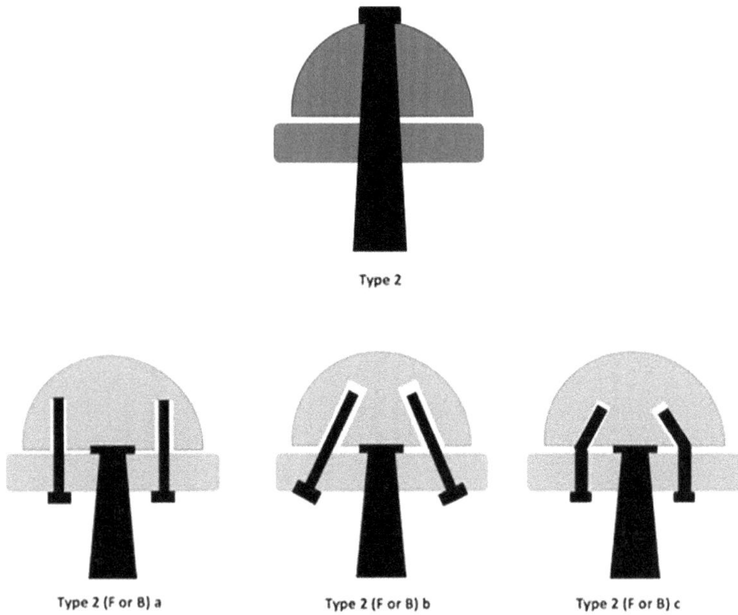

Figure 3 Type 2 pommels (F or B) a–c.

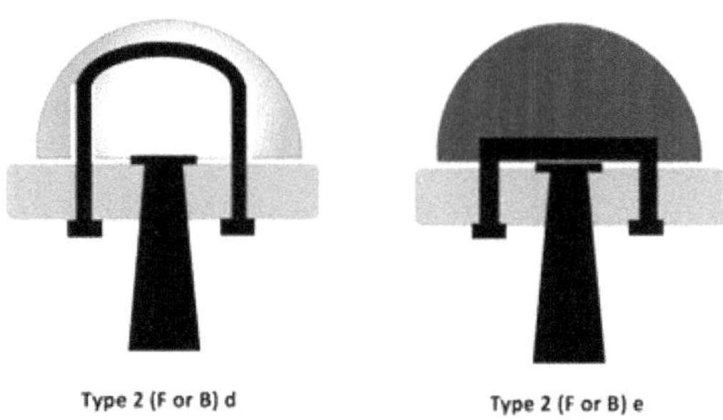

Figure 4 Type 2 pommels (F or B) d–e.

- *Type 2a* Upper guard attached to the pommel by straight rivets (this corresponds to Geibig IIb).
- *Type 2b* Upper guard attached to the pommel by angled rivets.
- *Type 2c* Upper guard attached to the pommel by bent rivets.
- *Type 2d* Upper guard attached to the pommel by means of a U-rivet or ring. Pommels attached in this way seem to be hollow or at least filled with radiologically low density material (this corresponds to Geibig Type IIa). Perhaps this is some form of resin.
- *Type 2e* Upper guard attached to the pommel by means of a staple-shaped rivet which might be welded to the base of a solid pommel.

In Type 2a–e pommels, the tang is peened over the top of the upper guard and the pommel then attached to the upper guard by some form of riveting, with varying degrees of complexity, as described above. There is a tendency for this type of pommel to be decorated. It is unclear what advantages the Types 2d and 2e might have had over the simpler and more common Type 2a but this may be related to the sequence of the attachment and decorating process because it does not seem likely that a finely decorated pommel would then be heated up in a forge. Type 2d would seem to have been less secure than some of the other types, and it is noteworthy that some swords have been found in which the U-rivet is still present on the upper guard while the pommel is missing (as illustrated in the Petersen classification *aertype* 20, no. 135). This raises the possibility that in this type the pommel was designed for easy removal with a valuable, prized pommel being removed or exchanged, for example, for a plain pommel for battle or even attached to a new sword blade. It should also be remembered that the vast majority of Viking-era swords have either been found in graves or as sacrifices to the Pagan gods and the decoration of the pommel also had to be appropriate for this purpose.

All the Type 2 pommels in this series were ferrous with one exception (39), which had a separately cast bronze pommel and upper guard, but with no visible rivets to connect the two (Type 2B). The style is Curonian and there are several known examples of swords from this area with cast and decorated bronze pommels, although many are Type 1c. Viking 37 is a Type 2F of known provenance from a single warrior burial in Latvia. A very similar sword to 37 is described by Svarane from Saraji, Latvia.[9] Viking 37 was found in pieces and very poor condition, requiring extensive conservation. Despite, or perhaps because, of this, it was possible to obtain a good deal of scientific information on the sword. Careful conservation revealed the presence of silver decoration on the pommel, and metallurgical analysis of a complete section of blade showed an iron core with steel edges welded on. Strictly speaking, a CT scan was not needed as the pommel construction became apparent during conservation. As this is a Type 2F there were no rivets and the tang was peened over the top of the pommel. For this type of construction, the tang would have to be trapezoidal in shape otherwise the guard would slip down. This should be compared with Viking 39, described above, an equivalent in bronze of Type 2B, also with a trapezoidal tang and again from the Baltic area.

CT scanning is also very useful in detecting repairs or rehilting, as shown in Viking sword 18: either the slot in the pommel was made too large and it was necessary to insert a metal dowel to secure the pommel to the tang or the sword was rehilted with a new pommel. In Viking sword 32, the tang is clearly broken just above the upper guard where it enters the pommel – what may have occurred is a matter of conjecture but perhaps the pommel was originally Type 1, then the tang broke off inside, so the pommel came off and was subsequently attached to another blade, thereby converting it into a Type 2.

In Medieval sword 3 the pommel is bronze, and while visual inspection might suggest that the tang was peened over the top of the pommel and then capped, the CT scan shows this was not the case. The tang enters the base of the pommel then seems to bifurcate into a void with some dense material (perhaps some form of solder) but never passes the mid-point of the pommel, which is solid at this level. This seems to have been a rehilting, possibly for reasons of fashion. These three examples show that when rehilting has been carried out it can be very difficult to detect without CT scanning because this technique has the advantage of being able to image the pommel in all three planes.

It is clear that similar pommels may be attached to the tang in different ways, which would support the hypothesis that hilting was a separate activity to blade manufacture and that this was a geographically widespread practice. Viking swords 17, 33, 41 and 30 have K style pommels (Petersen classification) although the photographs show that there are subtle differences not only in shape but in surviving decoration (see the appendix). Sword 41 in particular exhibits a radical difference from the other three: the CT scans not only show a different technique of attachment but also that this has a one-piece pommel compared to two-piece pommels on the other three. This demonstrates the value of assessing pommel construction as sword hilts that appear very similar may in fact be made quite differently. Indeed, anomalies such as these might raise concerns over provenance and metallurgical testing may then be considered necessary.

In addition to the decorative and constructional aspects, it is also important to appreciate the function of the pommel. In swords of this era the pommel served to prevent the hand slipping off the end of the grip and losing the pommel in battle: this could potentially be disastrous and whatever method of construction was used had to be able to prevent this from happening. In the more complex Type 2 pommels there were various ways of using rivets to attach the pommel to the guard and if the pommel did detach then the upper guard would remain with the blade. It is possible that in this group, the prepared upper guard was attached first with the heated tang firmly peened over the top of the guard and then, in a second stage, the cold, decorated pommel was attached with heated rivets in Types 2a–c. Of these three types, Type 2a would have been the least secure although the simplest. In Types 2d and e, the U-ring or staple may have been welded or somehow attached to the pommel which was decorated to match the upper guard. The ends of the rivet were then passed through preformed holes in the upper guard, heated to softness and peened under the guard. These pommels had the potential to be exchanged, particularly Type 2d. The weight of the pommel is also important as it balances the sword. Some pommels appear hollow or possibly filled with radiologically dense material, probably for sword balance as well as to facilitate exchange. Large pommels can also be used as a punching weapon but this seems less likely in this particular group.

Blade geometry by CT

In medical use, the initial CT topogram (a projection radiograph) is used to predict the current settings needed to produce the desired image quality and the scanner can then self-calibrate. In this study, the settings were entered manually but the CT topogram is still very useful as an object up to 2 m long can be fitted onto one image without any divergence of the X-ray beam, thus facilitating accurate measurements without any magnification factor. The topogram can then be used for a number of purposes:

- Accurate measurement of the length and width of a sword or part of a sword.
- Easier identification of the extent and shape of fuller in a corroded blade.
- Measurement of the taper of a blade.

There are several classifications of blade shape of which perhaps that of Geibig,[10] with 14 types, is the most useful. While length measurements may be included, there is a tendency to be merely descriptive when it comes to the shape or taper of the blade and the extent of the fuller. The illustrations of these typographies have ideal diagrams clearly showing the taper and the fuller. However, in practice, with a recently excavated rusty sword, it is often difficult to see the fuller or describe its shape. A topogram greatly facilitates assessment of the blade type as the fuller can be seen more clearly in a manageable-sized image and a variety of accurate measurements is possible including blade taper.

The CT software used is equipped with Cobb angle measurement. The Cobb angle is used in orthopaedics to measure a spinal curve (scoliosis), but it can also be employed to measure the taper of the sword blade on a CT scan (Figure 5). The coronal or profile shape of non-curved sword blades has evolved and changed from

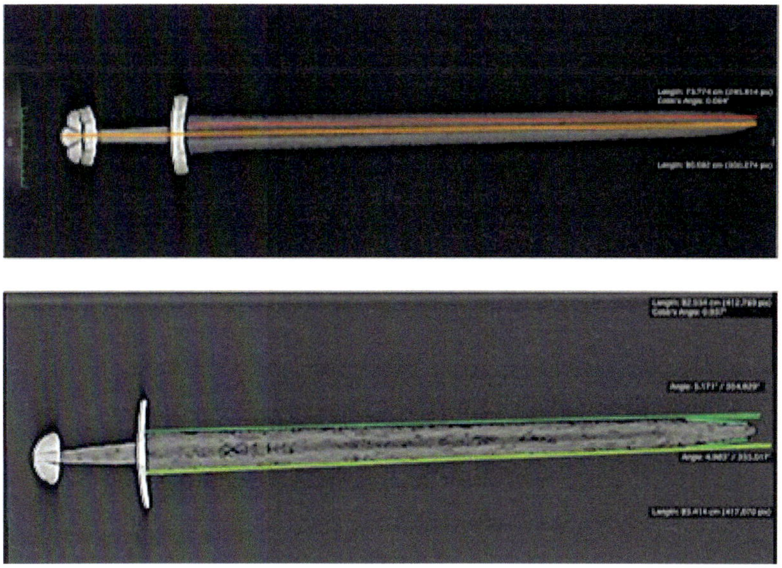

Figure 5 Blade measurements with topogram.

fairly straight-sided blades with a point of varying sharpness to blades that are more acutely tapered although there is often a further taper towards the point, This can be seen in Figure 5. The angle of both tapers can be measured. In addition, if a sword blade is viewed side on (saggital) there is also a further taper from the lower guard down to the point. This can also be measured by CT but this was not done in this study.

Examination of 'Vlfberht' swords

There has been considerable interest in 'Vlfberht'-inscribed swords of this era and therefore these were selected for closer analysis of their blade geometry (see the appendix, Viking nos. 4, 9, 13, 16, 24, 31, 36, 40). As well as interest there is controversy over the origin of these blades. There would seem to be agreement that these swords were produced sometime between AD 800 and 950,[11] and were the output of a workshop, rather than an individual, and it has been speculated that they had a Frankish connection. Frankish landowners raised armoured cavalry[12] and it has been suggested that the + preceding and following VLFBERHT might have had some ecclesiastical connotation.[13] A 'genuine Vlfberht' sword feels well balanced in the hand and is more than just a blade inscribed with the correct order of letters: these swords have improved design and metallurgy with the use of consistently high quality steel. Williams reported a metallurgical study of specimens from 44 swords with 'Vlfberht' or related inscriptions and concluded, in the swords studied, that those with the inscription +VLFBERH+T were made of hypereutectoid steel (greater than 0.8% C) or eutectoid steel (0.8% C), whereas those with the inscription +VLFBERHT+ had hypoeutectoid (0.4–0.6% C) steel edges hardened by quenching.[14] Swords with the many other variants of the spelling had a variable metallurgy ranging from medium-carbon steel down to iron or very low-carbon steel. It has been suggested that some of these swords were contemporary copies made with inferior metal and not produced by the original 'Vlfberht' smiths. Forging high-carbon steel, quite apart from requiring a source of consistent quality, would have been difficult, since it has to be done at a lower temperature (as the melting temperature range of steel falls with increasing carbon content), and presumably such techniques were known to only a few smiths. The hilting of these swords probably took place elsewhere as a variety of hilt types are found on 'Vlfberht' blades, doubtless to suit the taste of the owner.

Table 3 Blade geometry of 'Vlfberht' swords.

Reference number and notes	Pommel construction type	Blade length lower guard to tip (in cm)	Fuller length (in cm)	% Fuller length to blade length	% Fuller width to blade width at lower guard	Shape of tip	Proximal blade taper (in degrees)
24. Index sword +VLFBERH+T ?IIXXIII	1Fb	81.15	69.36	86%	37%	Rounded	0.96
9. Contemporary ? Eastern copy. Metallurgy known +VLFBERH+?	2Fa	77.25	60.89 *	79% *	51%	Sharp	0.72
4. Corroded with some edge losses +VLFBERHT+	2Fd	80.17	61.12 *	76% *	32% *	Probably rounded	0.97*
16. Blade suspicious +VLFBERHT+	2Fa	74.1	71.21	96%	54%	Sharp	1.08
31. Bent rivet +VLFBERHT+	2Fc	74.81	69.58	93%	49%	Sharp – same as 13	1.08
13. Blade suspicious bent rivet +VLFBERTH+	2Fc	75.27	72.4	96%	42	Sharp – same as 31	0.92
36. Repaired cremation sword – very likely blade has been shortened +VLF......	2Fd	72.96	**	**	39%	Probably rounded	1.17*
40. +VLFBERHT+ Pattern welded letters IIIXXXIII	2Fe	73.77	65.89	89%	30%	Rounded	1.24

* Corrosion might have affected accuracy of measurement. ** Too corroded to measure accurately.

As noted by Williams,[15] there are many different spellings of 'Vlfberht', some of which are considered indicative of contemporary, lower quality, copies. However, over a period of 200 years there may have been some design changes in the 'genuine Vlfberht' swords as well. The design features of a 'Vlfberht' sword may include the width and extent of the fuller and the taper on the blade. While changes in design are more likely to have been evolutionary rather than sudden, it is possible that the shape of an early 'Vlfberht' blade differs from a late 'Vlfberht' blade. If these swords were indeed of Frankish origin then design changes might perhaps have been related to developments in the Carolingian and later the Ottonian army, and specifically the increasing importance of the mounted, armoured, soldier. Coupland mentions changes in design possibly to make swords more manoeuverable.[16] Very little attention has been paid to measuring these parameters except for an article by Jones in which some measurements on Geibig's blade Types I–V of the Viking Age are described.[17] As pointed out by Stalsberg, the Geibig blade classification is not very helpful in this context because of the overlap of date ranges; a more precise assessment would be desirable.[18]

In this study, measurements were undertaken on eight 'Vlfberht'-inscribed swords. The blade length from the lower guard to the tip was

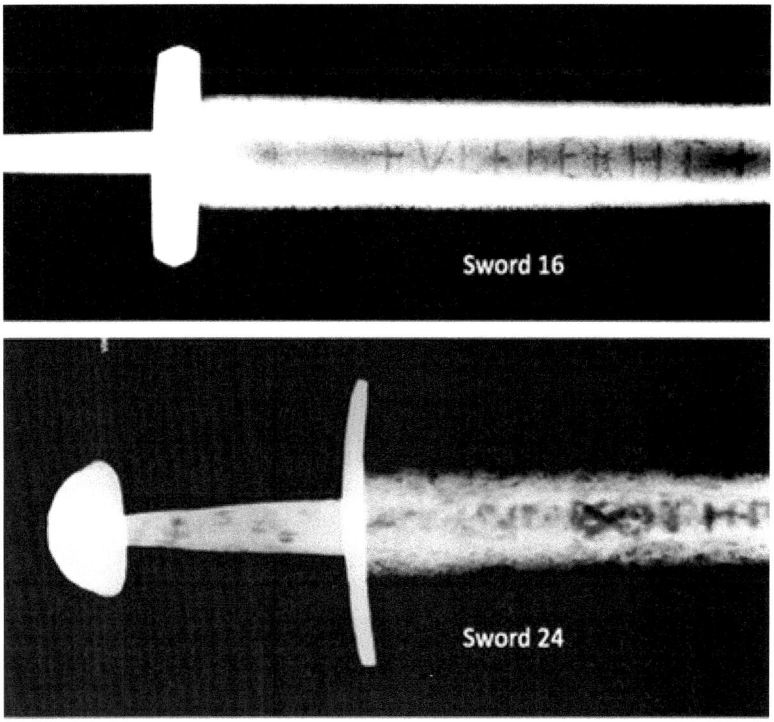

Figure 6 Digital X-rays of Swords 16 and 24. Sword 16 shows a homogenous uniform appearance with minimal corrosion whereas Sword 24 shows a streaky appearance with clear evidence of corrosion.

measured together with the fuller length, which was then expressed as a percentage of the blade length. The blade taper, the width of the blade and the fuller just below the lower guard were also measured. A CT topogram with post-processing software was used for these measurements (see Figure 5).

The results are presented in Table 3. Some allowance has to be made for the inaccuracy of measurements on corroded blades. Two of these eight swords (13 and 16) had suspect features, i.e. no heterogeneity of the blade on X-radiography (see below and Figure 6), while three of the other six have reasonable provenances and no suspect features. One was found in a cremation burial in pieces and repaired (36), another on the basis of its pommel shape might be a contemporary Eastern copy (9) and the third (24) came from an important private collection with good provenance including a known find site. The latter, which is perhaps the most reliable sword for the purposes of comparison, was taken as an index, and its blade taper and the extent of its fuller compared to those of the other six. As any 'Vlfberht' sword may have had a date range of up to 200 years, we did not expect to find identical blades and hilts but believed that perhaps there would be some similarities in proportion.

'Streaking' of the blade on an X-ray suggests a heterogenous, laminar composition as might be expected from the forging of bloomery iron and steel, which would contain appreciable quantities of slag inclusions elongated by forging, as well as possibly corrosion products. A completely uniform appearance is what would be expected from modern (i.e. post-Bessemer) steels which have been melted to homogenise them (see Figure 6 for an example of this finding).

With respect to pommel classification, of the two swords with suspect features one had bent rivets (Type 2Fc) and the other straight rivets

(Type 2Fa). Of the other six 'Vlfberht' swords, one of which is probably a contemporary copy as its metallurgy was low-carbon steel (0.3–0.4% C), two had Type 2Fd pommels, one had a Type 1Fb, one a Type 2Fc, one a Type 2Fa pommel and the last a Type 2Fe. This variety of types supports the hypothesis (suggested above) that blade manufacture was a separate activity to hilting and that good blades may have been rehilted, perhaps in a different style. There is also the matter of balance and correlating this with pommel type. An analysis of pommels in a large group of 'Vlfberht' swords would be interesting. The blades of these swords are discussed below (see the appendix for the pommels and their corresponding CT scans).

Assessment of the blade construction and geometry of 'Vlfberht' swords

The sword used as an index (no. 24 in Table 3) has an inscription on the reverse of the blade which can be read as IIXII. While some 'Vlfberht' swords do have such an inscription or a similar one, others do not, further adding to the confusion outlined above. This sword has a blade length of 81.5 cm and a fuller length of 69.36 cm representing 86% of the length of the blade. At the lower guard, the fuller occupies 37% of the blade width and the profile taper on the blade is 0.94 degrees. There are obvious problems with only comparing this sword to the seven other 'Vlfberht' swords studied here, and further research on a larger sample would be of greater value.

The results that are of interest are the percentage fuller length and the blade taper. As noted above, two of these swords (13 and 16) were considered suspect on other grounds: they both had long fullers – 96% of the blade length compared to 86% in the index sword and 79%, 76% and 89% in three other 'Vlfberht' swords (4, 9 and 40). Sword number 36 is excluded because it was repaired and shortened. The final sword (31), which was not considered suspect, also had a very long fuller (93%) and interestingly the pommel was a Type 2Fc with bent rivets like the suspect sword number 13. The Type 2Fc pommel construction needs further evaluation.

With respect to blade taper, all the 'Vlfberht' swords with the exception of swords 9 (which might be contemporary Eastern copy) and 40, have a taper of about 1 degree. Sword 9 is more parallel-sided with a taper of 0.72 degrees. The 9th-century Arabic writer, Ya'cub Ishaq al-Kindī, describes Viking swords and distinguishes between Frankish swords and the swords of the Rus or Slimanish.[19] He observed that swords of the Rus are noted to taper less and have less rounded points than the Frankish swords, possibly supporting the view that sword 9 is of Eastern origin. Sword 40 is slightly shorter and has the most acute taper. It is said to have been found in East Anglia and has a pommel Type VII, based upon Wheeler's classification of sword hilt types associated with British find sites.[20] Its general form would suggest a 10th- or 11th-century sword, so it is a later 'Vlfberht'. It is very well balanced – perhaps late 'Vlfberht' blades were more acutely tapered.

In conclusion, based on this very small sample, it is evident that 'Vlfberht' swords had a variety of pommel types, both by design and construction, reflecting individual ownership. Their edges are not parallel-sided but have a taper approaching 1 degree. The fullers are probably less than 90% of the length of the blade and less than 40% of the width of the blade just below the lower guard. These features might have contributed to the maneuverability of these swords.

Repaired and composite blades

X-radiography and CT scanning are useful in identifying blade repairs which might have been contemporary or modern. A sword might be broken in battle and then repaired, perhaps only for use as a burial offering or possibly deliberately broken for burial and repaired after excavation. A good example of both these sorts of repair is shown in Figure 7, a 'Vlfberht' sword from a cremation burial which was found in pieces and a modern reconstruction carried out by the current private owner. The repair is very skilful and not easily detectable by the naked eye. Interestingly, there appears to be a further repair in the blade (nearer the hilt) using a different technique with

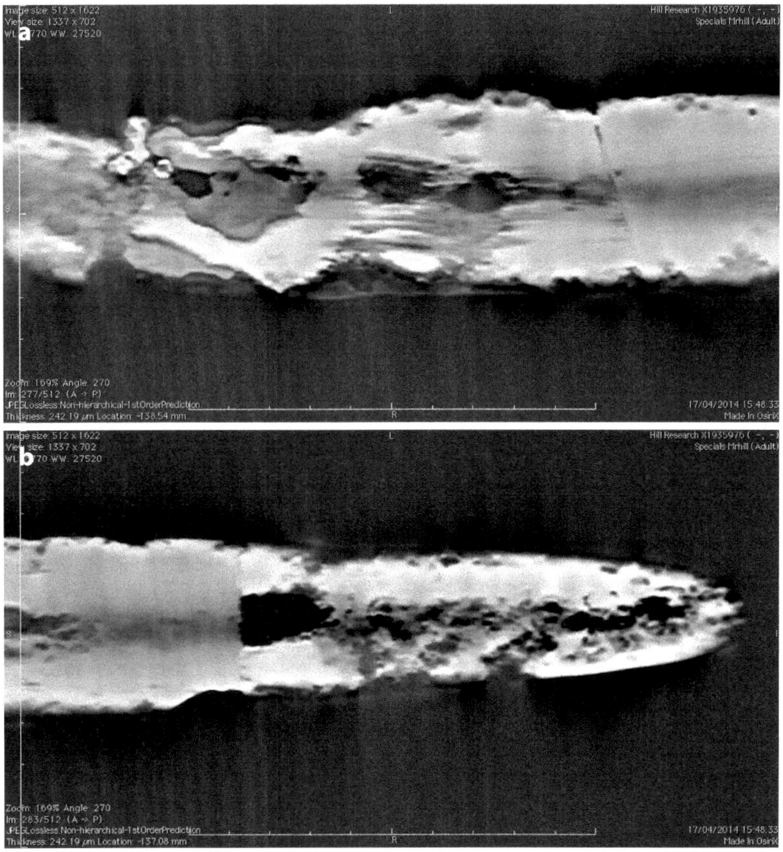

Figure 7 Sword 36: (a) mid-blade repair and (b) tip repair.

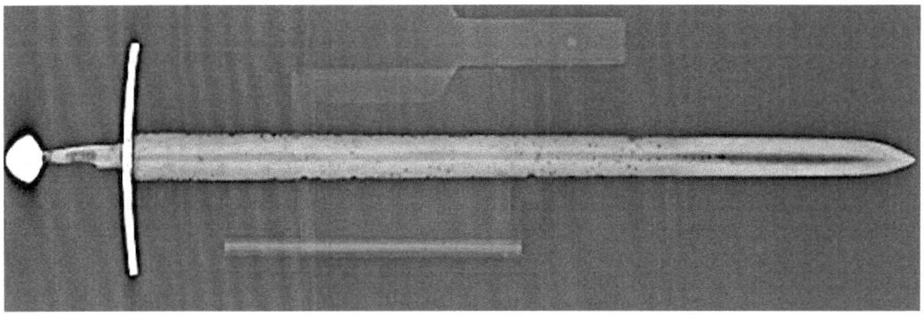

Figure 8 Topogram of composite blade.

some form of riveting, which might have been contemporary. A composite sword is made up of parts from different swords or perhaps an original part might be incorporated into a modern copy. The suspicious collector (or curator) may take a small specimen for metallurgical analysis but unless this is removed from the right area or areas, incorrect conclusions may be drawn.

Figure 8 illustrates another medieval sword (not part of the group studied here) in which the CT topogram clearly showed a composite blade: it was then possible to take specimens from the two separate areas for metallography which, as expected, showed quite different appearances under the microscope. One sample consisted of ferrite only with a little grain boundary cementite and no slag, consistent with modern metal. The other sample comprised ferrite and some pearlite, which was partly divorced, and some oxides, consistent with a low-carbon steel (0.2% C), possibly medieval. In medical parlance this would be termed a CT guided biopsy!

Conclusions

The assessment and investigation of archaeological artefacts should be multidisciplinary both in concept and scope. There is increasing use of scientific techniques to aid the understanding of not just what a sword looks like but what it is made of and how it was put together. This is a small study of 50 swords from private collections but it has yielded some very interesting results, most importantly using non-destructive analysis. Unfortunately, the provenance of many swords in private collections is unknown which does devalue the interpretation of the results, and vigilance is needed to guard against the possibility of copies and composites. It would be very useful to extend this study to a larger number of swords ideally of known provenance, as this would enable analysis with less speculative conclusions. Industrial CT scanners may be more accurate than medical CT scanners but this advantage is to some extent outweighed by the availability of the latter.

Appendix: Catalogue of pommels

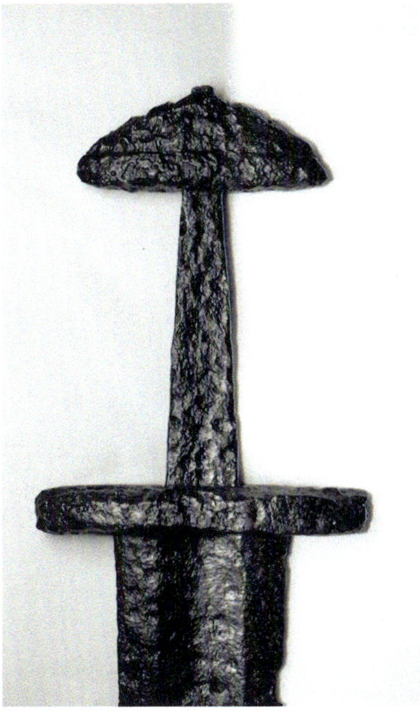

Viking 1: hilt

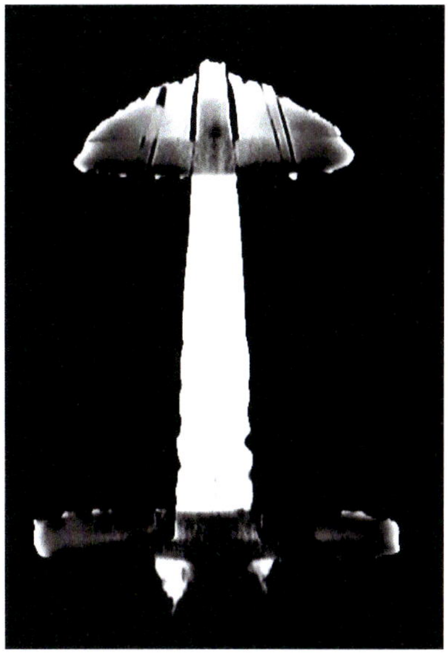

Viking 1: coronal CT scan (pommel Type 2Fb)

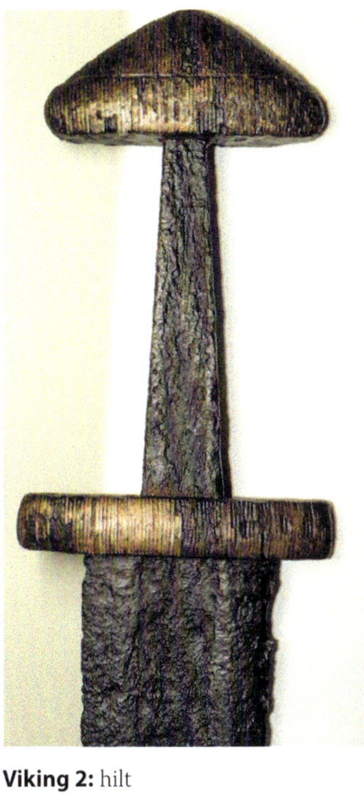

Viking 2: hilt

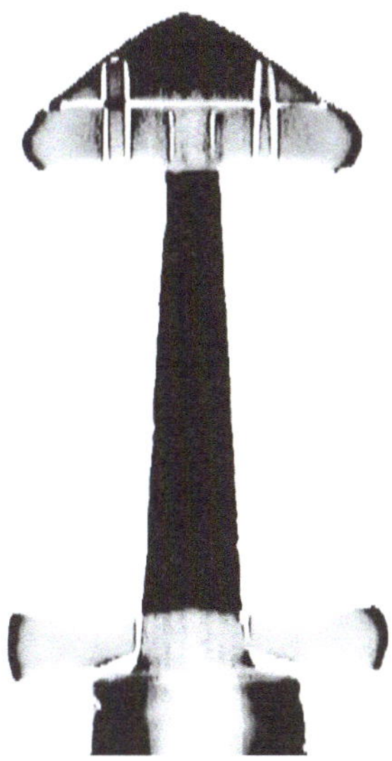

Viking 2: coronal CT scan (pommel Type 2Fa)

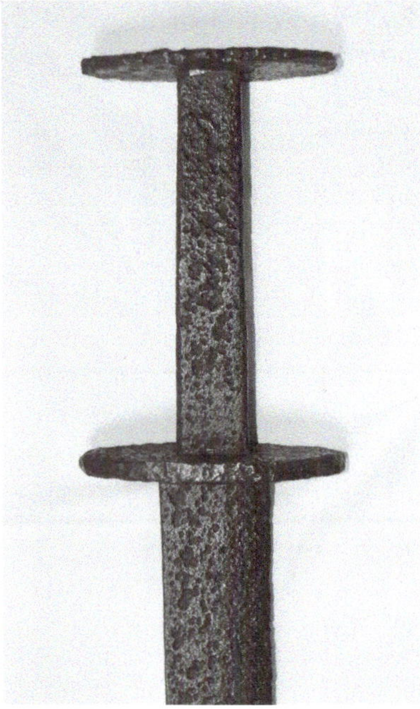

Viking 3: hilt

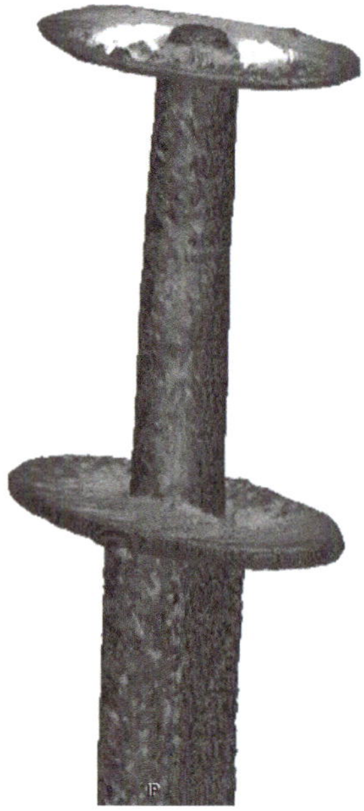

Viking 3: 3D image from CT scan (pommel Type 1Fb)

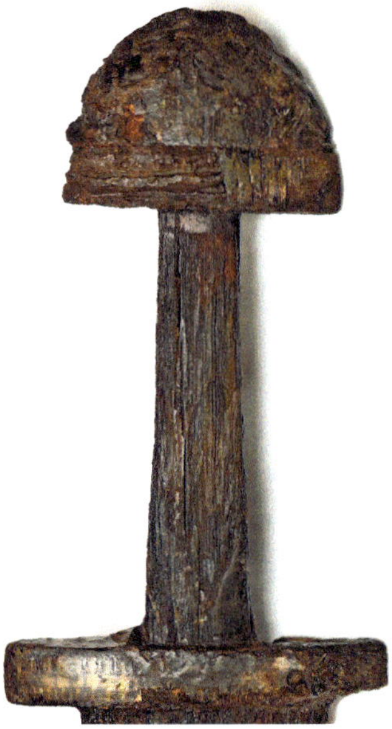

Viking 4: hilt (Vlfberht)

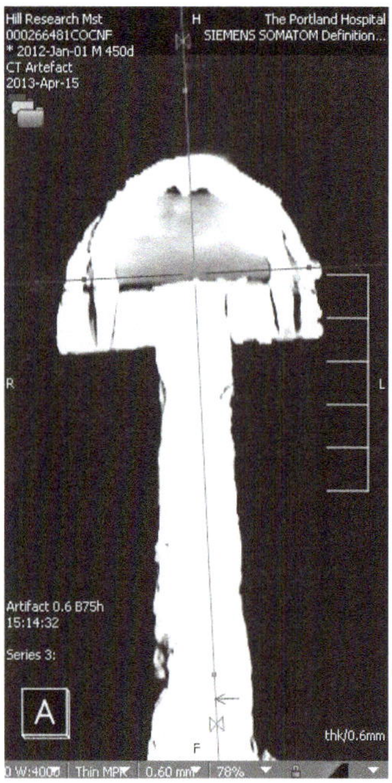

Viking 4: coronal CT scan (pommel Type 2Fd)

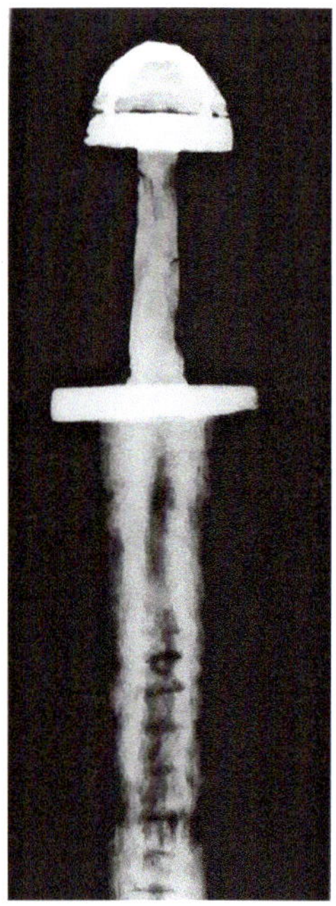

Viking 4: X-radiograph image

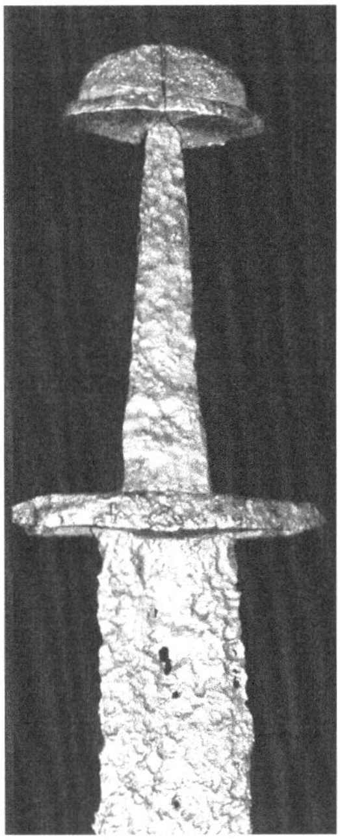

Viking 5: hilt

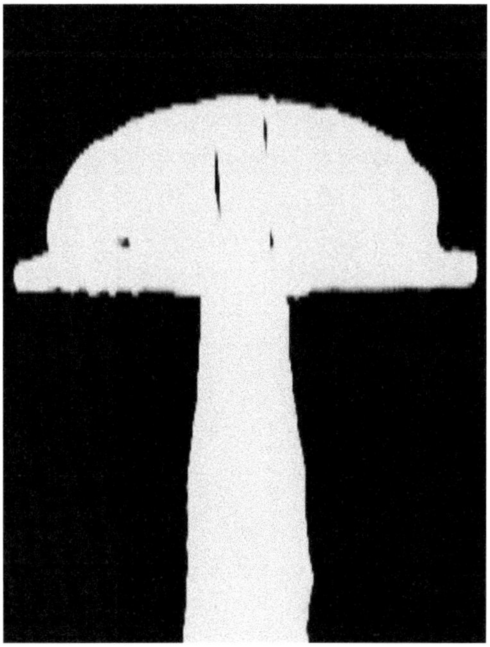

Viking 5: coronal CT scan (pommel Type 1Fc)

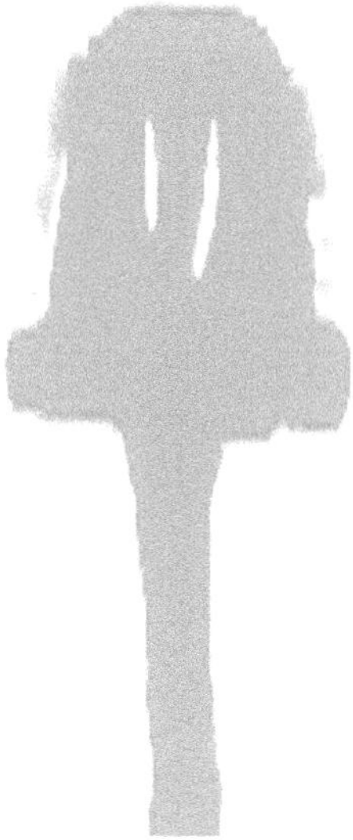

Viking 5: sagittal CT scan

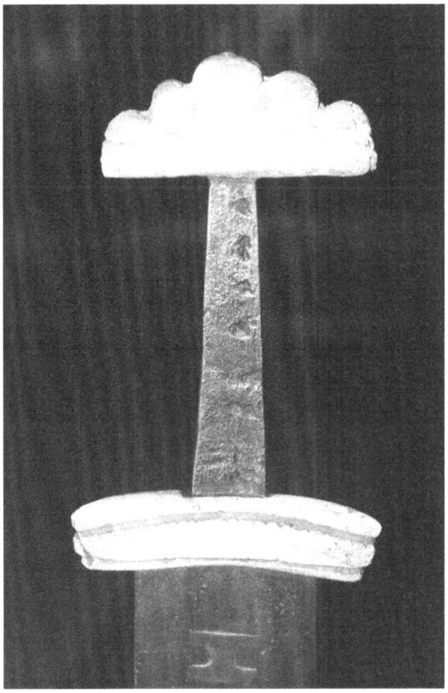

Viking 6: hilt

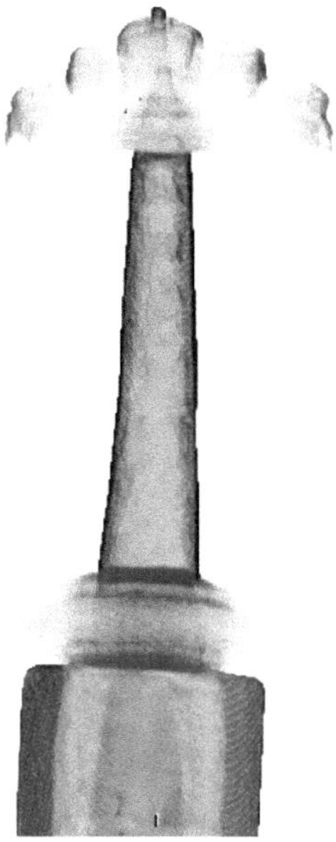

Viking 6: coronal CT scan (the ivory pommel is a possible replacement Type 1)

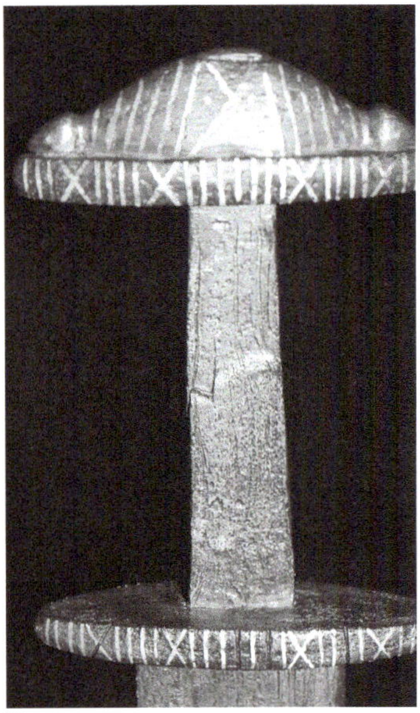

Viking 7: hilt

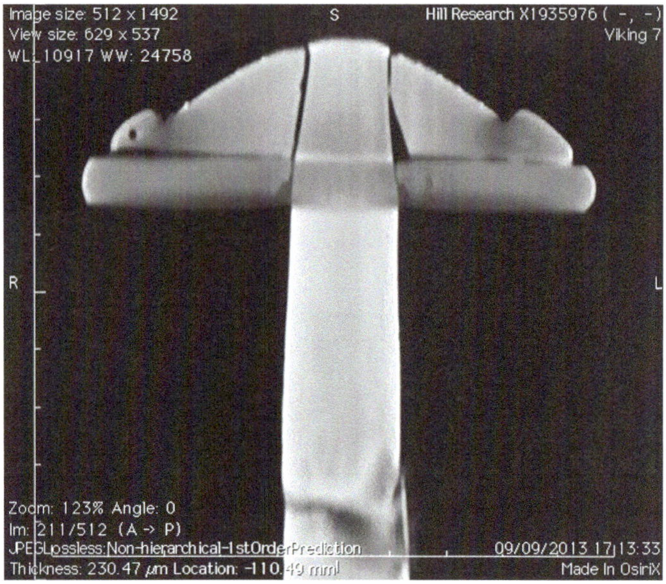

Viking 7: coronal CT scan (pommel Type 2F)

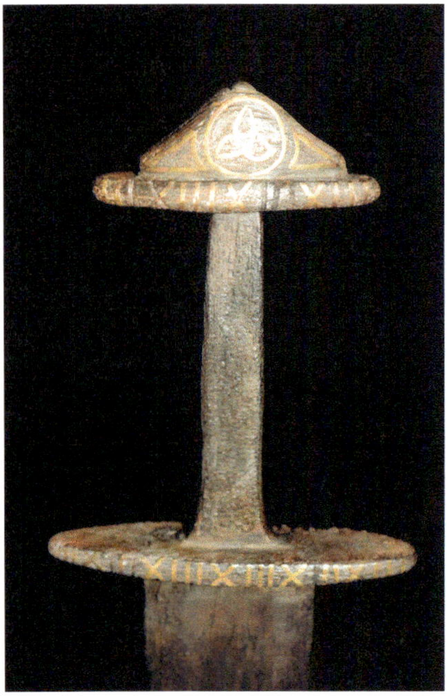

Viking 8: hilt

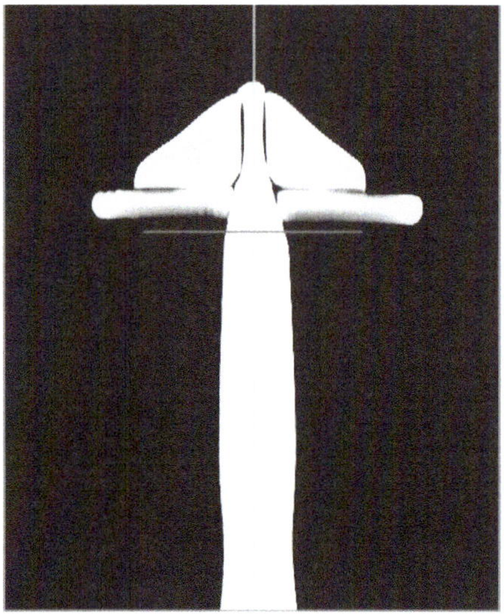

Viking 8: coronal CT scan (pommel Type 2F)

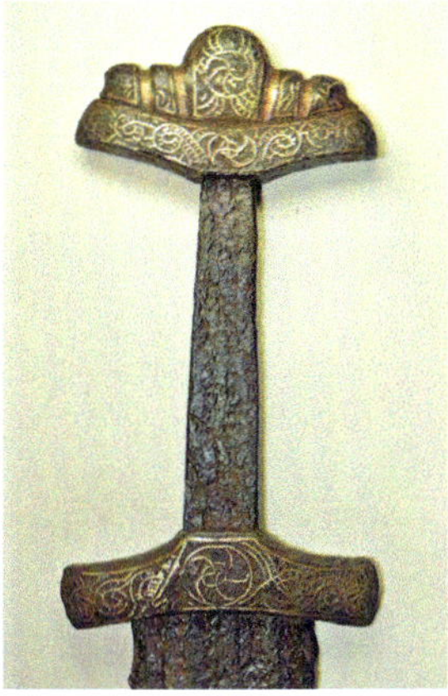

Viking 9: hilt (Vlfberht)

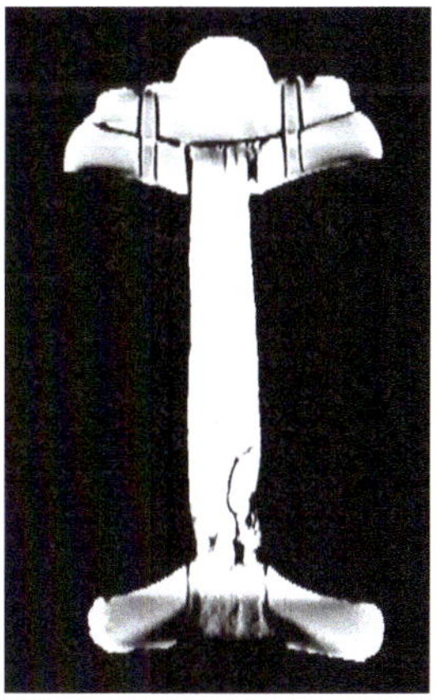

Viking 9: coronal CT scan (pommel Type 2Fa)

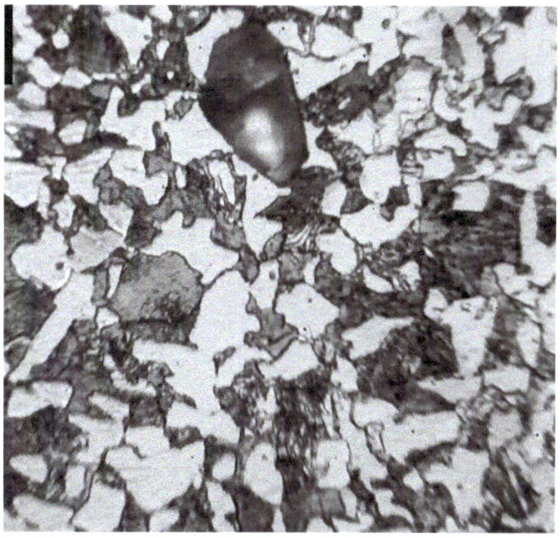

Viking 9: photomicrograph of blade (metallurgy is pearlite, ferrite and slag inclusion), scale bar = 50 μm

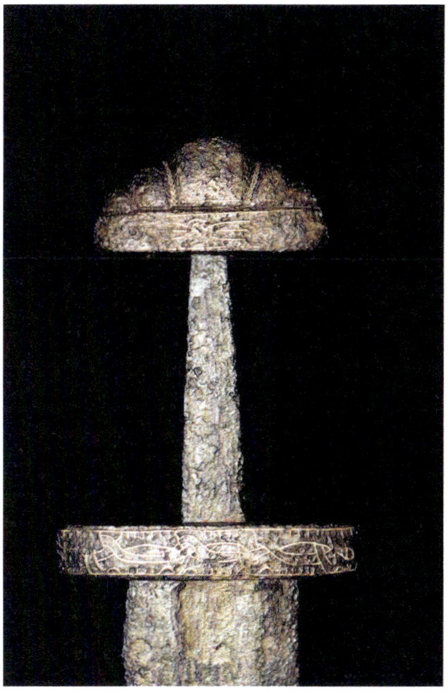

Viking 10: hilt

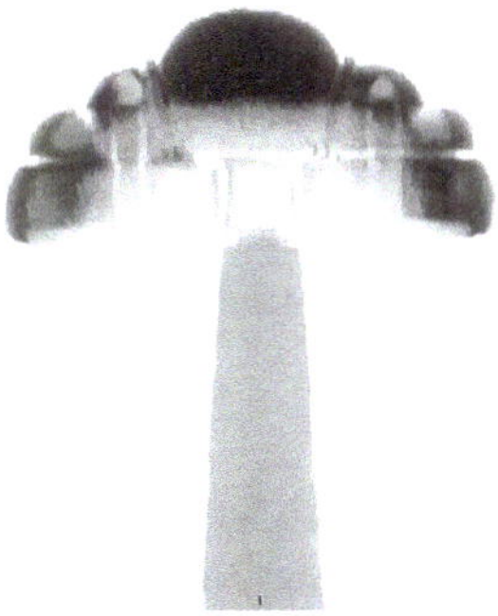

Viking 10: 3D image from coronal CT scan (pommel Type 2Fa)

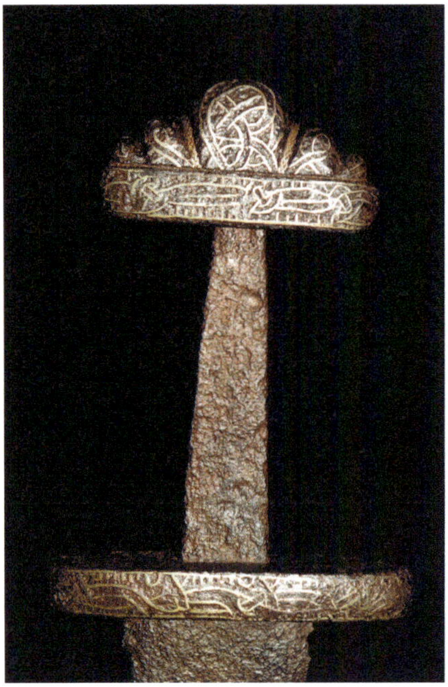

Viking 11: hilt

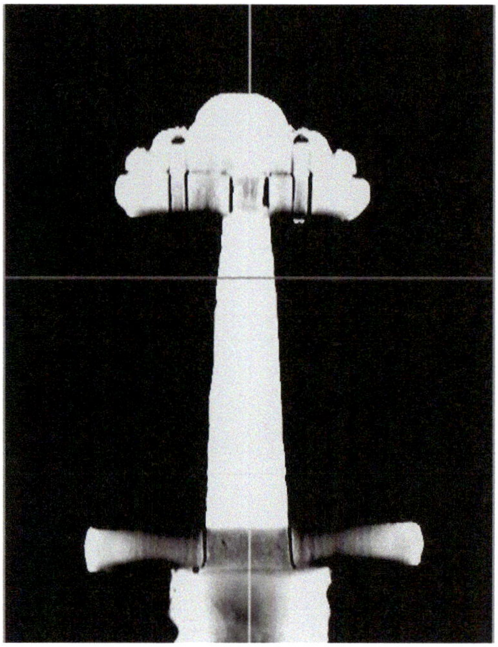

Viking 11: coronal CT scan (pommel Type 2Fa)

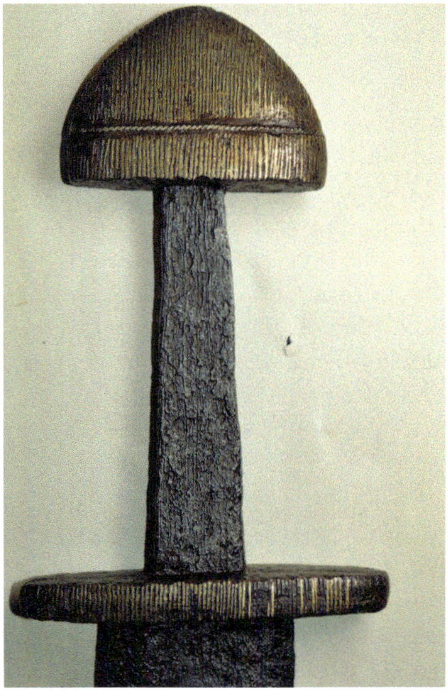

Viking 12: hilt

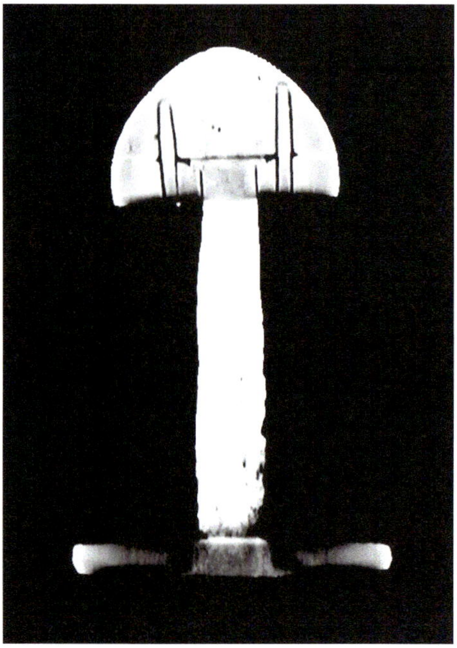

Viking 12: coronal CT scan (pommel Type 2Fa)

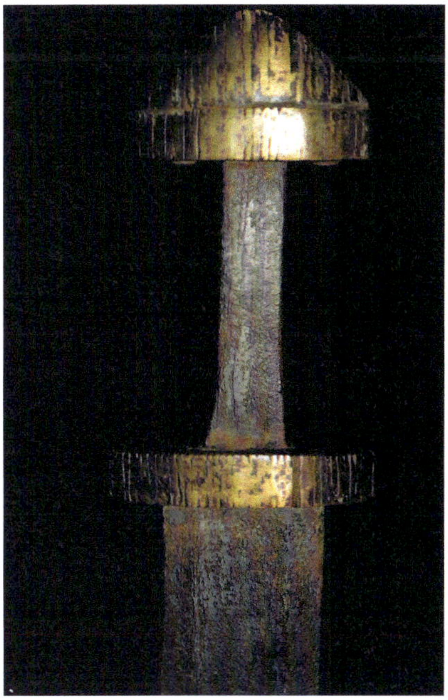

Viking 13: hilt (Vlfberht)

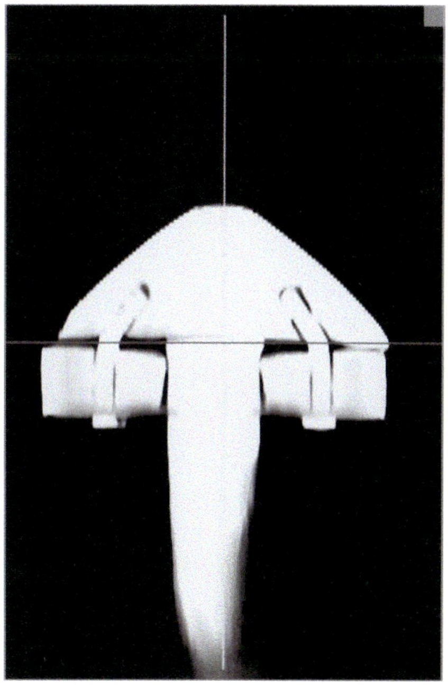

Viking 13: coronal CT scan (pommel Type 2Fc)

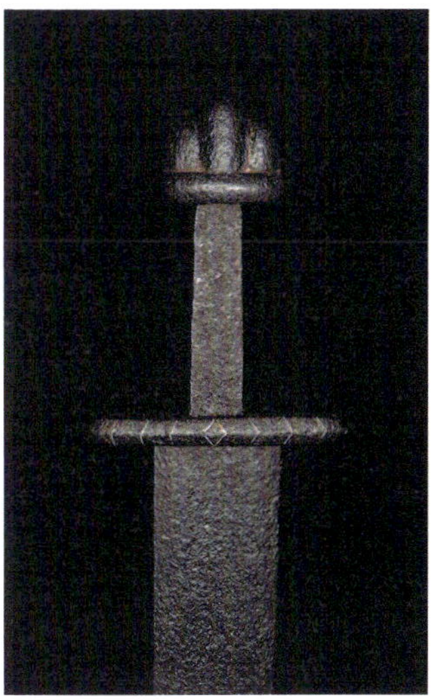

Viking 14: hilt

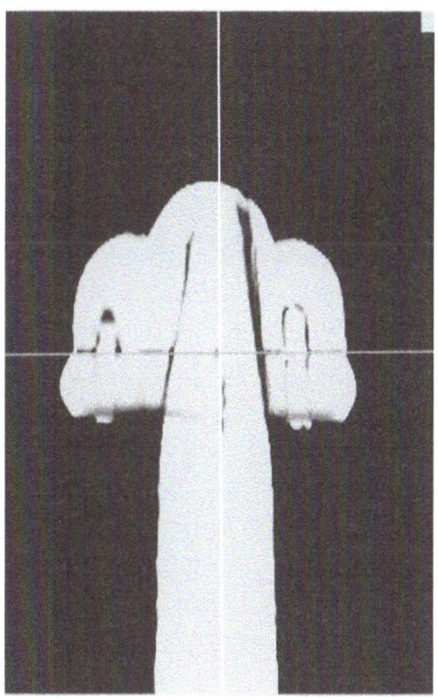

Viking 14: coronal CT scan (pommel Type 1Fc)

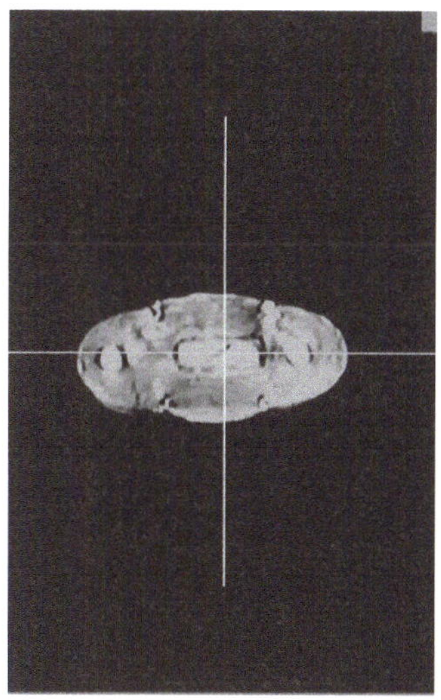

Viking 14: CT cross-section of pommel

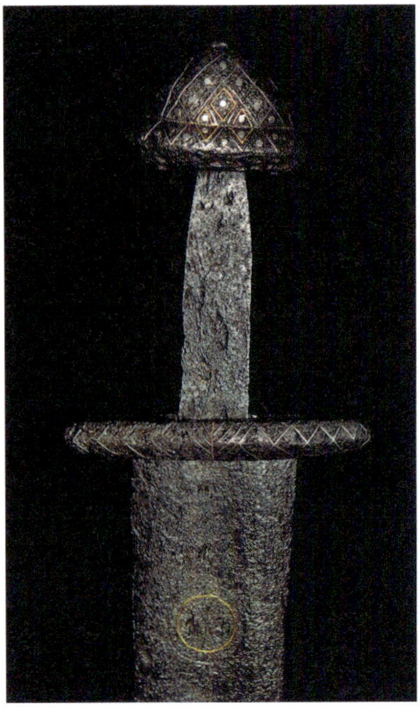

Viking 15: hilt (with suspicious embellishment of decoration)

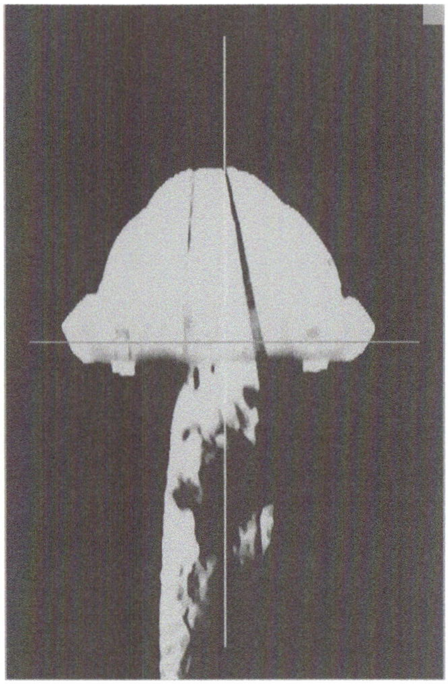

Viking 15: coronal CT scan (pommel Type 1Fc)

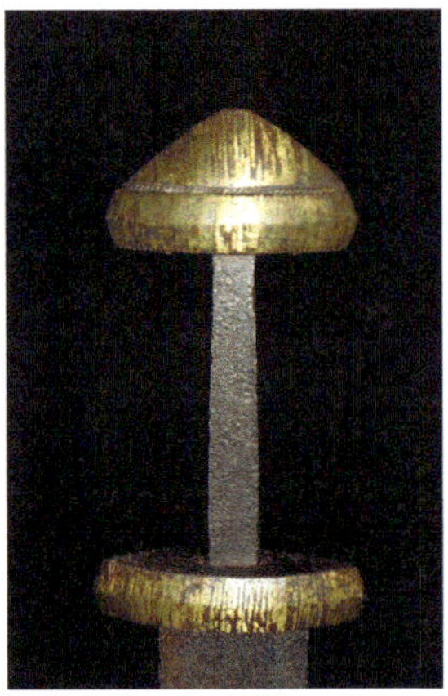

Viking 16: hilt (Vlfberht)

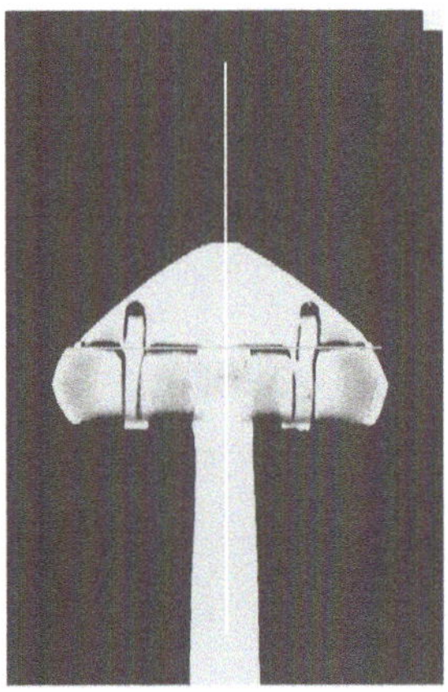

Viking 16: coronal CT scan (pommel Type 2Fa)

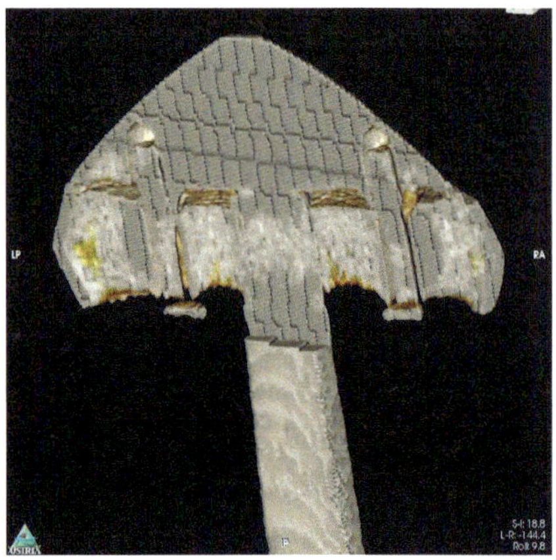

Viking 16: oblique 3D CT scan: cross-hatching is artefactual

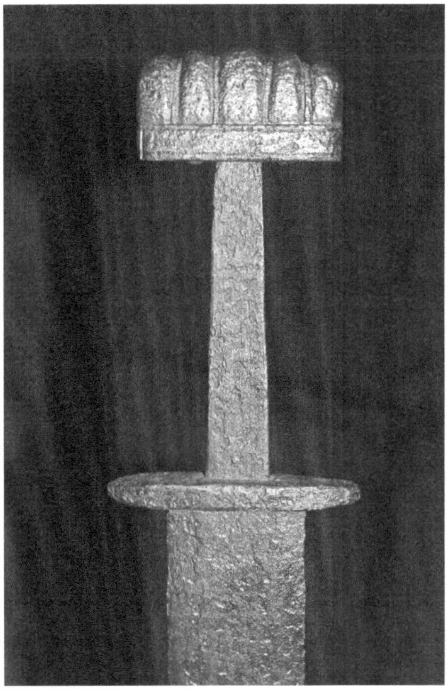

Viking 17: hilt

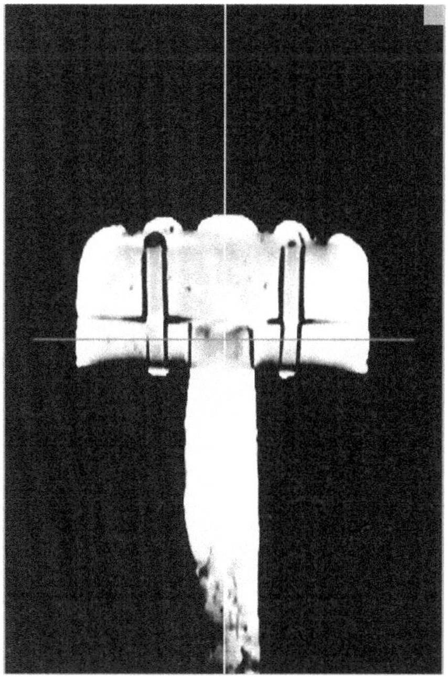

Viking 17: coronal CT scan (pommel Type 2Fa)

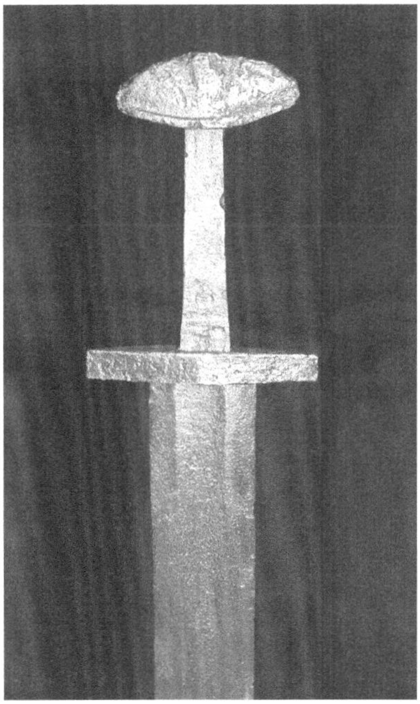

Viking 18: hilt

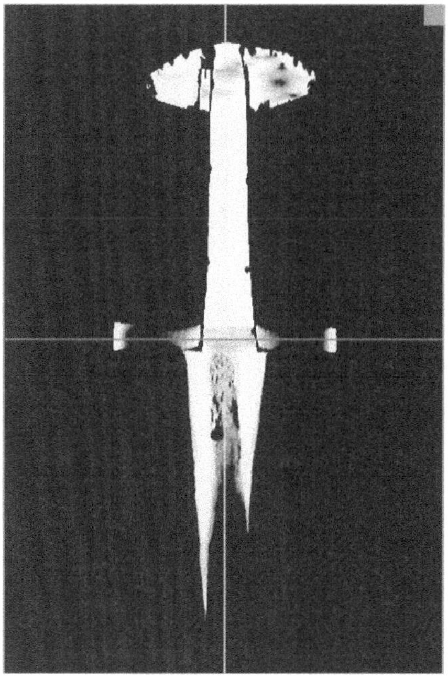

Viking 18: coronal CT scan (pommel Type 1Fb)

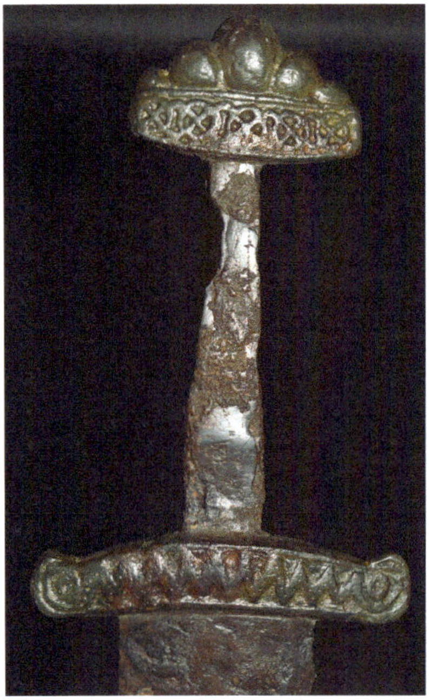

Viking 19: hilt (bronze pommel and crossguard)

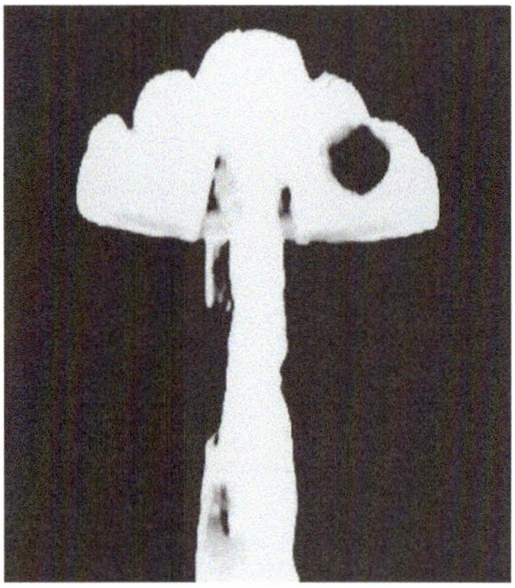

Viking 19: coronal CT scan (pommel Type 1Bc – note void)

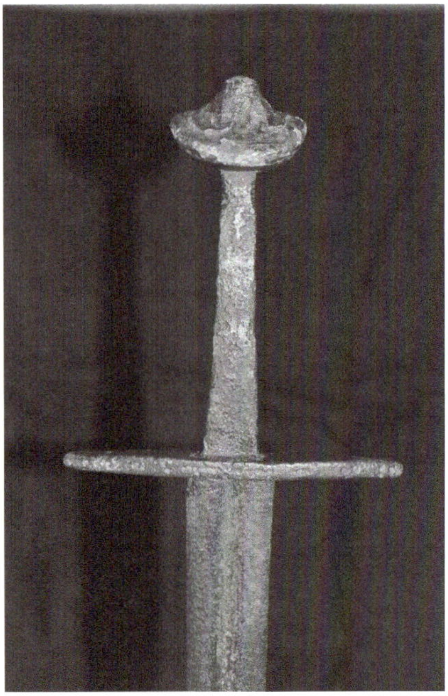

Viking 20: hilt

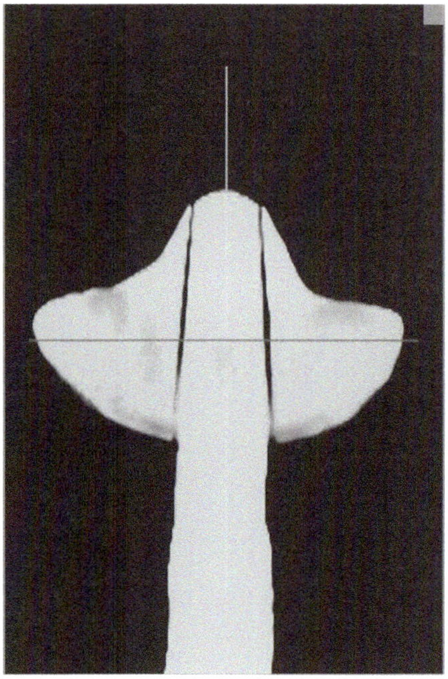

Viking 20: coronal CT scan (pommel Type 1Fb)

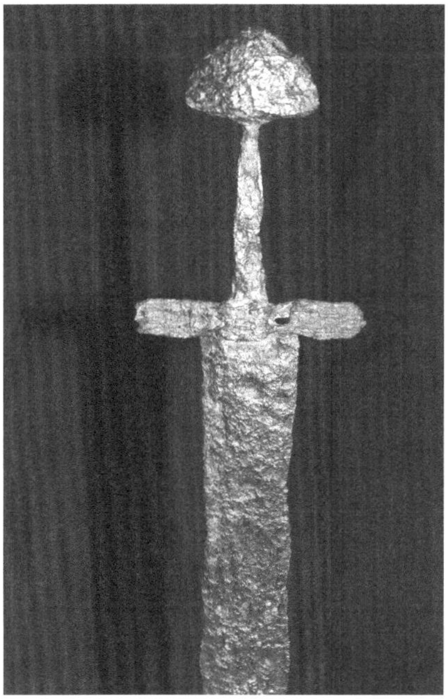

Viking 21: hilt

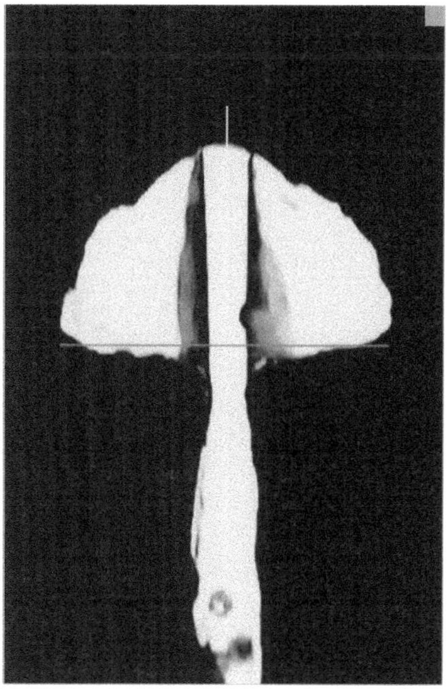

Viking 21: coronal CT scan (pommel Type 1Fb)

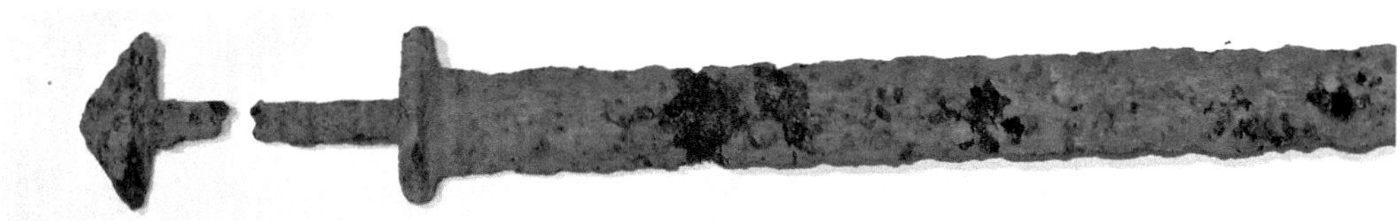

Viking 22: hilt

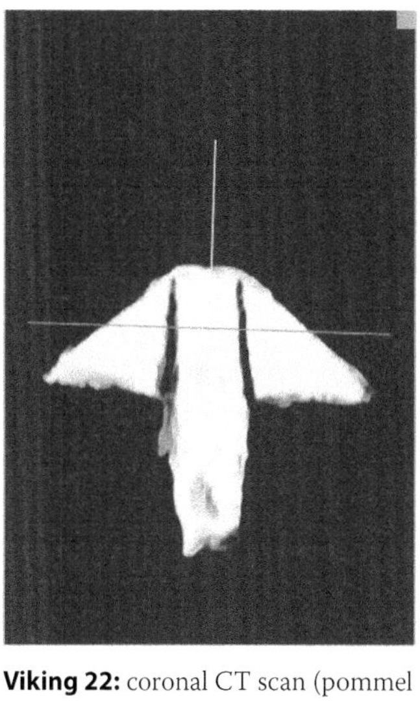

Viking 22: coronal CT scan (pommel Type 1Fb)

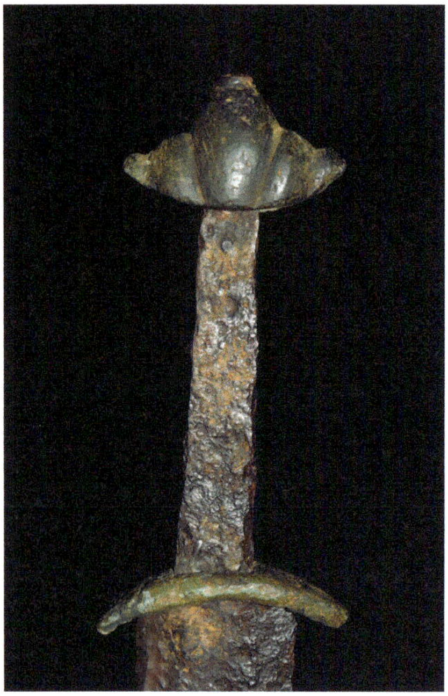

Viking 23: hilt (bronze pommel and crossguard)

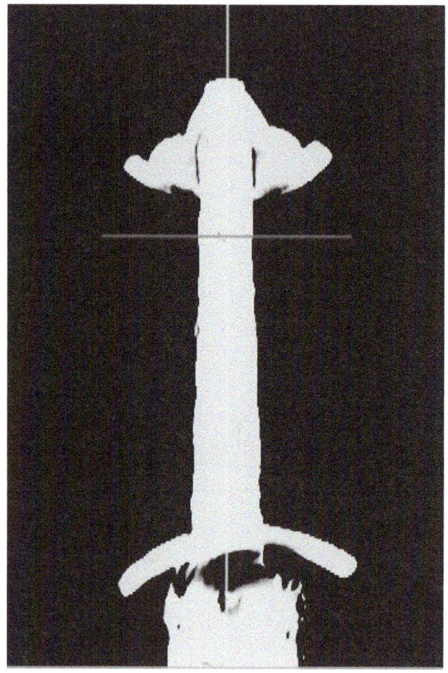

Viking 23: coronal CT scan (pommel Type 1Bb)

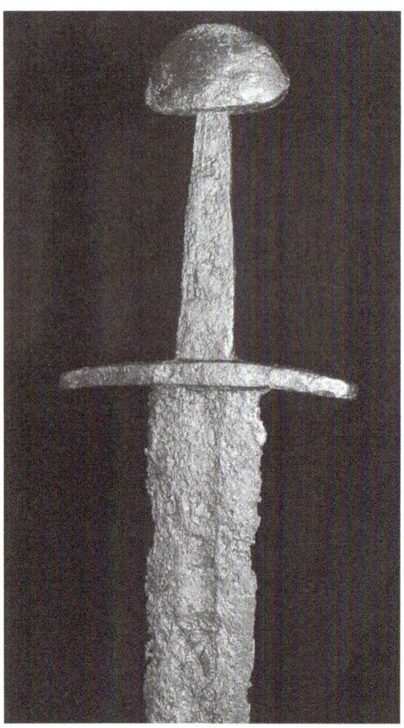

Viking 24: hilt (Vlfberht)

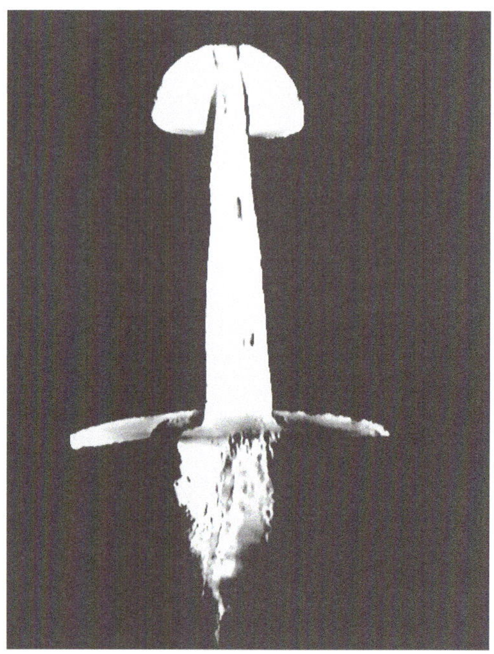

Viking 24: coronal CT scan (pommel Type 1Fb)

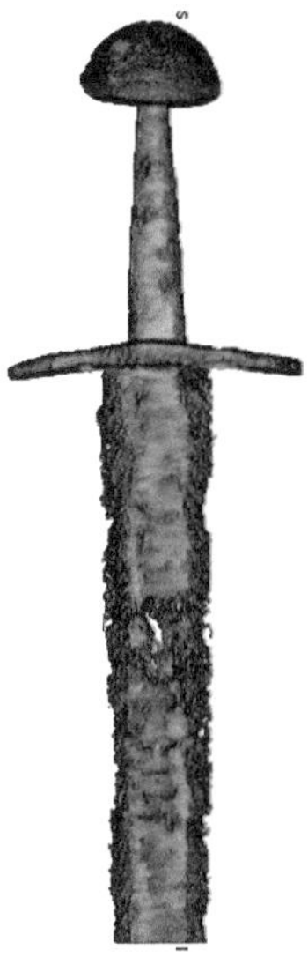

Viking 24: VR image from CT

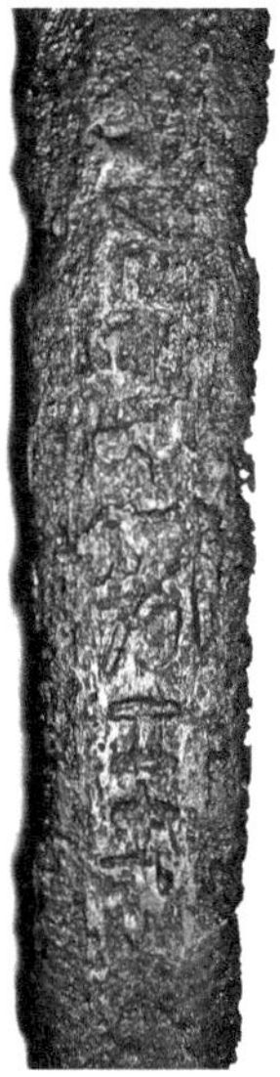

Viking 24: Vlfberht inscription

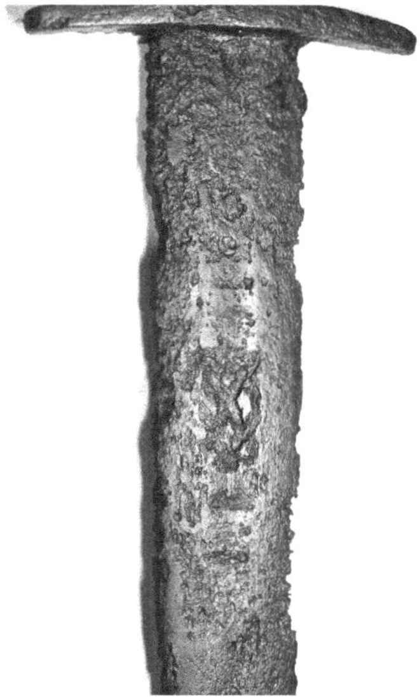

Viking 24: reverse of inscription

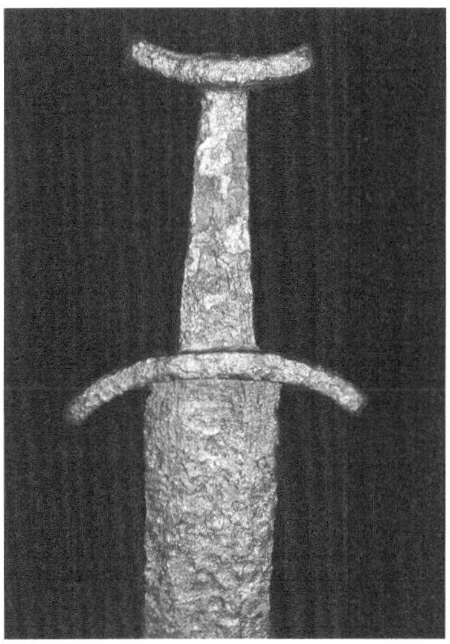

Viking 25: hilt

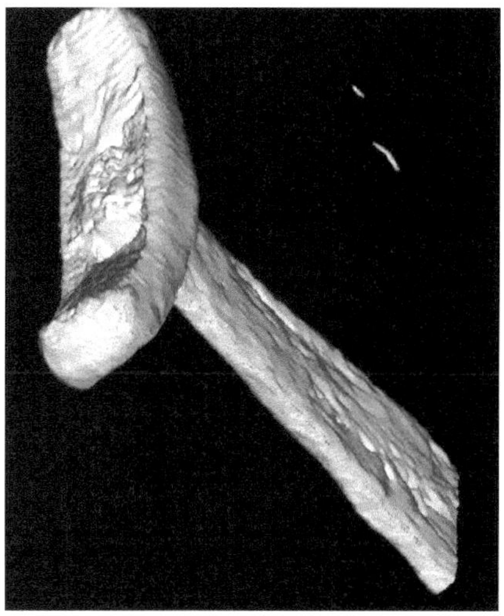

Viking 25: 3D image from CT scan (pommel Type 1Fb)

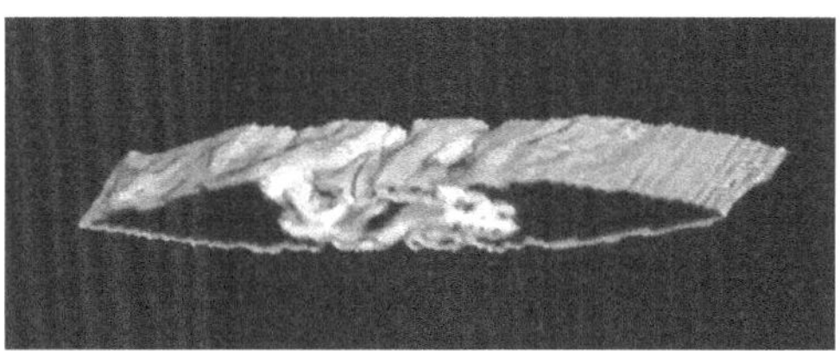

Viking 25: 3D image from CT of cross-section (showing central panel of pattern welding)

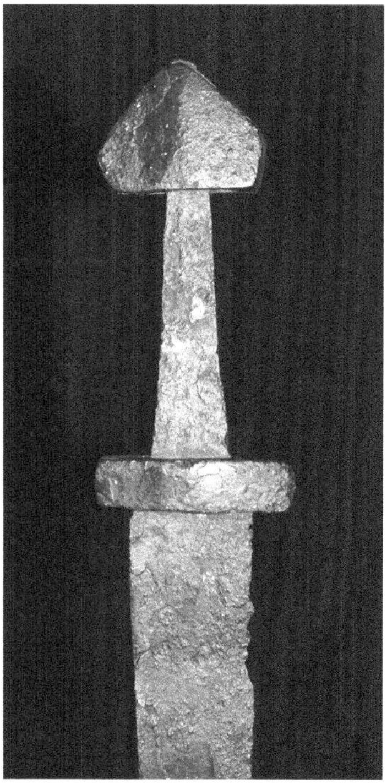

Viking 26: hilt (of single-edged sword)

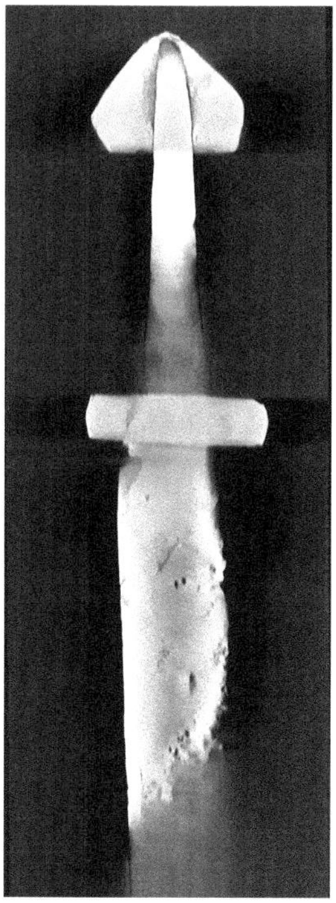

Viking 26: coronal CT scan (pommel Type 1Fb)

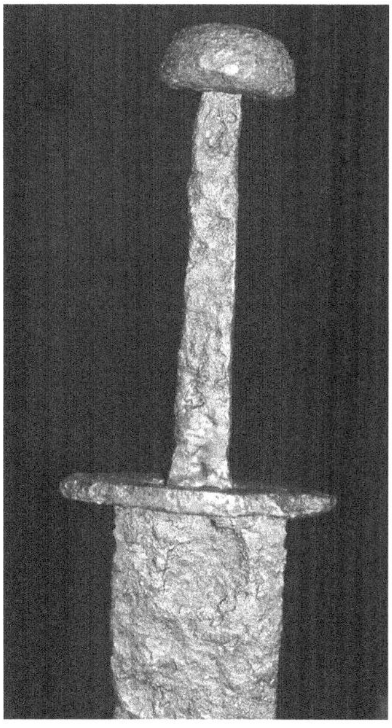

Viking 27: hilt (Merovingian sword?)

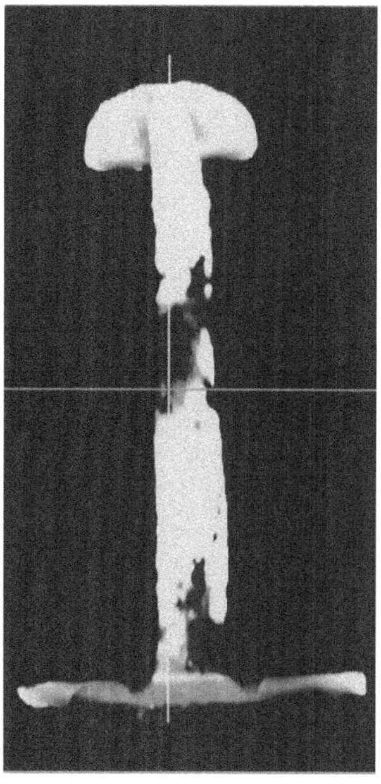

Viking 27: coronal CT scan (pommel Type 1Fa)

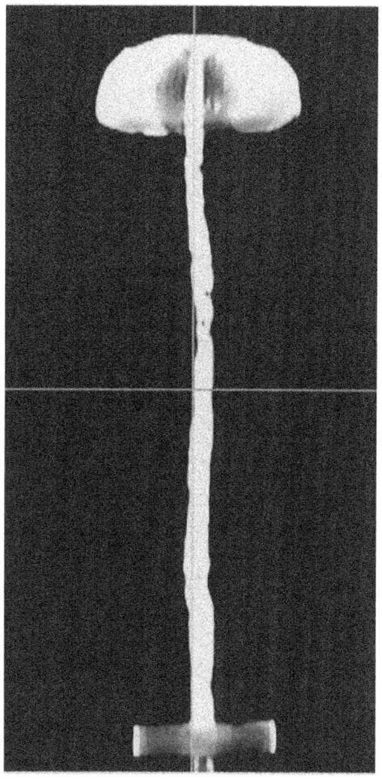

Viking 27: sagittal CT scan

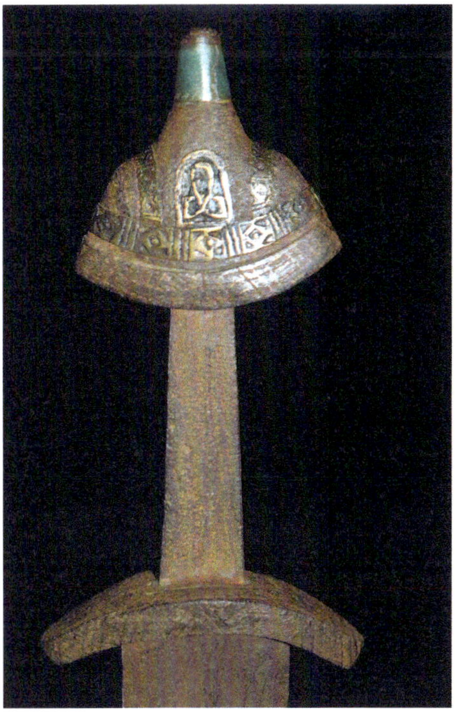

Viking 29: hilt

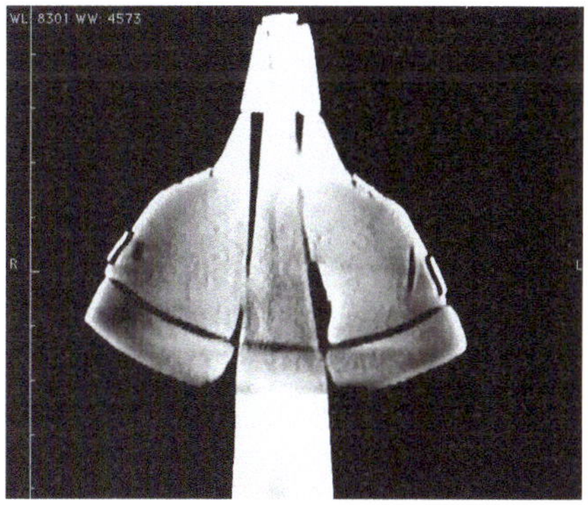

Viking 29: coronal CT scan. Note: the tang does not have a defect – this is a superimposition of the pommel at the level of the CT cut (pommel Type 2Fa)

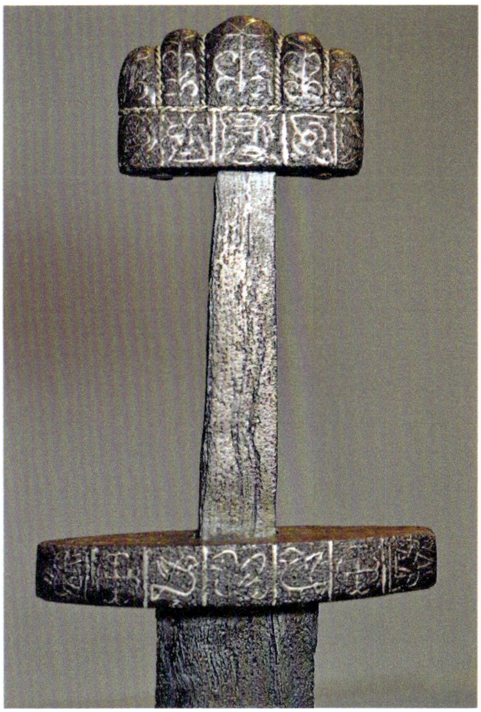

Viking 30: hilt

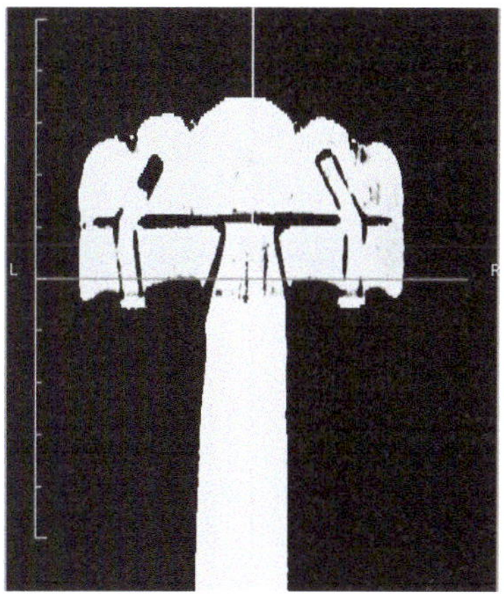

Viking 30: coronal CT scan (pommel Type 2Fc)

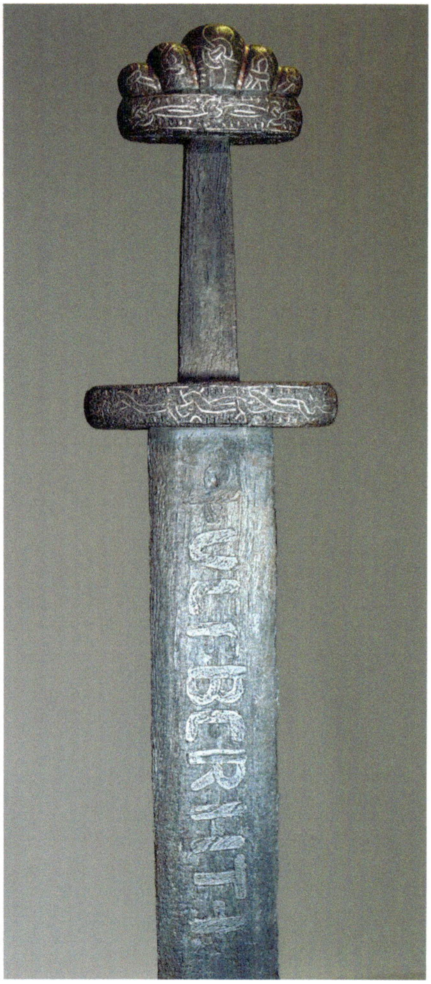

Viking 31: hilt

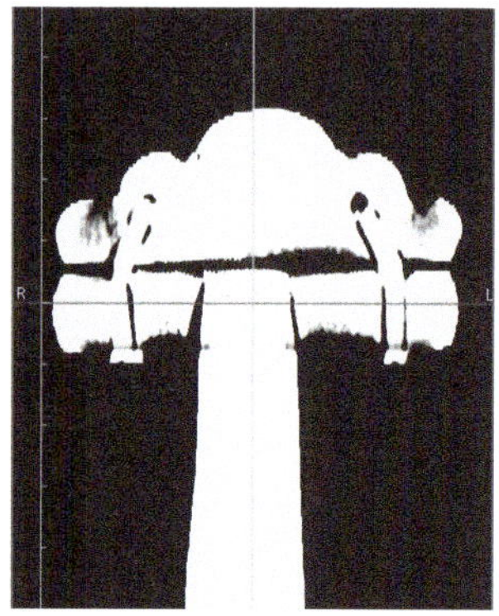

Viking 31: coronal CT scan (pommel Type 2Fc)

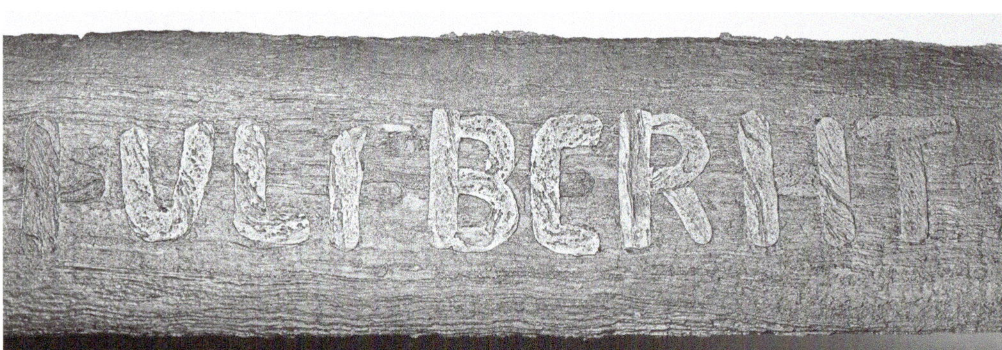

Viking 31: inscription

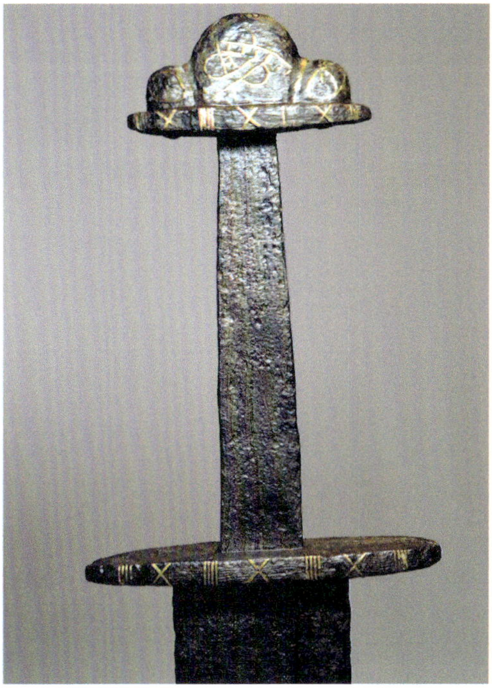

Viking 32: hilt

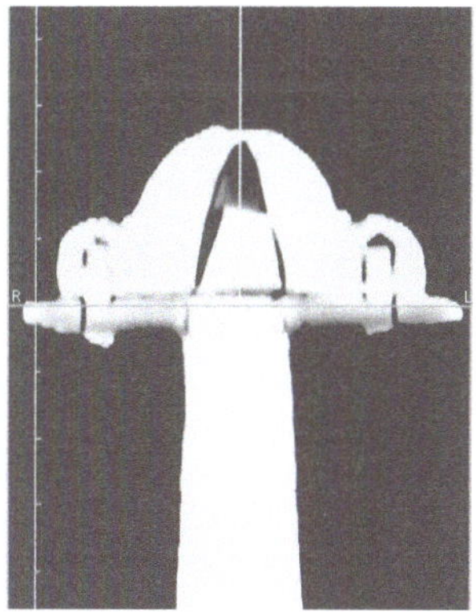

Viking 32: coronal CT scan (pommel Type 2Fa)

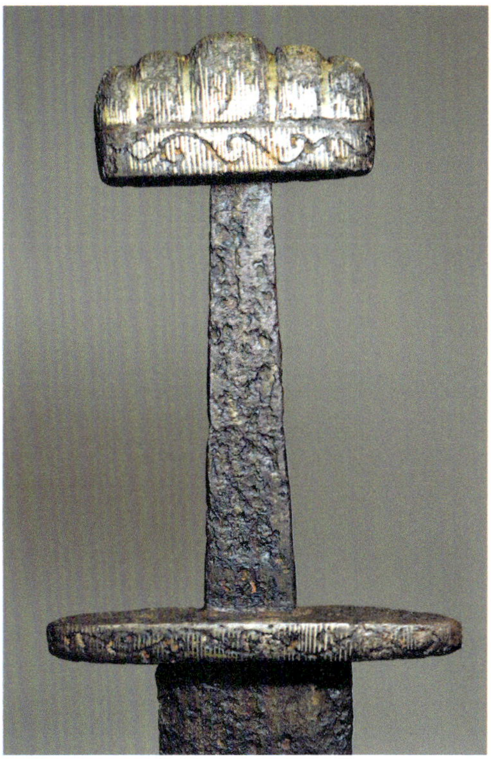

Viking 33: hilt

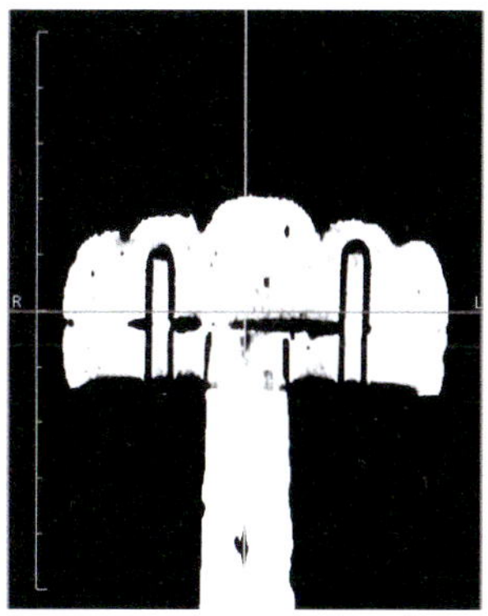

Viking 33: coronal CT scan (pommel Type 2Fa)

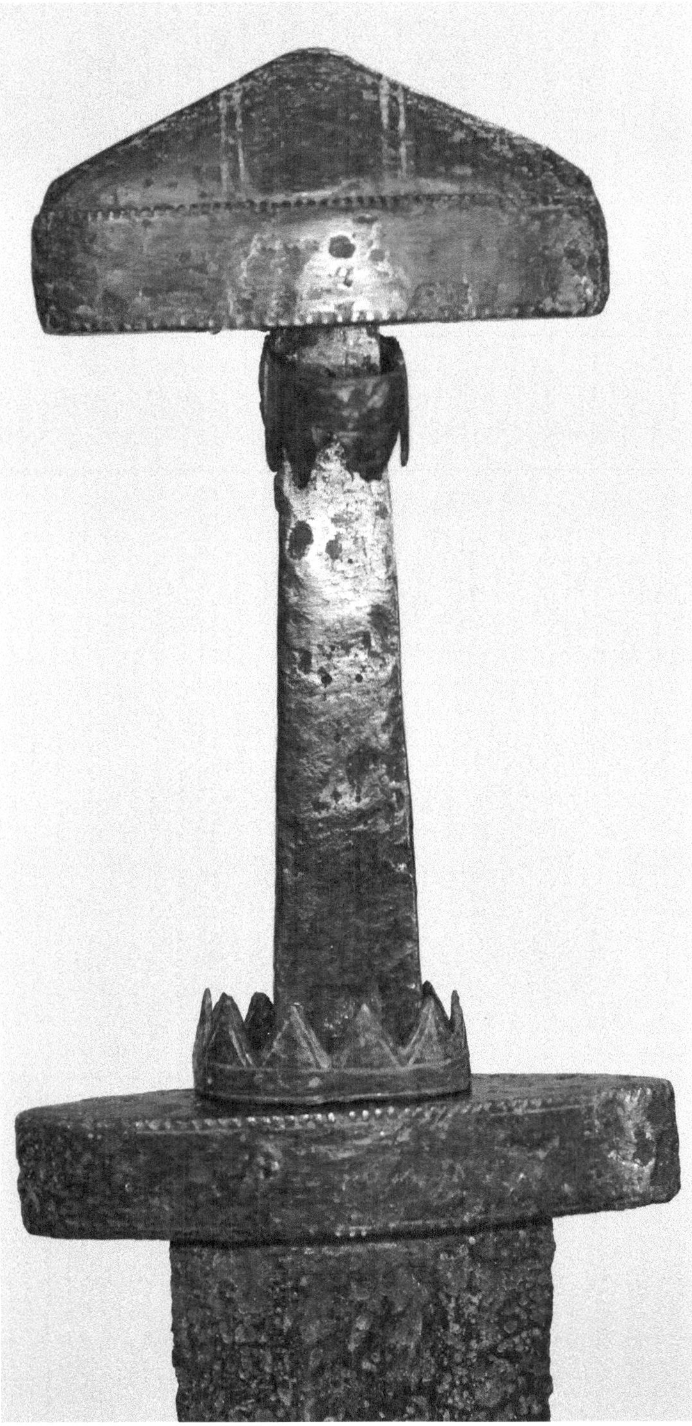

Viking 34: hilt

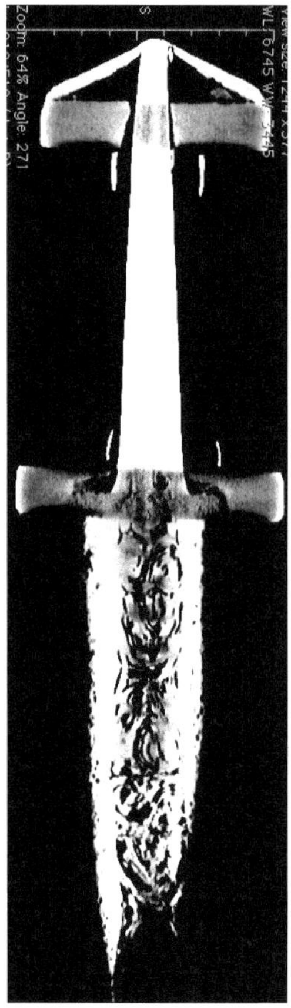

Viking 34: coronal CT scan also showing part of the pattern-welded blade (pommel Type 1Fc). Note pommel may be hollow

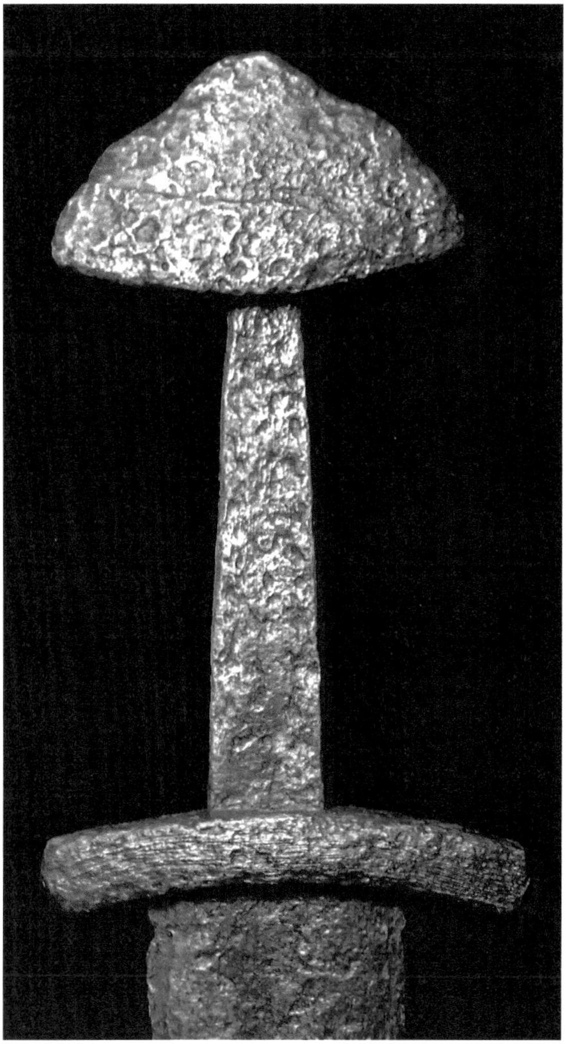

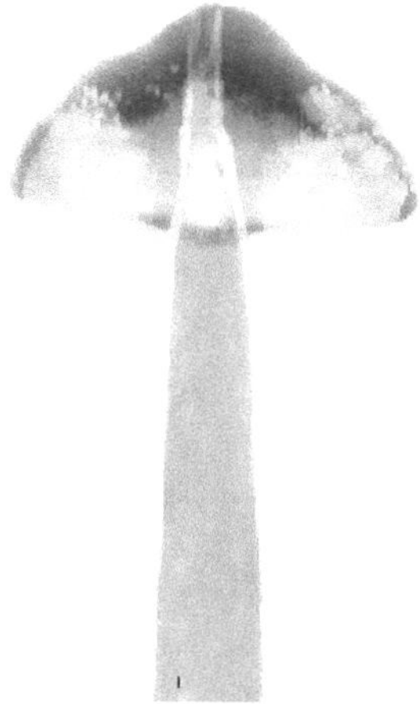

Viking 35: coronal CT scan (pommel Type 1Fc)

Viking 35: hilt

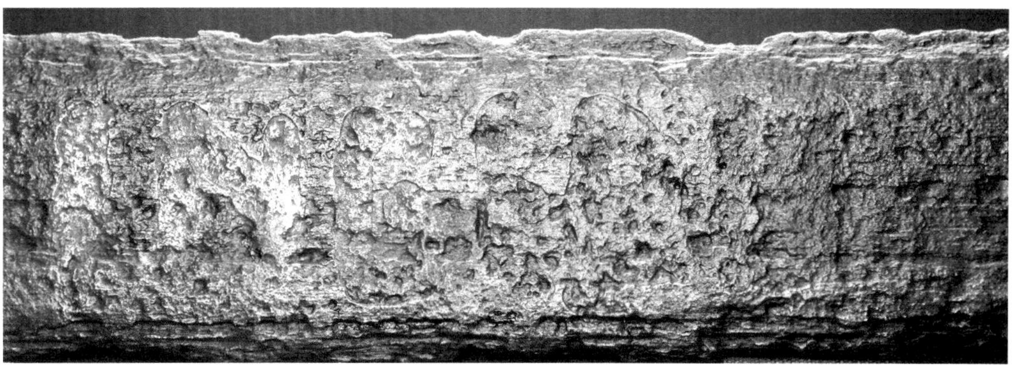

Viking 35: Ingelrii inscription

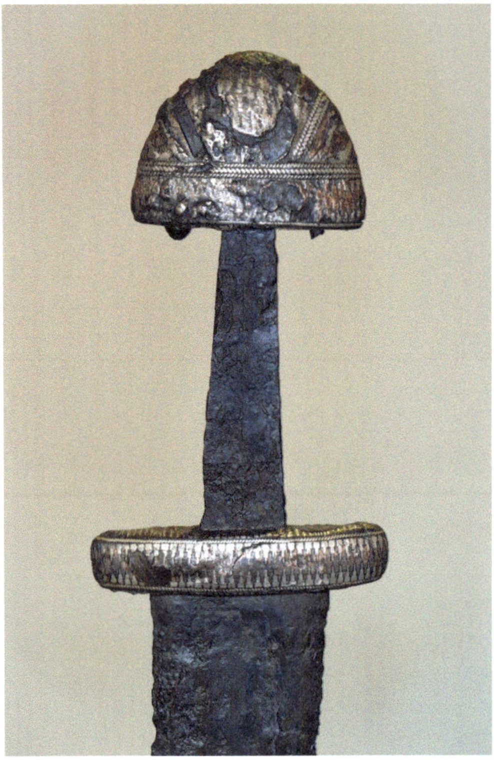

Viking 36: hilt (Vlfberht)

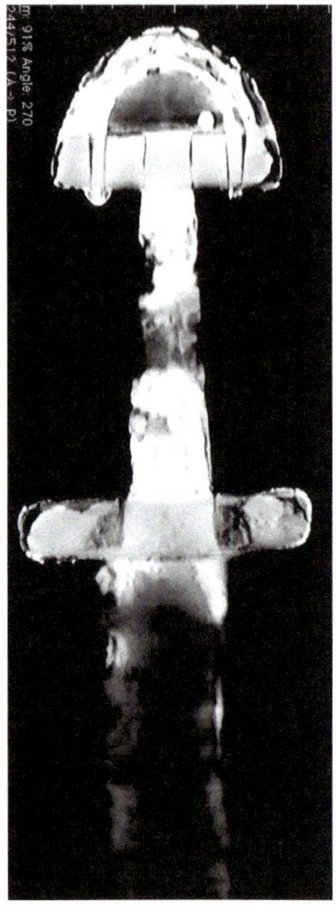

Viking 36: coronal CT scan (pommel Type 2Fd)

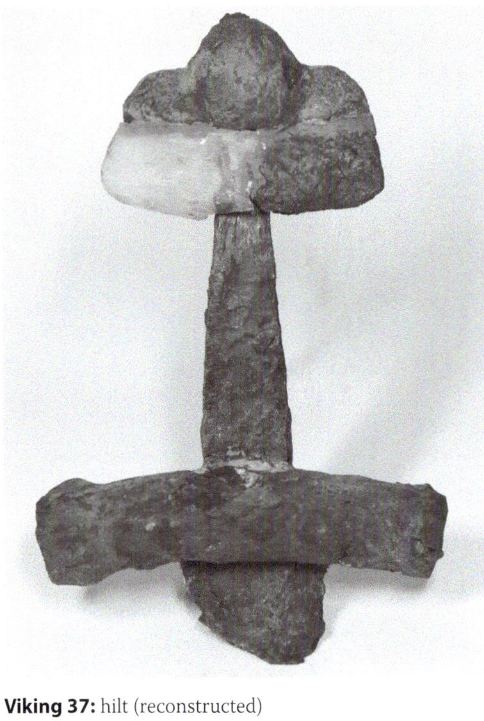

Viking 37: hilt (reconstructed)

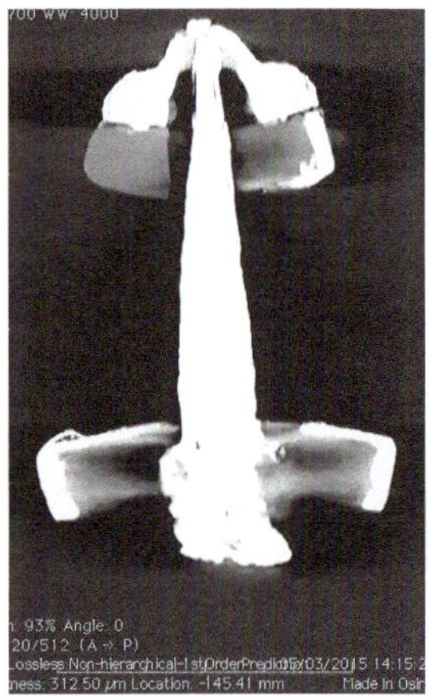

Viking 37: coronal CT scan (pommel Type 2F)

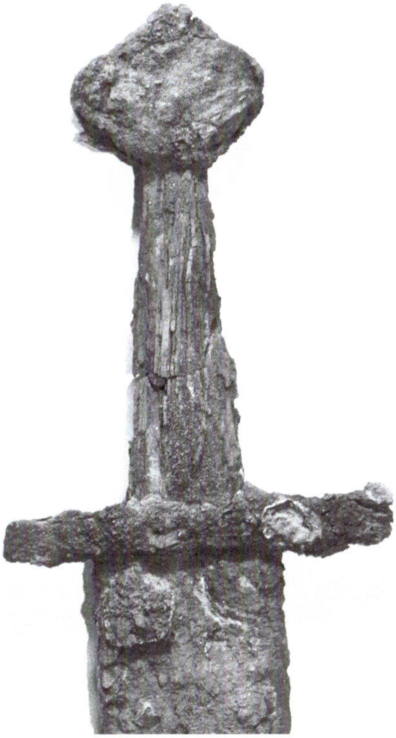

Viking 38: hilt

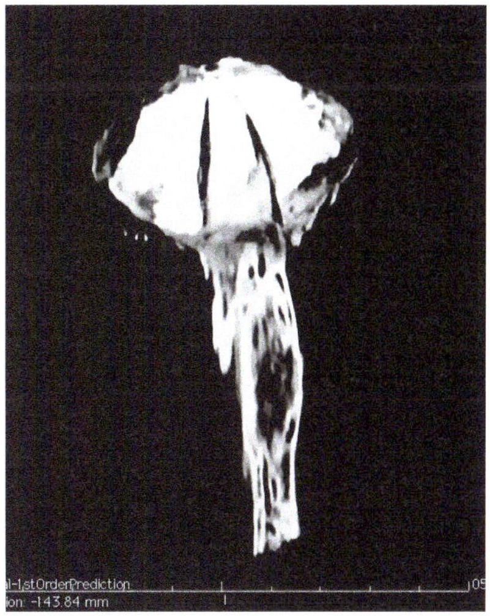

Viking 38: coronal CT scan (pommel Type 1Fb)

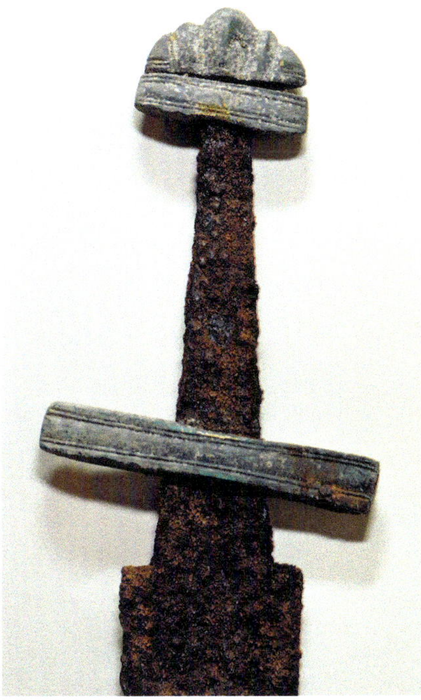

Viking 39: hilt (bronze pommel and crossguard)

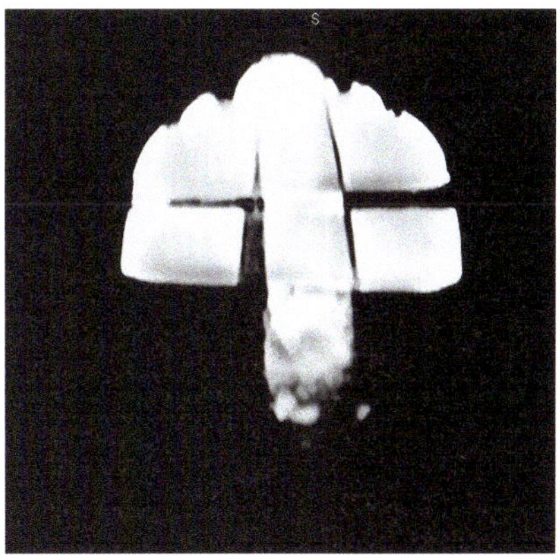

Viking 39: coronal CT scan (pommel Type 2B without rivets, note triangular tang)

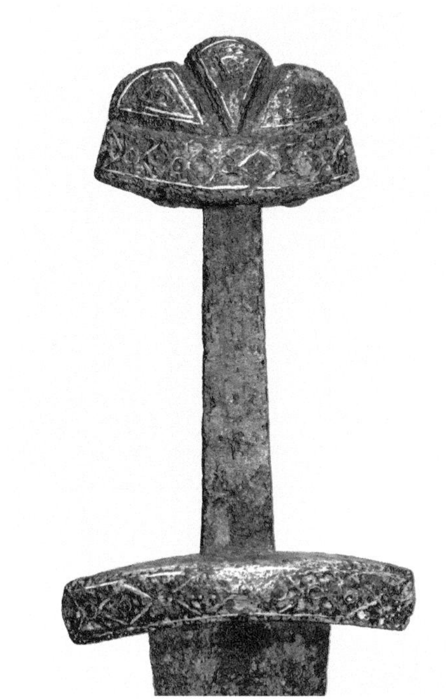

Viking 40: hilt (Vlfberht)

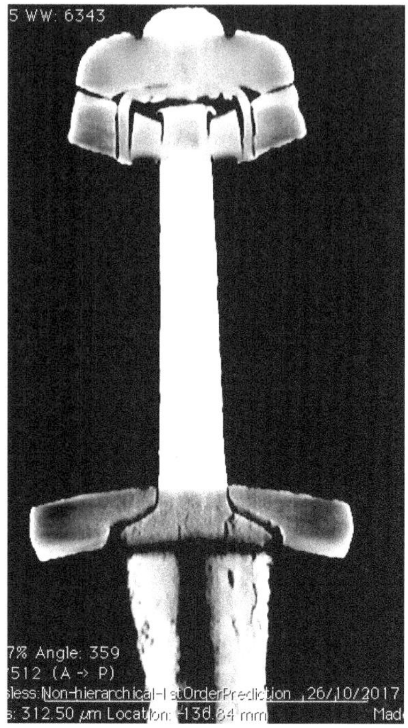

Viking 40: coronal CT scan (pommel Type 2Fe)

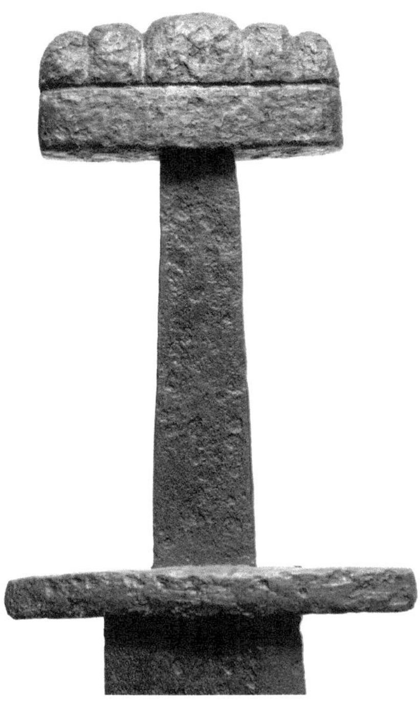

Viking 41: hilt

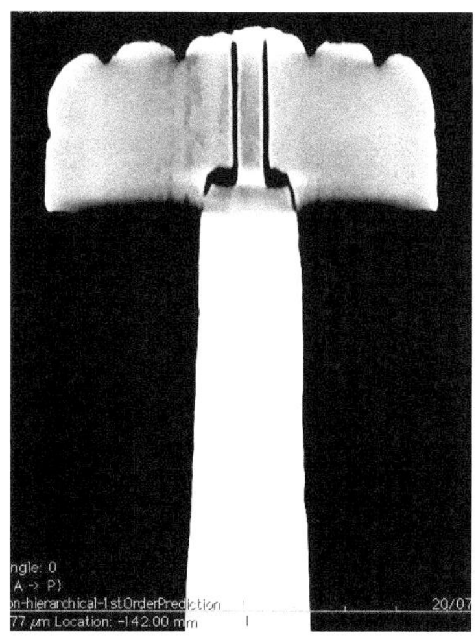

Viking 41: coronal CT scan (pommel Type 1Fc)

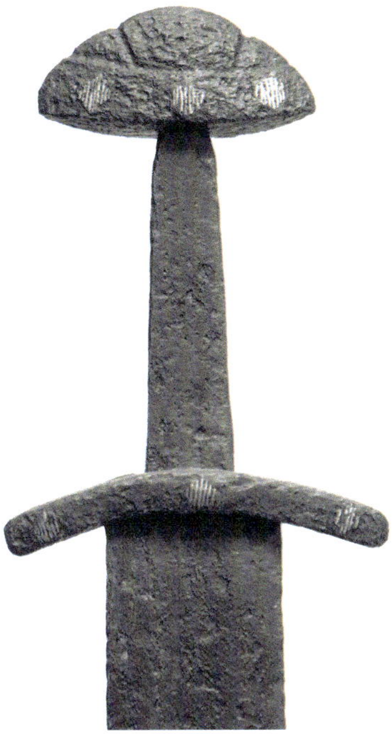

Viking 42: hilt (with silver decoration)

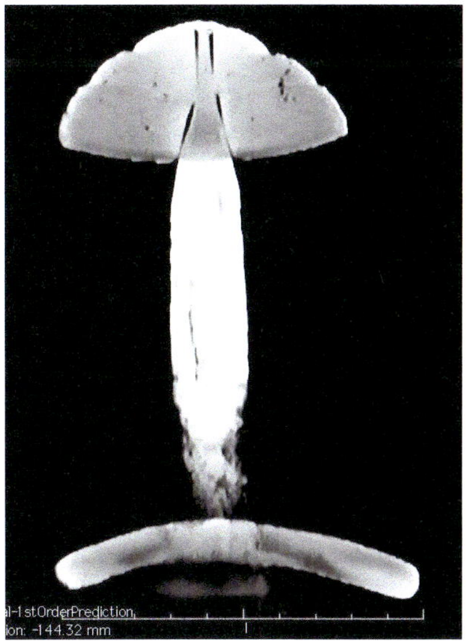

Viking 42: coronal CT scan (pommel Type 1Fc)

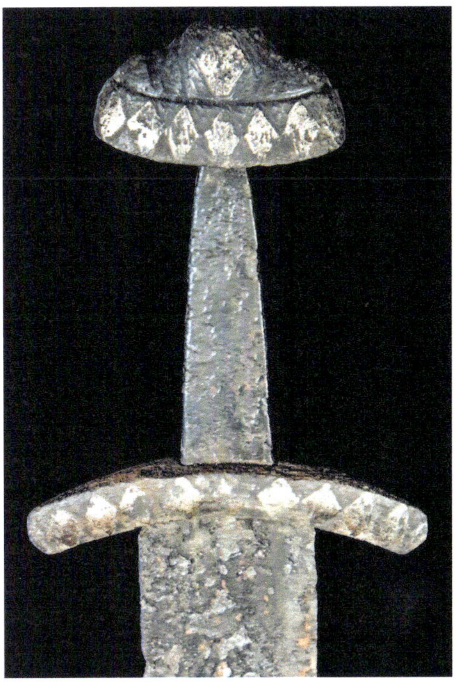

Viking 43: hilt (Leutril inscription on blade)

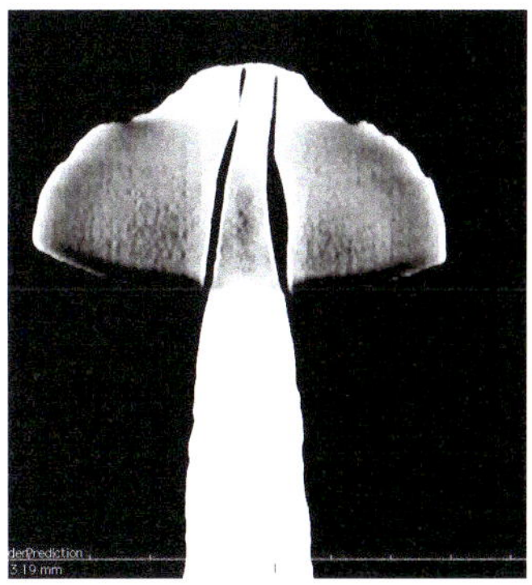

Viking 43: coronal CT scan (pommel Type 1Fc)

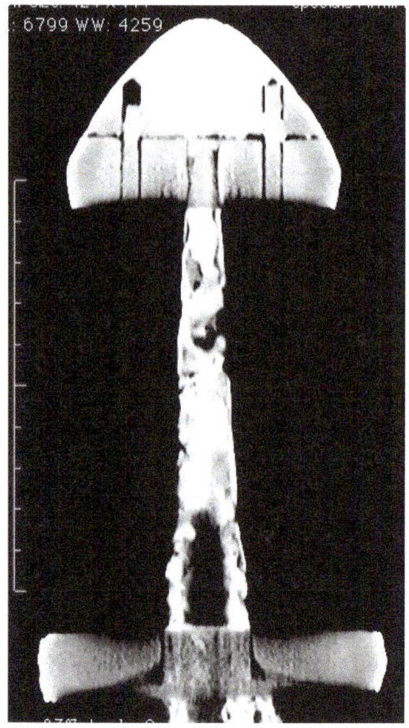

Viking 44: coronal CT scan (pommel Type 2Fa)

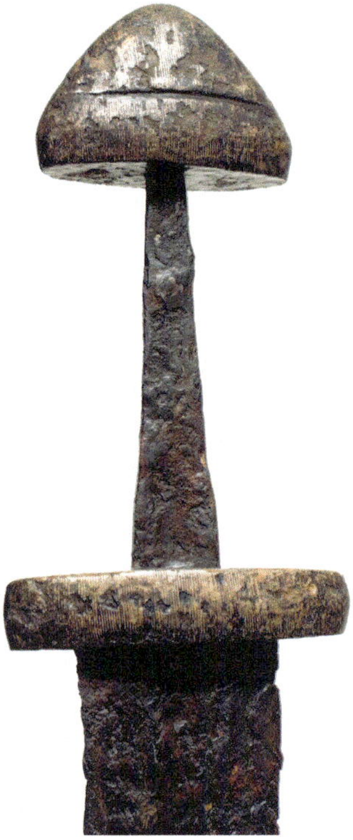

Viking 44: hilt

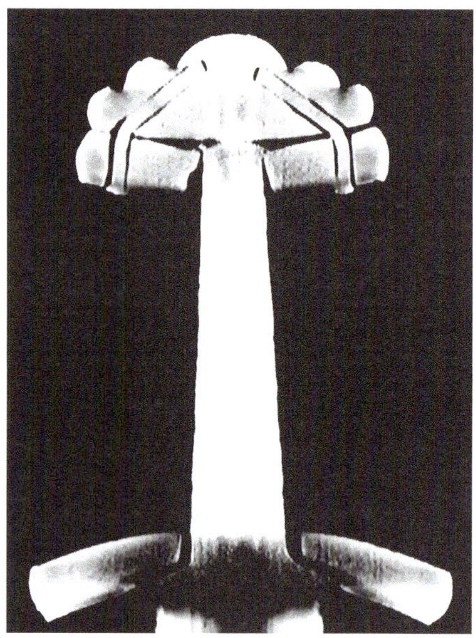

Viking 45: coronal CT scan (pommel Type 2Fc)

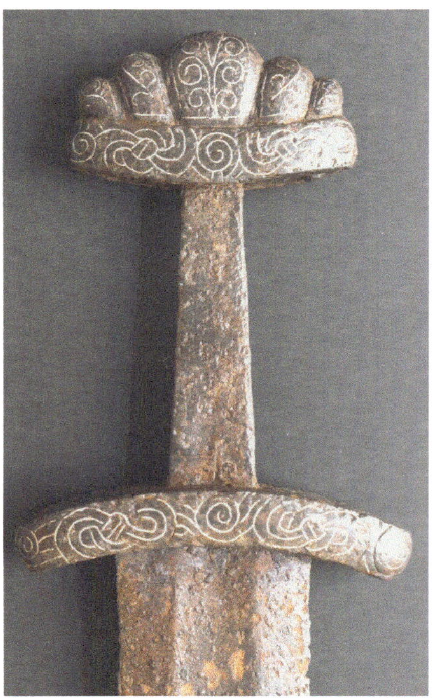

Viking 45: hilt

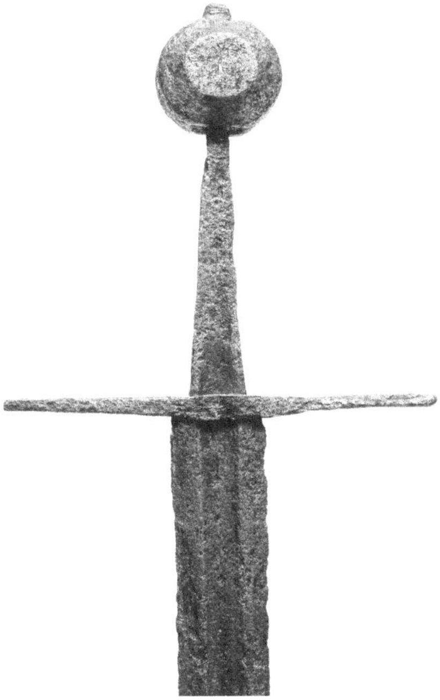

Medieval 1: hilt

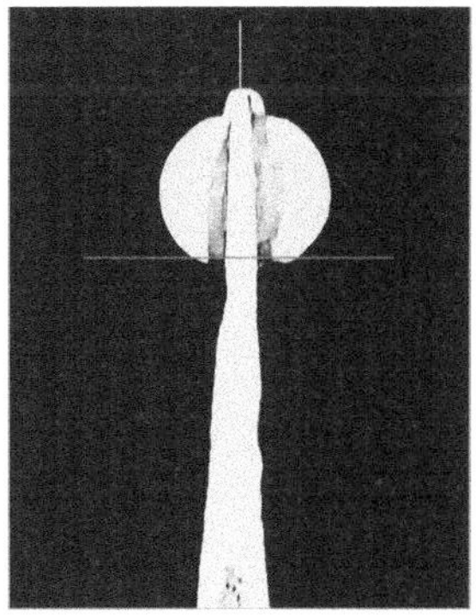

Medieval 1: coronal CT scan
(pommel Type 1Fb)

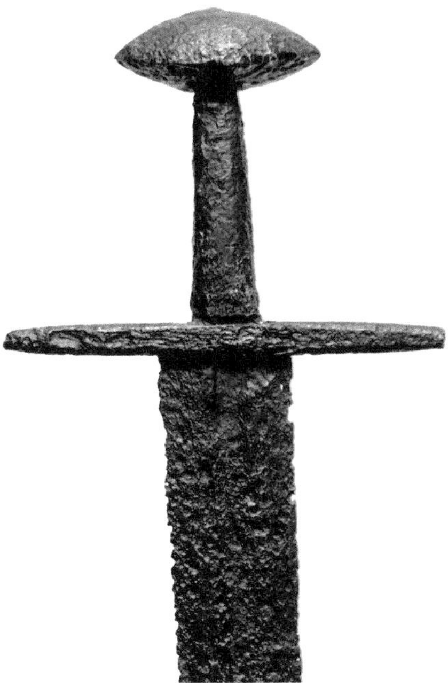

Medieval 2: hilt

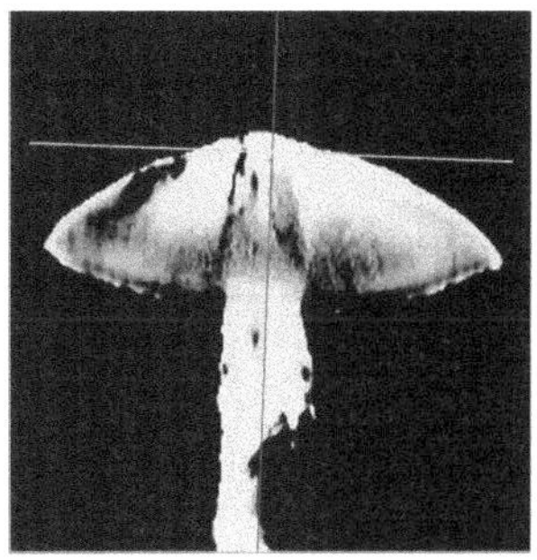

Medieval 2: coronal CT scan
(pommel Type 1Fb)

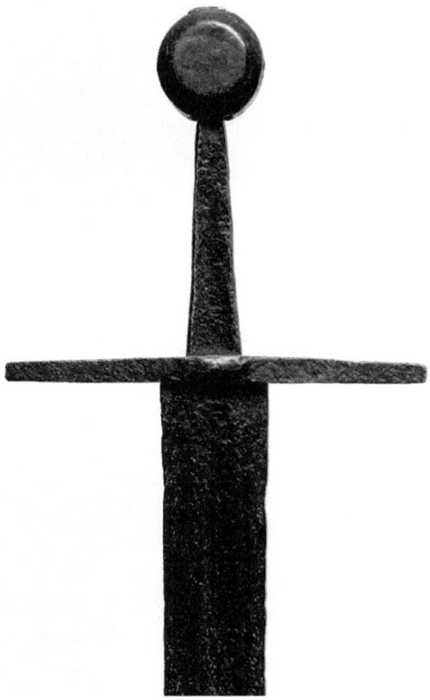

Medieval 3: hilt (bronze pommel)

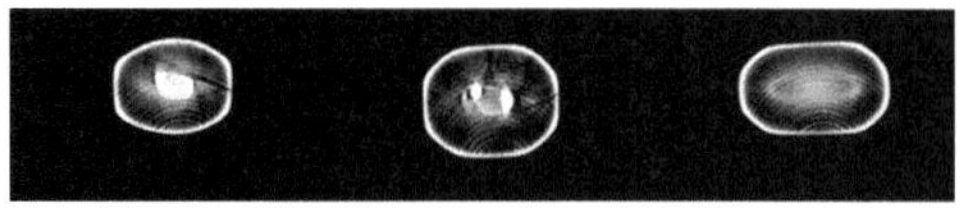

Medieval 3: CT cross-sections of pommel

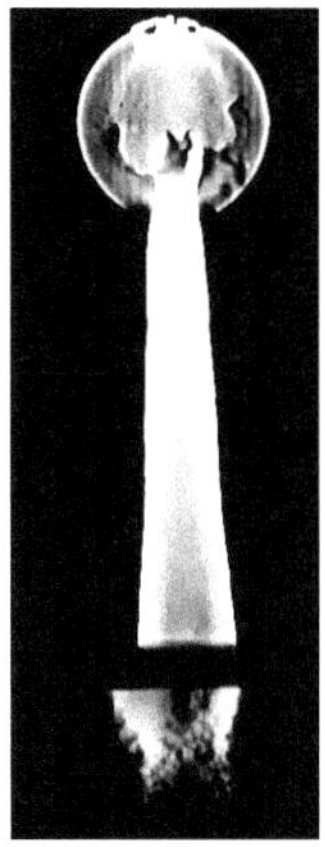

Medieval 3: coronal CT scan (pommel Type 1Ba)

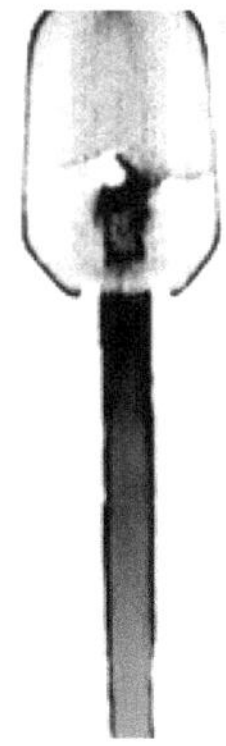

Medieval 3: sagittal CT scan, clearly showing that the tang only extends as far as the middle of the pommel

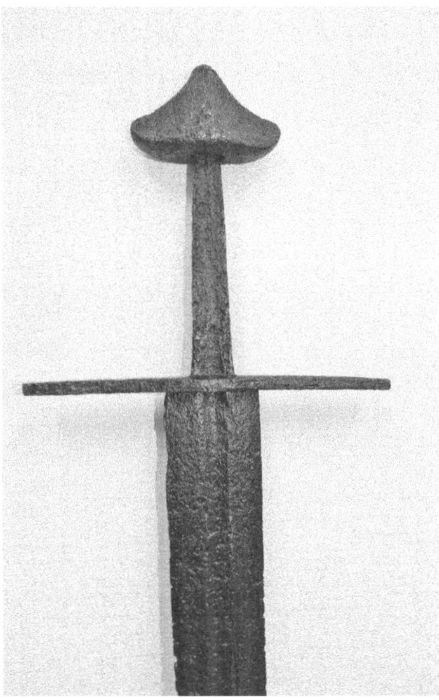

Medieval 4: hilt

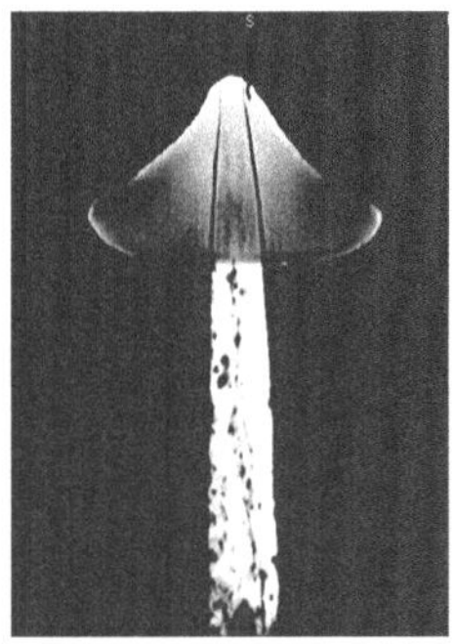

Medieval 4: coronal CT scan (pommel Type 1Fb)

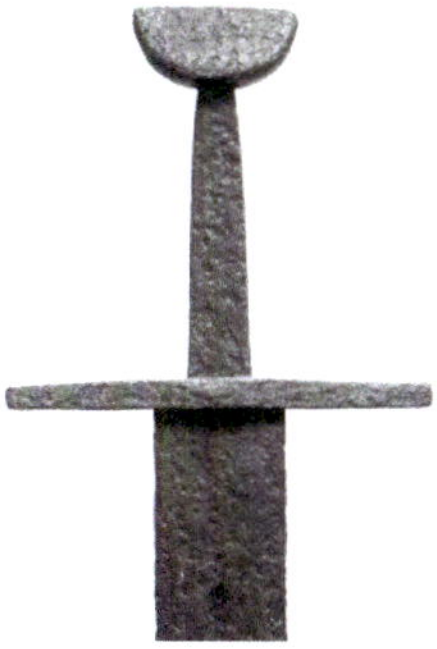

Medieval 5: hilt

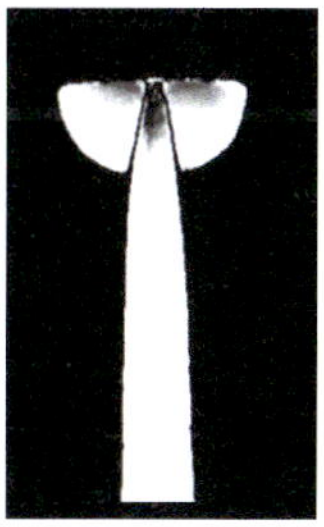

Medieval 5: coronal CT scan (pommel Type 1Fb)

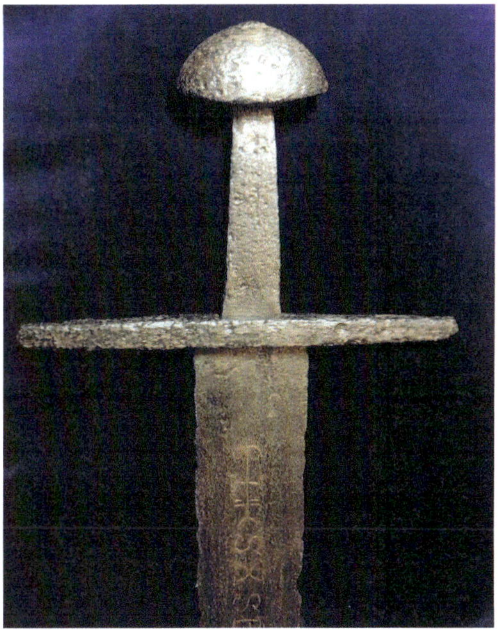

Medieval 6: hilt

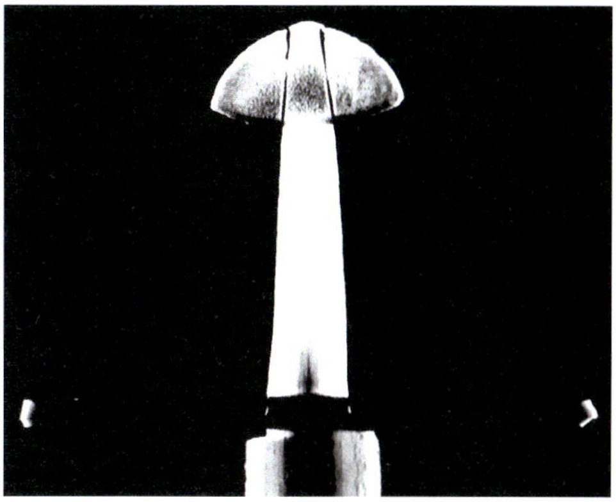

Medieval 6: coronal CT scan (pommel Type 1Fb)

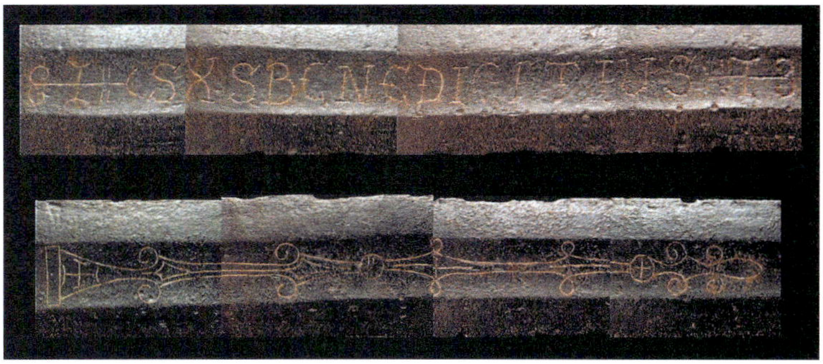

Medieval 6: composite photograph of inscription

Acknowledgements

I would like to thank the anonymous private owners for permission to scan and X-ray their swords and hope that their lead will be followed by museum curators. Thanks are also extended to the staff and management of the Imaging Department of the Portland Hospital, London, for their enthusiasm and assistance.

Notes and references

1. Fell, V., Mould, Q. and White, R. (2006) *Guidelines on the X-radiography of Archaeological Metalwork*. Swindon: English Heritage.
2. Lang, J. and Ager, B. (1989) 'Swords of the Anglo-Saxon and Viking periods in the British Museum: a radiographic study', in S.C. Hawkes (ed.), *Weapons and Warfare in Anglo-Saxon England*. Oxford: Oxford University Committee for Archaeology, Monograph 21, pp. 85–122.
3. Ebinger-Rist, N., Peek, C., Stelzner, J. and Gauß, F. (2011) 'Computed tomography: a powerful tool for non-destructive mass documentation of archaeological metals', in P. Mardikian, C. Chemello and C. Waters (eds), *Metal 2010: Proceedings of the Interim Meeting of the ICOM-CC Metal Working Group, 11–15 October 2010*. Charleston, SC: Clemson University, pp. 458–65.
4. Hill, R., Williams, A. and Owens, C.M. (2014) 'Preliminary results of medical computerised scanning (CT) of swords and other edged weapons', *Journal of the Arms and Armour Society* XXI(4): 141–55.
5. Williams, A., Edge, D., Grazzi, F. and Kardjilov, N. (2019) 'A new method of revealing armourers' marks', *Studies in Conservation* 64(1): 10–15.
6. Petersen, J. (1919) *De Norske Vikingesverd*, Oslo.
7. Geibig, A. (1991) *Beiträge zur morphologischen Entwicklung des Schwertes im Mittelalter. Neumünster*: Wachholtz Verlag; Oakeshott, E. (2008) *The Archaeology of Weapons: Arms and Armour from Prehistory to the Age of Chivalry*. Woodbridge: Boydell Press, p. 133; Oakeshott, E. (2009) *Records of the Medieval Sword*. Woodbridge: Boydell Press, p. 10 and pp. viii–ix; see also Pierce, I. (2002) *Swords of the Viking Age* (Woodbridge: Boydell Press) with an introduction by E. Oakeshott and a useful summary by L.H. Jones on pp. 15–24 ('Overview of hilt and blade classifications) and pp. 14–151 ('Blade construction and pattern welding').
8. Wilson. D.M. (1965) 'Some neglected Late Anglo-Saxon swords', *Medieval Archaeology* 1(9): 32–54.
9. Svarane, D. (2007) 'The use of inlay technique in the production of medieval Curonian and Semigallian military equipment (12th to 14th century)', *Archaeologica Baltica Volume 8: Technological Aspects of Weaponry*, pp. 368–76.
10. Geibig 1991 (cited in note 7), passim.
11. Gorman, M.R. (1999) 'ULFBERHT: innovation and imitation in early medieval swords', in *Catalogue of 16th Park Lane Arms Fair, London*, pp. 7–12.
12. Nicolle, D. (1984) *The Age of Charlemagne*. Oxford: Osprey, p. 27.
13. Stalsberg, A. (2008) 'Herstellung und Verbreitung der ULFBERHT-Schwertklingen. Eine Neubewertung', *Zeitschrift für Archäologie des Mittelalters* 36: 89–118.
14. Williams, A. (2012) *The Sword and the Crucible: A History of the Metallurgy of European Swords up to the 16th Century*. Leiden and Boston: Brill, esp. Ch. 8.
15. Ibid.
16. Coupland, S. (1990) 'Carolingian arms and armor in the ninth century', *Viator: Medieval and Renaissance Studies* 21: 29–50.
17. Jones in Pierce 2002 (cited in note 7).
18. Stalsberg 2008 (cited in note 13).
19. Davidson, H.E. (1998) *The Sword in Anglo-Saxon England*. Woodbridge: Boydell Press, pp. 114–15.
20. Wheeler, R.E.M. (1927) *London and the Vikings*. London Museum Catalogues 1.

10

The helmet of the Black Prince analysed by neutron diffraction

David Edge, Francesco Grazzi, Antonella Scherillo and Alan Williams

Introduction

The Conservation Department of the Wallace Collection, London, has a small laboratory that has been operating for 20 years in which plate armour and excavated swords are examined by metallography, that is, the microscopic examination of a prepared surface. If the armour plate has a cut straight edge, or the sword is broken, this is a non-invasive technique.[1] However, when the armour plate has a rolled edge or the sword is on display in the museum because it is not broken or damaged, this method of analysis is difficult to employ without causing some harm. Accordingly, the department has been seeking to identify other techniques in order to analyse difficult objects such as helmets and swords.[2]

One promising technique is neutron diffraction. Unlike X-rays, neutrons generally pass almost unperturbed through metals since they do not interact with electrons but only with nuclei, an interaction that is much less intense. Depending on the spacing between the atoms and the velocity (and therefore the wavelength) of the neutrons, some may be diffracted and generate interference (observing Bragg's Law). Crystals of ferrite (iron) will diffract them through a different angle to cementite (iron carbide) and other phases, allowing them to be detected and measured separately. A 'white' neutron beam contains neutrons with a mixture of velocities and therefore a mixture of energies (and equivalent wavelengths λ). Neutrons acting as waves can be diffracted by layers of atoms within a crystal. For certain wavelengths ($n\lambda$), positive interference leads to peak readings on the detectors. According to the Bragg equation:

$$n\lambda = 2d \sin \theta$$

where θ is the scattering angle and d the interatomic spacing, so the different phases in the metal (ferrite, cementite, martensite, wuestite, perhaps goethite and magnetite) can be identified and quantified. In a polycrystalline metal, diffraction from all possible planes will be included in the results thereby enabling quantification and characterisation of the phases in the microstructure.

As well as determining the proportions of the different phases present, the shape of the diffraction peaks and their relative intensities can be used to obtain information on the presence and intensity of texture, the residual strain distribution and size of the grains. Smaller grains will have more grain boundaries and therefore more disordered atoms, so the peaks will be broader (with a shape following the Lorenz mathematical function) than the same phase with a larger grain size.

Cold working may lead to plastic deformation of the ferrite and the cementite whereas hot working in a hypereutectoid steel (above Acm,

727 °C) will induce texture in the cementite only. Texture, whether induced by cold working (which would include battle damage) or asymmetrical hot working (e.g. forging hypereutectoid steels above Acm) will lead to an anomalous variation in the intensity of the Bragg peaks in any specific phase, with respect to that expected from an isotropic homogeneous powder (a condition similar to that found in normalised steels). This occurs because the spatial distribution of some planes of atoms has been altered.

Ferrite has a strong peak at 2.02 Å and two weaker ones at 1.43 Å and 1.17 Å with many more at smaller distances. Cementite has two significant peaks at 1.85 Å and 1.87 Å with a very similar intensity (as well as numerous smaller ones). The ratio between these two peaks can vary if texture is induced in the cementite because they are formed by different planes of atoms, which are quite differently oriented in the crystalline structure. Texture alters the relative intensity of the planes measured in the same direction.

Untempered martensite (formed by quenching without tempering, but seldom found in historic specimens) has a structure very similar to ferrite – a cube with one somewhat elongated side – so it has the same peaks as ferrite with notable shoulders on the sides of the ferrite peaks, on the side of the longer wavelengths. Essentially, martensite is a highly supersaturated solid solution of carbon in iron which, on tempering, rejects carbon in the form of finely divided carbides, dispersed in a ferrite matrix. The cementite grows in globule size with prolonged tempering, resulting in a decrease in hardness but generally an increase in toughness. The cementite peaks will become narrower as a result of this process. Slack quenching may produce a mixture of several phases, tempered martensite, bainite and pearlite. Since these are all different arrangements of ferrite and cementite, deducing the precise nature of the microstructure may be difficult because all these microconstituents engender diffraction peaks in only those two phases.

Bloomery iron will always contain some slag (mostly fayalite Fe_2SiO_4) and wüstite (FeO) if the original bloom was not completely reduced. In addition, historical specimens usually contain some corrosion products such as goethite (FeO(OH)).[3]

Experimental methods

The helmet from Canterbury Cathedral as well as some others from the Wallace Collection were analysed using the time-of-flight diffractometer INES (Italian Neutron Experimental Station) at the ISIS pulsed neutron spallation source of the Rutherford Appleton Laboratory, Harwell. In the present work, a systematic study of the helmets was undertaken to produce a microstructural and compositional mapping of their different areas. Where possible, several cross-sections for each sample were scanned; these areas are described in the results. The helmets were mounted on an aluminium support frame clamped on an x-y table in order to change the measuring points.

The average flux on the INES sample position is of the order of 10^6 n/cm^2/s^{-1}. The neutron beam cross-section was adapted to the size of the desired parts of the helms using ceramic boron-enriched jaws at the entrance window of the instrument to minimise data superimposition and background signal. The neutron flux, measured through the integrated synchrotron proton current, was 600 μAh for the average measurements, corresponding to about 4 hours, assuming an average synchrotron current of 150 μA. After irradiation, the activation of the sample was measured using the Geiger counter provided on the beamline: the samples showed very low activation (less than 0.5 μSv.h^{-1}).

The phases searched for were those related to metallic iron and steel such as ferrite, cementite, martensite and austenite, and the mineralisation and smelting-related phases such as wüstite, goethite, magnetite, hematite, fayalite and troilite. Data analysis was carried out through the use of the GSAS program. The microstructural properties were determined by analysing the shape and relative intensity distribution of the diffraction peaks, and the size of the crystallographic domains was identified using the Lorenzian broadening of the peaks.[4]

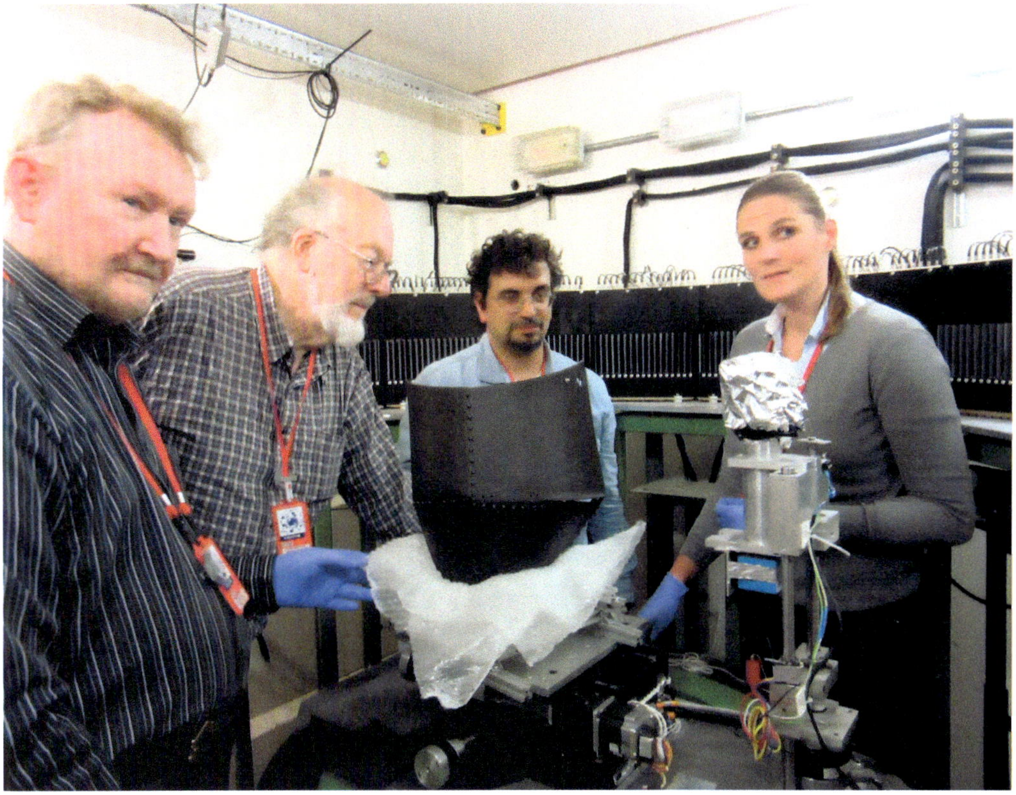

Figure 1 The helm of the Black Prince from Canterbury Cathedral during analysis. *From left to right*: David Edge, Alan Williams, Francesco Grazzi and Ariane Langreder.

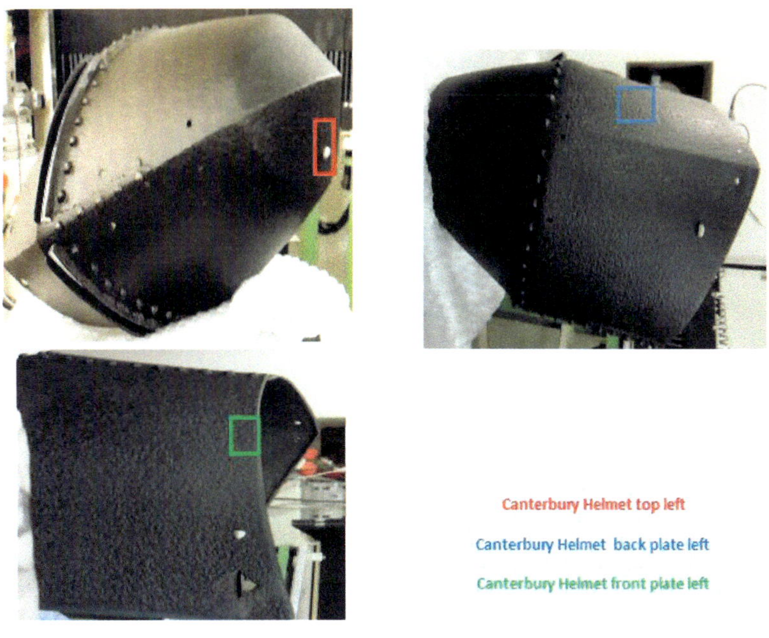

Figure 2 The boxes show the areas analysed of the Black Prince's helm.

Experimental results

The helmet of the Black Prince

The great helm was part of the funeral achievements of the Black Prince, Edward (1330–1376), the eldest son of King Edward III, and suspended over his tomb in Canterbury Cathedral (Figures 1 and 2). It comprises three plates riveted together with a total weight of 3.24 kg. All three plates were made of low-carbon steel with traces of fayalite which would be expected from a bloomery steel. Traces of microstrain were detected (by the broadening of the diffraction peaks) in the front plate, possibly the result of a blow which, having dented the plate, was later repaired by a cold hammering. This would have left some of the metal strained, although the surface might have appeared undamaged. The modest quality of the metallurgy suggests that it may have been a local product, rather than an import of potentially higher quality. By the end of the 14th century, plate armour was being made of steel in some northern Italian centres, although their export trade to England does not seem to have developed until the 15th century. The carbon contents of the three plates following measurement were:

- Top plate (left) 0.216% C (+/-0.007%)
- Front plate (left) 0.186% C (+/-0.007%)
- Back plate (left) 0.089% C (+/-0.007%)

In all the measured areas there is cementite but the amount is variable. No wüstite was found so the ore reduction was completed up to the point that all the iron was reduced. Usually cementite and wüstite cannot coexist for this reason and the presence of cementite and the absence of wüstite mean that the helm parts were manufactured using a good quality bloom. There is a fixed amount of fayalite (0.2%) so the iron silicate (slag) amount is constant; there is also an almost constant amount of goethite due to the external layer of oxides.

There is little texture in the whole helm so it was hot worked during manufacture. The grains are quite large indicating that the helm was not rapidly cooled as a final treatment: there was no quenching. The stress distribution is not constant: it is weak everywhere but has a larger value in the front plate (0.9) than in the other plates (around 0.4–0.45), therefore it is possible that the front was subjected to some cold working, probably the result of battle damage, that was repaired by hammering back without subsequent annealing.

Analyses of other helmets from the Wallace Collection

A tournament helm, A186 is made in two pieces, weighing 7.40 kg (Figures 3 and 4). Mann[5] pointed out that the thickness varied from 7.6 mm to 3.0 mm to maximise protection. It is possibly English, however a resemblance has been noted to a French jousting helm of the early 15th century.[6] Three areas were examined, all of which were steels, with no trace of quenching: the front plate (0.44% C), the top plate (0.41% C) and the neck (0.50% C) – all +/-0.007%. This tallies with the results of earlier metallography, which showed a pearlitic steel.[7]

A sallet, probably made in Germany in the mid-15th century, A79 is made in one piece and repaired, with a weight of 2.92 kg (Figures 5 and 6).[8] Analysis of three areas of this skull confirmed that all were low-carbon steels: the left side (0.121% C), the central keel (0.175% C) and the right side (0.155% C), the side, protecting the left cheek (0.115% C) and a repair within the skull – all +/-0.007%. The repair consisted of a patch of iron (0% C) secured by 'brazing' but in this case a 3% tin bronze was used rather than a brass. All the areas contained some fayalite.

A80 is a sallet probably made in South Germany in the later 15th century and apparently used for a tournament (Figures 7 and 8). It weighs 3.62 kg and has a roller in the centre of the lower edge, the *bundrennen*, which was a fitting for the joust. The two holes on each side of the brim are for bolting the helmet to upright supports on the mechanical breastplate. This was triggered by a direct hit on the shield, which was then ejected upwards. The roller was to prevent the shield from catching on the helmet.[9] The components of this helmet were examined in five places:

- The right side of the helmet around the cheek 0.134% C (+/-0.007%)
- The left side of the helmet around the cheek 0.175% C (+/-0.007%)

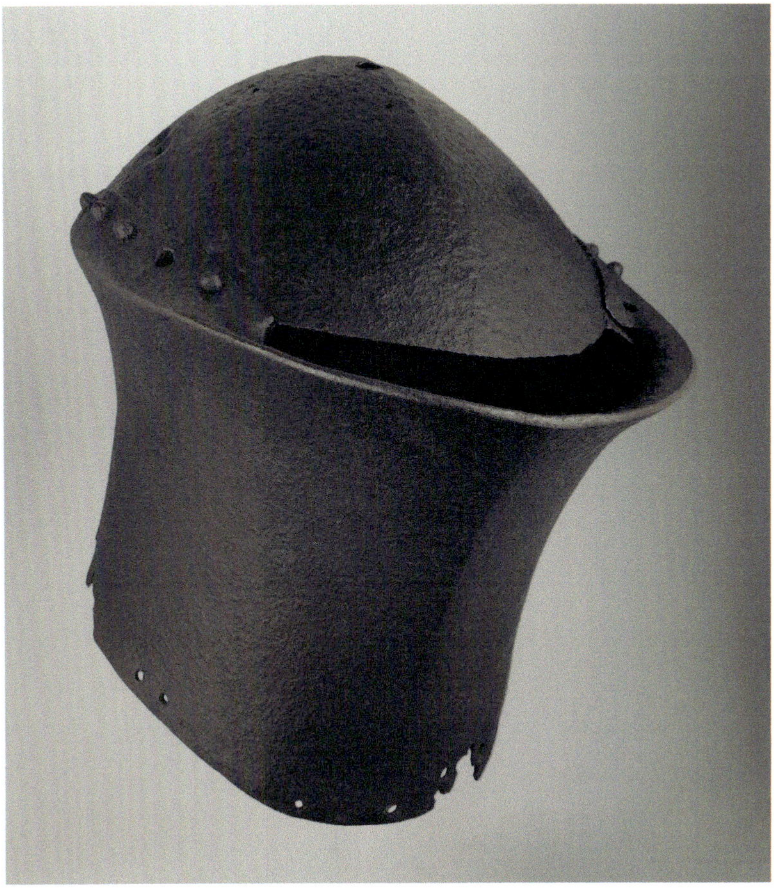

Figure 3 The helm A186. © Trustees of the Wallace Collection.

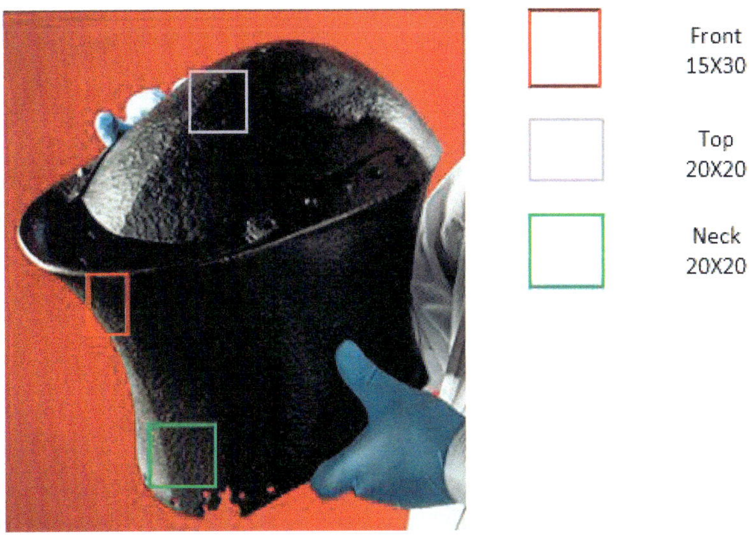

Figure 4 The helm A186 showing the areas analysed: top of the skull; the front at the lowest part protecting the neck; and the front below the sight.

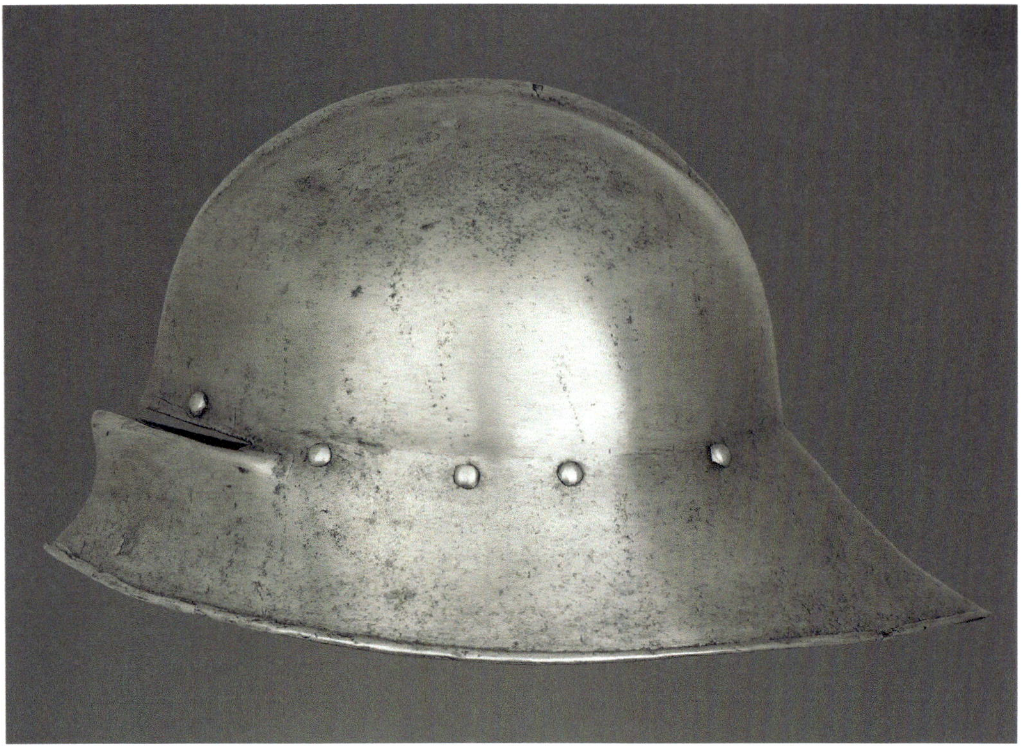

Figure 5 The sallet A79. © Trustees of the Wallace Collection.

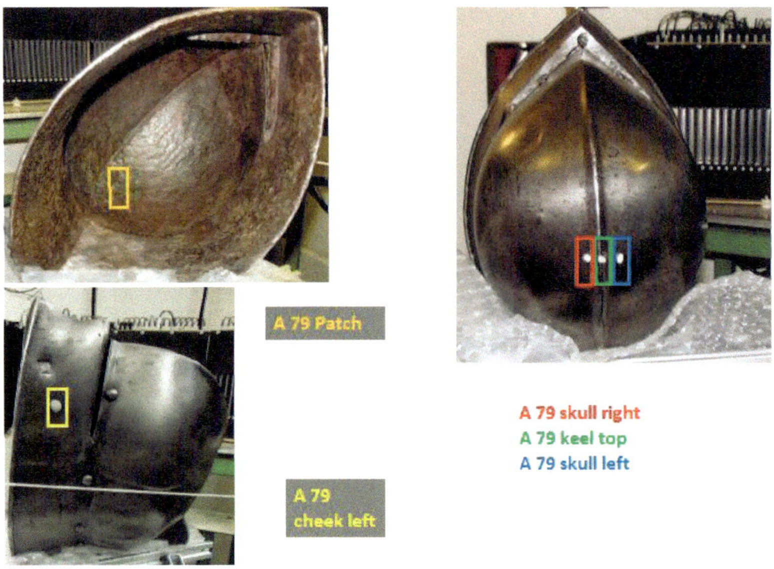

Figure 6 The sallet A79 showing the areas analysed: a patch inside the skull; the top of the skull on the proper left side; the keel on the top of the skull; the skull on the right side; and the lower part of the front on the left side, protecting the chin.

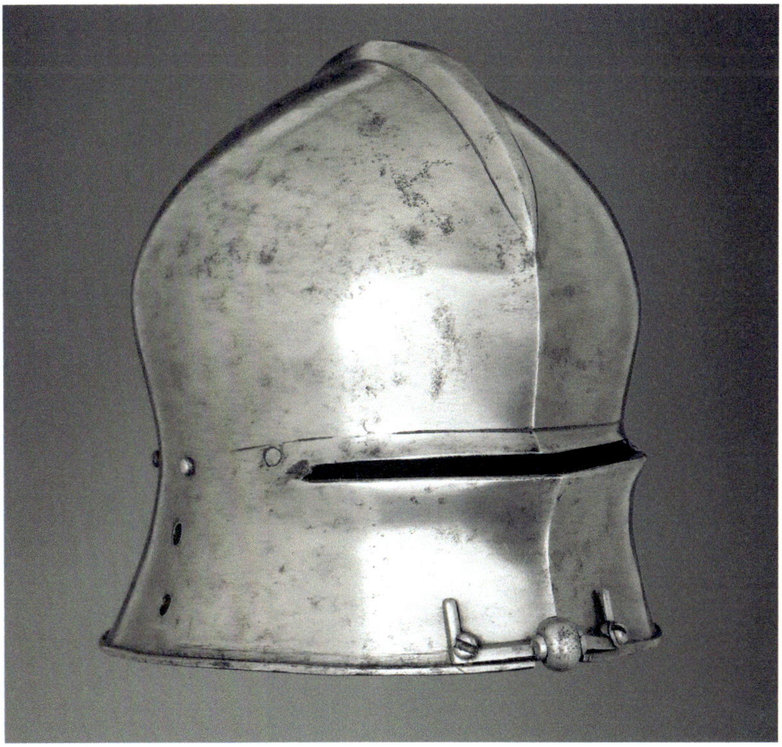

Figure 7 The sallet A80. © Trustees of the Wallace Collection.

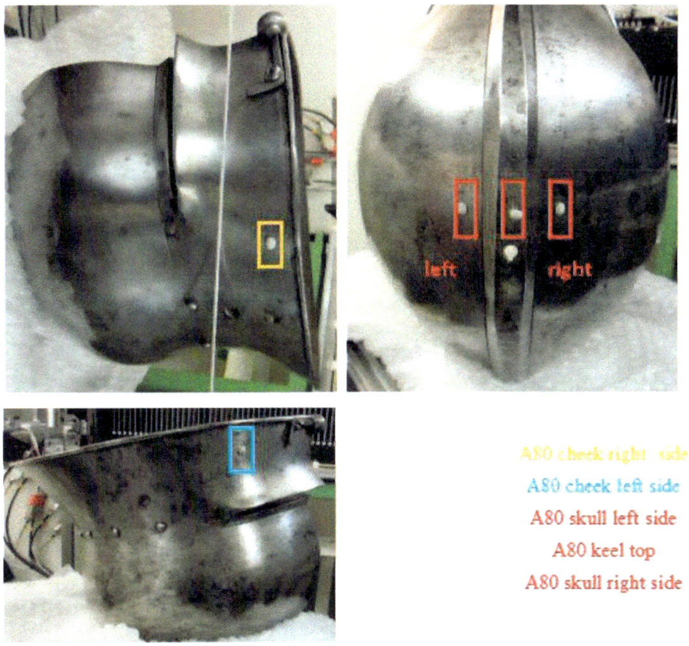

Figure 8 The sallet A80 showing the areas analysed: the lower front right side; the top of the skull on the proper left side; the top of the keel; the top of the skull on the right side; and the lower front left side.

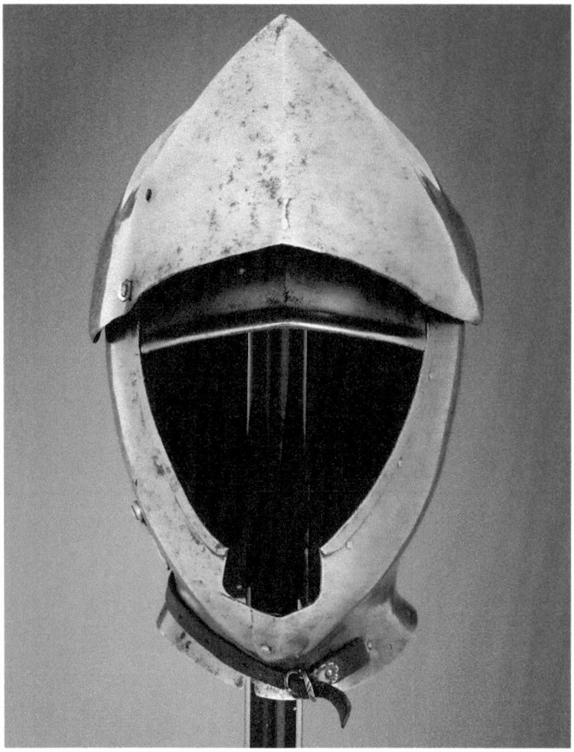

Figure 9 The armet A151 from the front with the visor raised. © Trustees of the Wallace Collection.

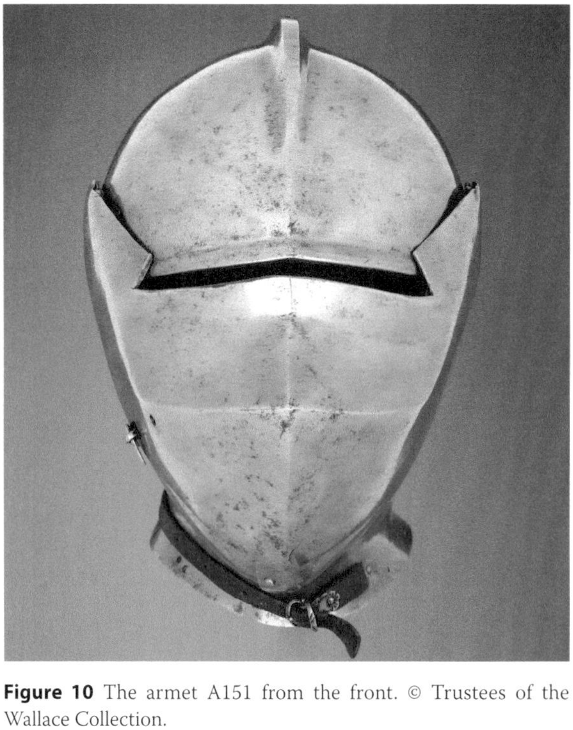

Figure 10 The armet A151 from the front. © Trustees of the Wallace Collection.

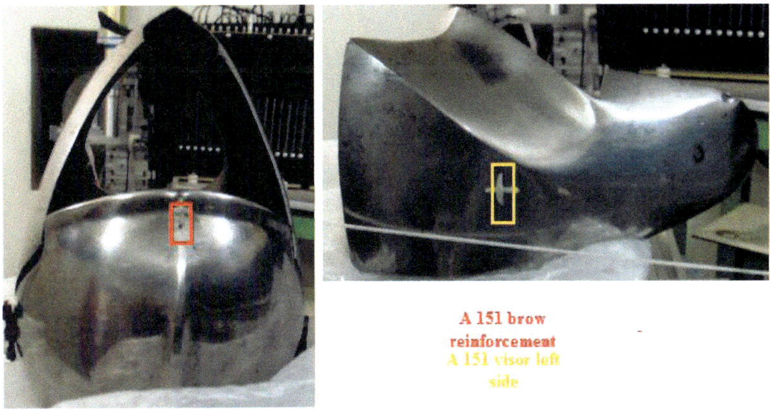

Figure 11 Areas of A151 analysed: the brow reinforcement in the centre and the visor on the left side.

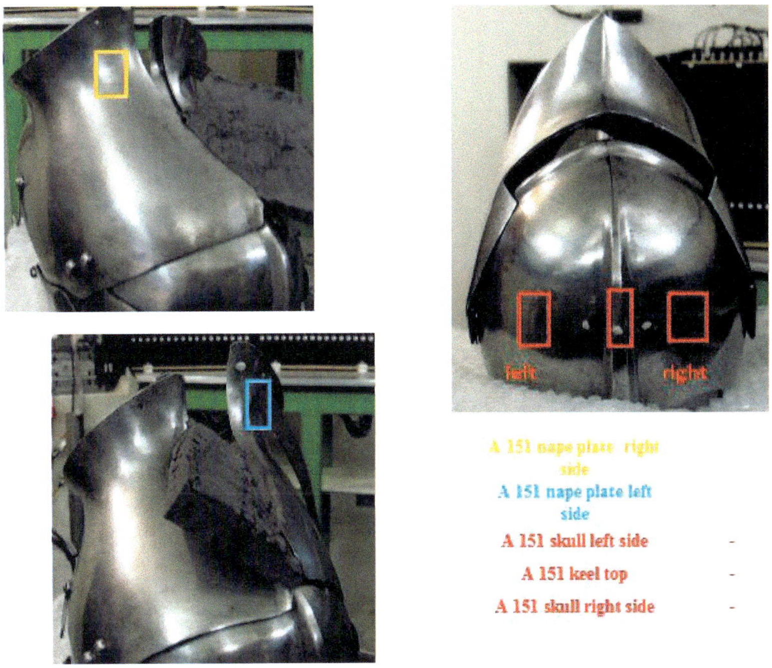

Figure 12 Additional areas of A151 analysed: the nape plate on the right side; the top of the skull on the proper left side; the top of the keel; the top of the skull on the right side; and the nape plate on the left side.

- The skull on the left side 0.362% C (+/- 0.007%)
- The keel on the top of the skull 0.276% C (+/- 0.007%)
- The skull on the right side 0.323% C (+/- 0.007%)

Martensite was not found in any of these components but a very small domain size in each case suggests the possibility of a quenched and then fully tempered microstructure of ferrite and cementite. There has also been some cold working after the final heat treatment possibly due to tournament usage.

On armet A 151 (Figures 9–12), probably made in Italy in the early 15th century, there is an armourer's mark which has apparently been

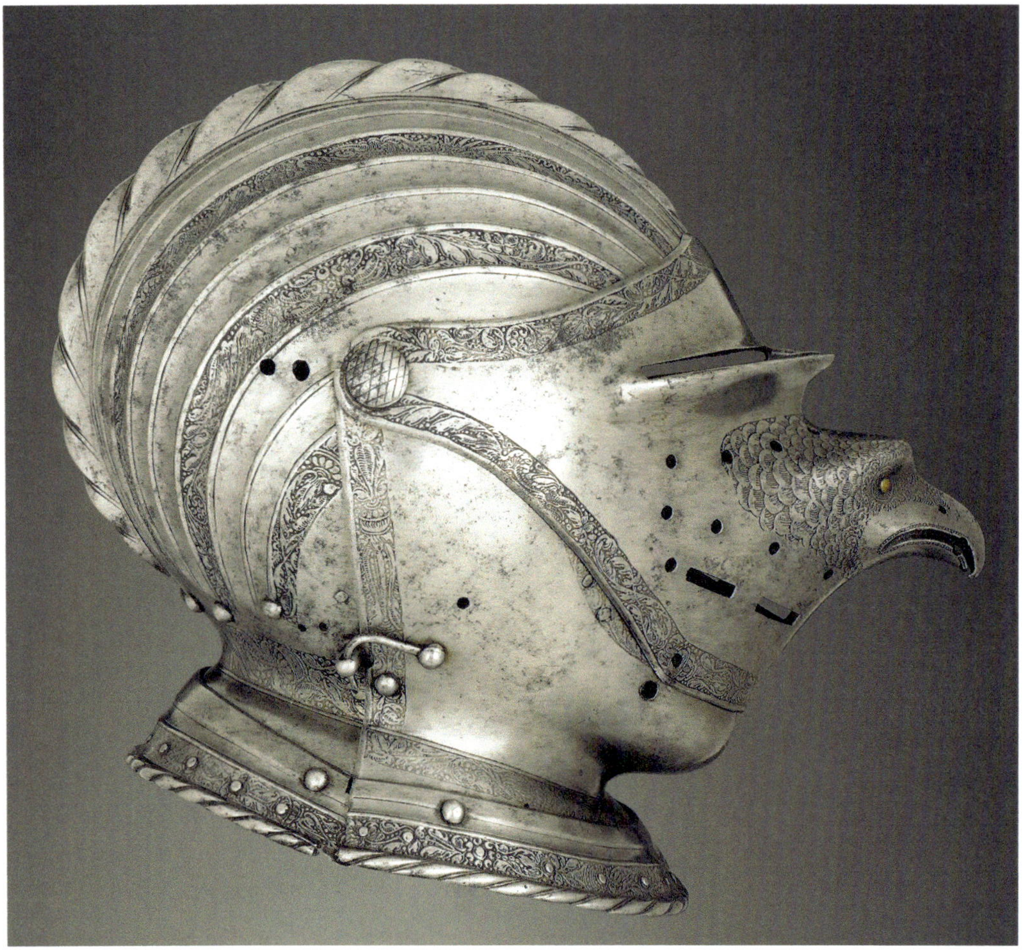

Figure 13 The close helmet A158. © Trustees of the Wallace Collection.

struck twice on the back of the skull. It weighs 3.38 kg.[10] The components of this helmet were examined in six places. The cheek plate which extends to cover the nape of the neck on the right side contained 0.147% C (+/-0.007%). As this contained no fayalite it may be a modern repair. The cheek plate which extends to cover the nape of the neck on the left side contained 0% C. It contained 0.4% fayalite but showed considerable texture, consistent with cold working. This might also be a repair, if a contemporary one, especially as the patination on the inside of these two plates seems to be very different. No cementite was detected in the skull on the left side, the keel of the skull and the skull on the right side but these three specimens from the skull showed the presence of martensite, indicating that the skull had been quenched in a futile attempt to harden it. The brow reinforce contained 0.242% C (+/- 0.006%) but no martensite.

A158 is a close helmet with a visor in the form of an eagle's beak (Figures 13–16). Made in South Germany, possibly Augsburg, around 1530, its weight is 3.36 kg.[11] The components of this helmet were examined in seven places:

- The right cheek plate 0.275% C (+/-0.009%)
- The right side of the skull on a polished rib 0.108% C (+/-0.007%)
- The right side of the skull on an etched rib 0.129% C (+/-0.008%)

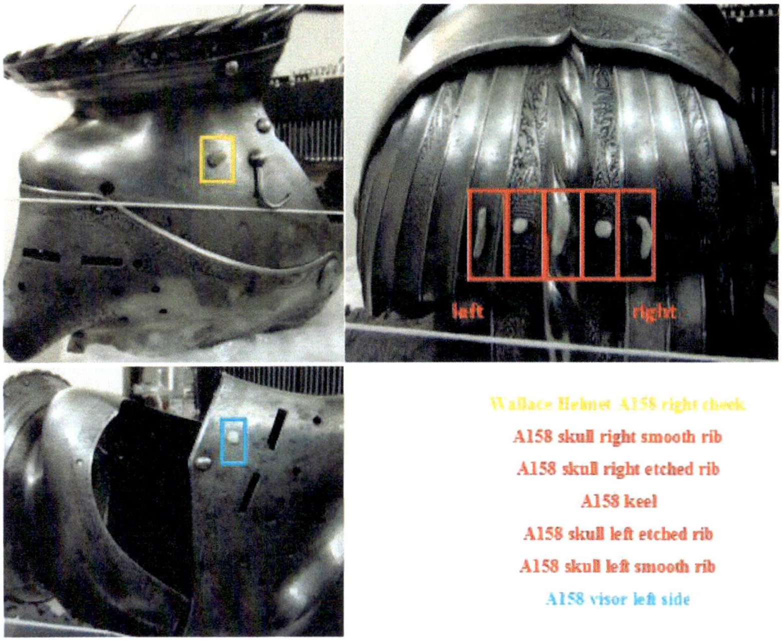

Figure 14 Areas of A158 analysed: the right cheek piece; the proper left side of the skull, polished flute, etched band, top of the keel, etched band on the right side, polished band; and the lower part of the visor on the left side.

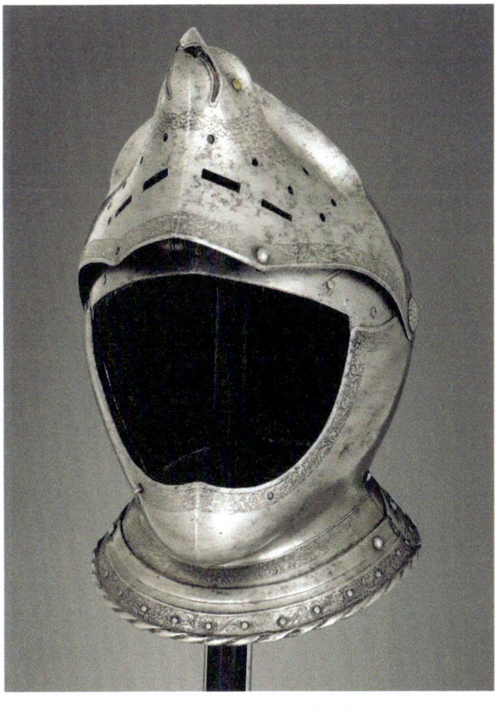

Figure 15 The helmet A158 with the visor raised. © Trustees of the Wallace Collection.

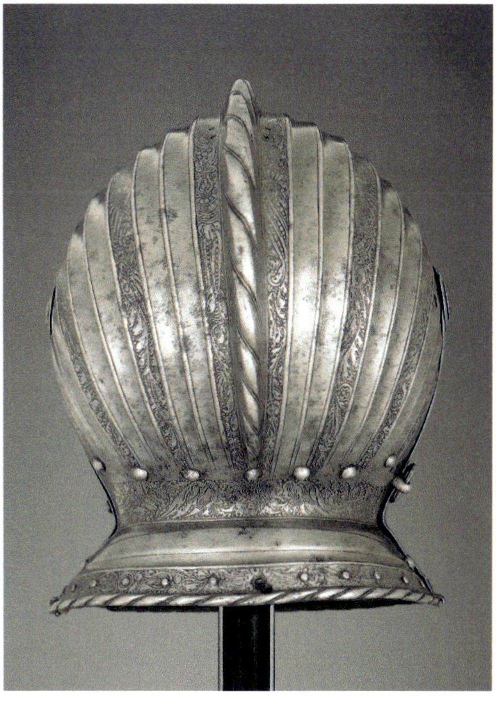

Figure 16 The back of the helmet A158. © Trustees of the Wallace Collection.

- The keel of the skull 0.136% C (+/-0.013%)
- The left side of the skull on an etched rib 0.116% C (+/-0.008%)
- The left side of the skull on a polished rib 0.061% C (+/-0.007%)
- The visor on the left side 0.130% C (+/--0.008%)

Untempered martensite was not found in any of these, but a very small domain size in each case suggests the possibility of a quenched and fully tempered microstructure of ferrite and cementite. The surface hardness (measured with a Branson Sonodur electronic hardness tester) was around 240–260 VPH, which would agree with such a microstructure.

Acknowledgements

We are very grateful to the Dean and Chapter of Canterbury Cathedral for their permission to allow the Black Prince helm to be brought to the Rutherford Appleton Laboratory for analysis. The vital assistance of the cathedral staff, Heather Newton (Head of Conservation), Sarah Turner (Collections Manager) and Ariane Langreder (Conservator) is much appreciated. We also acknowledge gratefully the support of STFC for experiment RB 1761003.

Notes and references

1. Williams, A. and Edge, D. (2003) 'Some early medieval swords in the Wallace Collection and elsewhere', *Gladius* 23: 191–210; Williams, A. and Edge, D. (2004) 'Great helms and their development into helmets', *Gladius* 24: 123–34; Williams, A., Edge, D. and Atkins, T. (2006) 'Bullet dents – "Proof marks" or battle damage?', *Gladius* 26(1): 175–209. Available at: http://gladius.revistas.csic.es/index.php/gladius/article/view/8/8 (accessed 6 June 2021); Williams, A. and Edge, D. (2007) 'The metallurgy of some Indian swords from the Arsenal of Hyderabad and elsewhere', *Gladius* 27: 149–76.
2. Williams, A., Edge, D., Kasztovszky, Z., Kis, Z., Kovács, I., Rosta, L., Szőkefalvi-Nagy, Z. and Káli, G. (2015) 'Nondestructive methods of analysis applied to oriental swords', *Gladius* 35: 139–58.
3. Grazzi, F., De Francesco, A., Barzagli, E., Scherillo, A., Williams, A., Edge, D. and Zoppi, M. (2016) 'Determination of the manufacturing methods of Indian swords through neutron diffraction', *Microchemical Journal* 125: 273–8.
4. Larson A.C. and Von Dreele R.B. (1994) General Structure Analysis System (GSAS), Los Alamos National Laboratory Report LAUR, 86-748, pp. 1–224.
5. Mann, J.G. (1962) *European Arms and Armour. Volume I. Armour.* London: Wallace Collection Catalogues, p. 148.
6. Norman, A.V.B. (1986). *European Arms and Armour Supplement.* London: Wallace Collection Catalogues, p. 68.
7. Williams, A. (2003) *The Knight and the Blast Furnace.* Leiden: Brill, p. 747.
8. Mann 1962 (cited in note 5), p. 98; Norman 1986 (cited in note 6), p. 43.
9. Mann 1962 (cited in note 5), p. 99; Norman 1986 (cited in note 6), p. 43.
10. Mann 1962 (cited in note 5), p. 128; Norman 1986 (cited in note 6), p. 56.
11. Mann 1962 (cited in note 5), p. 132; Norman 1986 (cited in note 6), p. 59.

11

Shah Jahan and Mughal steel: recent studies on Indian blades by neutron diffraction

Arthur Bijl, David Edge, Francesco Grazzi, Antonella Scherillo and Alan Williams

Introduction

Among the most famous swords in history are those made of so-called 'Damascus steel' and much has been written about them.[1] They were made in India and Iran, have a 'watered-silk' or 'damask' pattern on their surfaces, and were renowned for their quality: no amount of sharpening would remove their edges. A unique method to prepare steel developed in the ancient Middle East called crucible steel: this was made by placing in a crucible lumps of iron with carbon-rich organic material (or even cast iron), which was heated for many hours or possibly even days. Once the crucible was broken, an ingot of cast steel was recovered from which a blade might be forged. Since the liquid metal had separated from the liquid slag, it was a homogeneous product with a high carbon content (1.2%–1.6% C) and virtually no slag. The earliest known artefacts made of crucible steel date from the 6th century: a 6th-century Sassanian blade in the British Museum was found to be made of a crucible steel.[2]

An important subgroup of crucible steel – called *wootz* in southern India but sometimes, misleadingly, 'Damascus steel' in the Western world – was used to make expensive swords of impressively high quality. Distinguished by a characteristic pattern (formed by the presence of large crystals of cementite, Fe_3C) on its surface, it was likened to 'watered silk'. Initially, *wootz* was made much like other crucible steels but forged in such a way that the cementite dendrites were not homogenised but retained as a network of visible crystals. This may have entailed a different forging procedure (perhaps a very slow forging at an extremely low temperature) to break up the cementite dendrites generated in the crucible into smaller (but still visible) particles, reducing brittleness and producing the characteristic pattern which became visible after polishing and etching the surface of the blade with a weak acid. Attempts by Verhoeven and Pendray to recreate this process have resulted in a series of detailed papers on *wootz* metallurgy.[3]

The Wallace Collection

The Wallace Collection in London is home to a 19th-century collection of Asian arms and armour which contains over 900 objects, including hundreds of swords and daggers. Many of these may have been made of patterned crucible steel (*wootz*) but the passage of time and inappropriate cleaning have obscured many of these patterns.

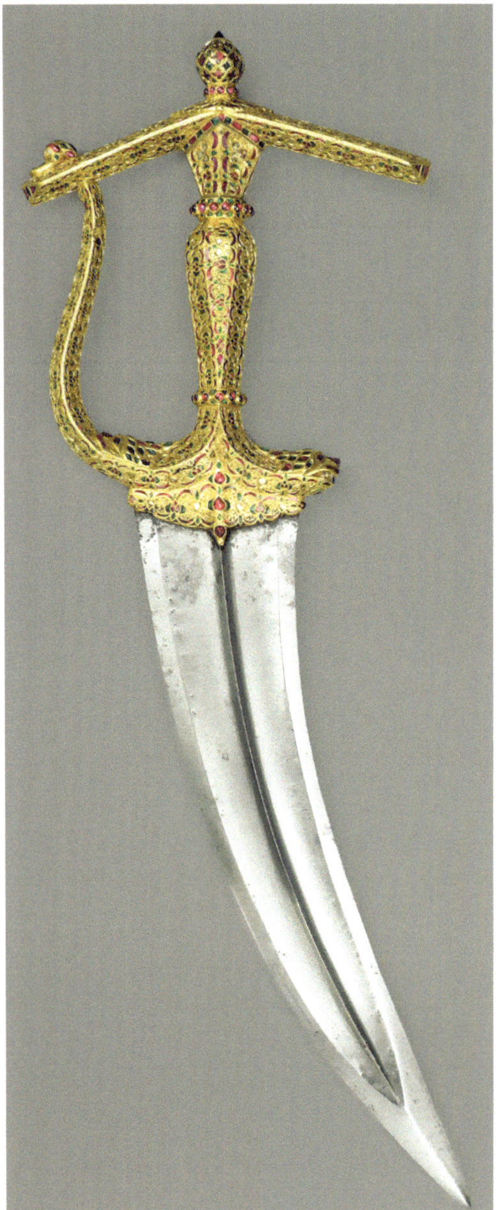

Figure 1 The dagger of Shah Jahan in the Wallace Collection (OA1409). © Trustees of The Wallace Collection.

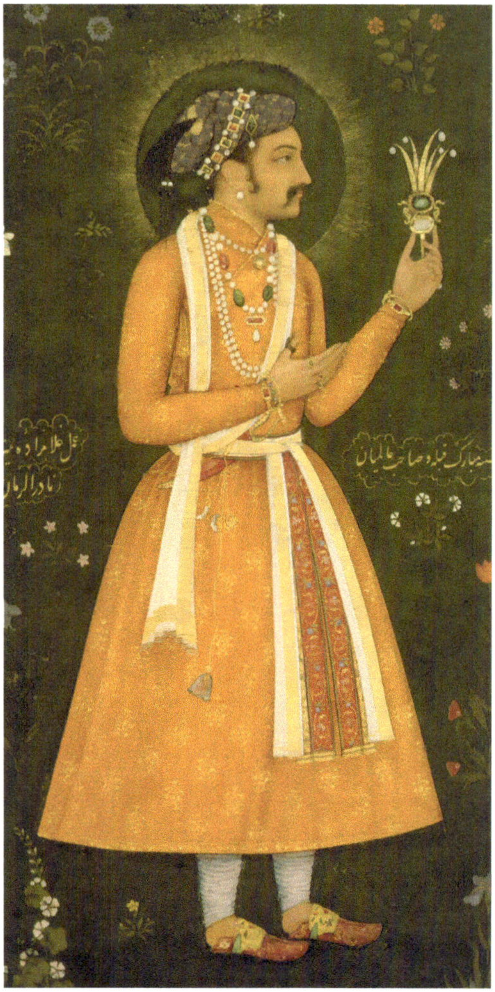

Figure 2 Shah Jahan, c.1620, illustrated in a manuscript from the Victoria and Albert Museum, IM.14-1925 © Victoria & Albert Museum, London.

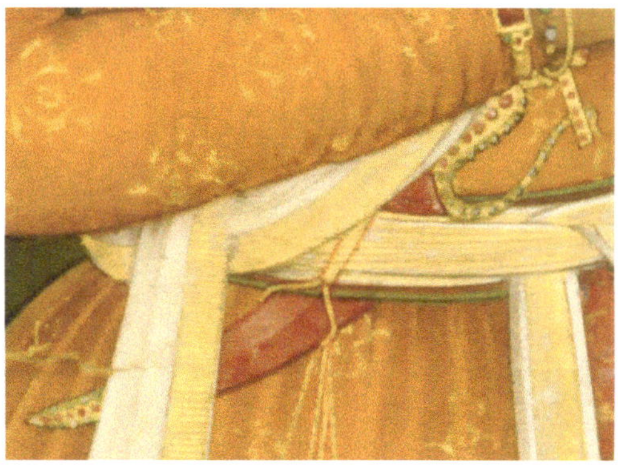

Figure 3 A detail showing the hilt of the dagger illustrated in Figure 2.

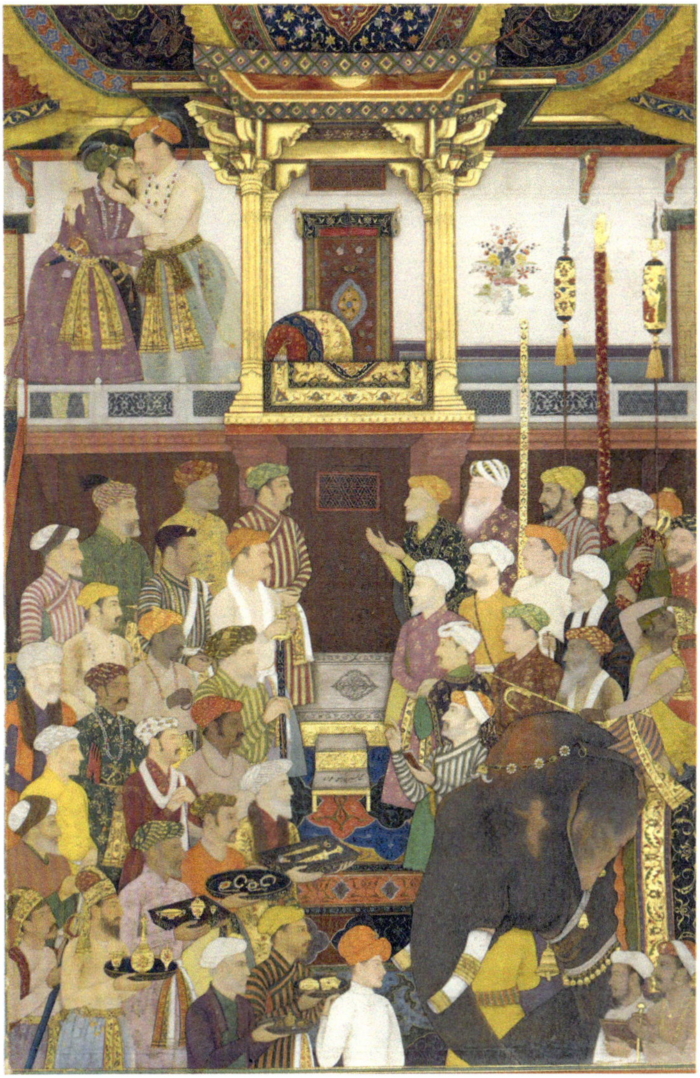

Figure 4 The Emperor Jahangir receives Prince Khurram on his return from the Deccan (10 October 1617), 1656–57, painting by Murar in a manuscript from the Royal Collection. RCIN 1005025.j f.49a, Royal Collection Trust © Her Majesty Queen Elizabeth II.

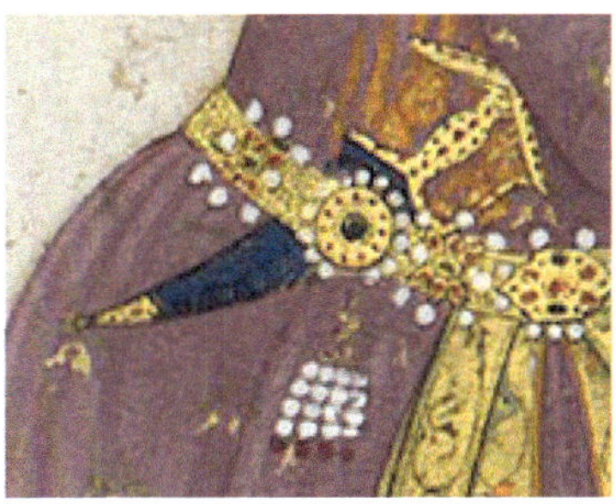

Figure 5 A detail showing the hilt of the dagger shown in Figure 4.

One of the highlights of the Wallace Collection is a jewel-studded dagger (Figure 1) attributed to the Mughal imperial workshop c.1615 (OA1409). Its likeness appears in two paintings, both showing Khurram, later known as Shah Jahan on his victorious return from campaigns in the Deccan to the court of his father, Jahangir, on 10 October 1617 (Figures 2–4).[4] Two rare features of the depicted dagger (Figure 5) suggest that it may be identified with the Wallace Collection object. Both paintings show the knuckle-guard connected to the pommel, a feature not commonly seen on this type of dagger hilt, and secondly, the dagger painted by Abu al-Hasan in the Victoria and Albert Museum depicts the head of a *haṃsa* or waterfowl on the pommel: no other dagger with these characteristics survives. Regardless of whether the dagger belonged personally to Khurram, it is without question a production of the imperial workshop during the reign of Jahangir, and it is therefore not surprising to find that it is made of an exceptionally high-carbon steel.

Analysis of OA1409 in the Wallace Collection

Analyses of museum objects, especially those of the quality in the Wallace Collection, must be completely non-destructive and the most suitable technique has proved to be time-of-flight neutron diffraction (TOF-ND). Obeying Bragg's Law, a neutron beam of known velocity behaves as a wave of known wavelength, and different crystalline phases (ferrite, cementite, fayalite, etc.) diffract the wave through different angles (see Figure 13). The proportion of cementite indicates the carbon content of a steel. Neutron diffraction can also reveal the presence of patterns by detecting anisotropy in the cementite. However, searching for a direct connection between the 'watered' pattern and the anisotropy has raised many more questions which further research is endeavouring to answer.[5] Examination of 10 areas of the blade found that the carbon content varies only slightly between 1.39% C and 1.59% C, confirmation that this is a high-carbon crucible steel, typical of the finest Indian and Iranian blades.[6]

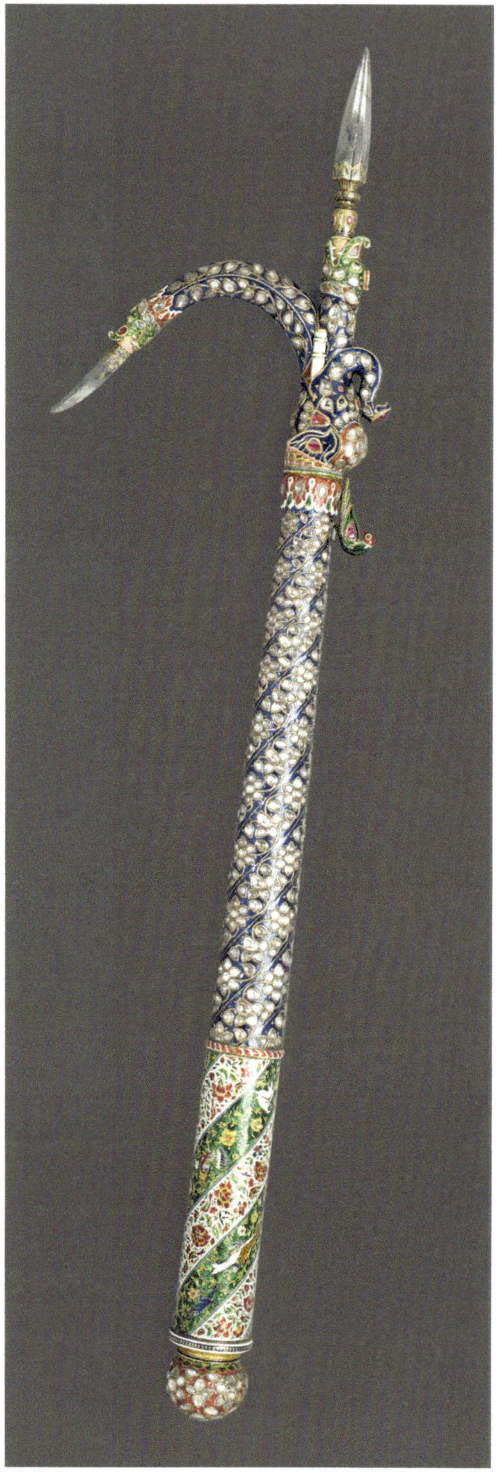

Figure 6 An elephant goad, Jaipur, c.1860–70 in the Wallace Collection (OA1382). © Trustees of the Wallace Collection.

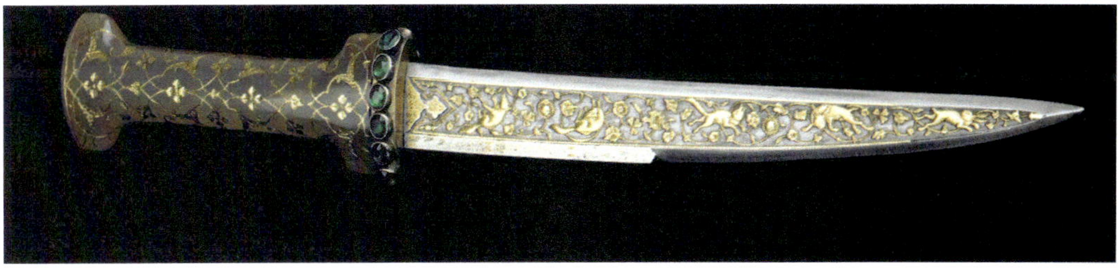

Figure 7 A dagger, Iran or Central Asia, late 15th century, in the Wallace Collection (OA1414). © Trustees of the Wallace Collection.

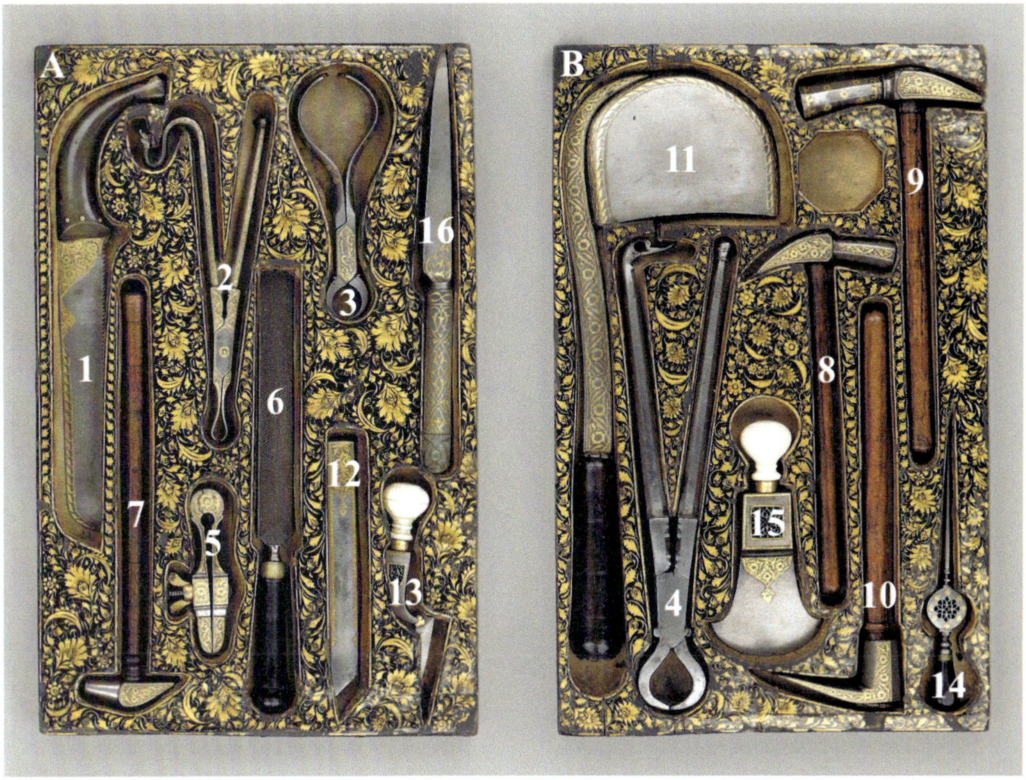

Figure 8 A set of tools from Iran, 19th century, in the Wallace Collection (OA1737). © Trustees of the Wallace Collection.

Further analyses

The results of examination of six blades of Indian and Iranian origin (five from the Wallace Collection and one from a private collection[7]) during the course of this experiment are included for comparison (Figures 6–12). All but one were made of similar very high-carbon steels with no martensite detected – the exception (OA1414) was produced in Iran or Central Asia during the late 15th century.[8]

- **Elephant goad, Jaipur, c.1860–70, OA1382** (Figure 6). The blade was examined in two places: the carbon content at the tip was 0.39% C and at the hook 1.12% C. This is also a high-carbon crucible steel.
- **Dagger, Iran or Central Asia, late 15th century, OA1414** (Figure 7). The blade was examined in six places: the carbon content varied between 0.30% C and 0.57% C. The tip was 0.18% C. This is a medium-carbon bloomery steel.

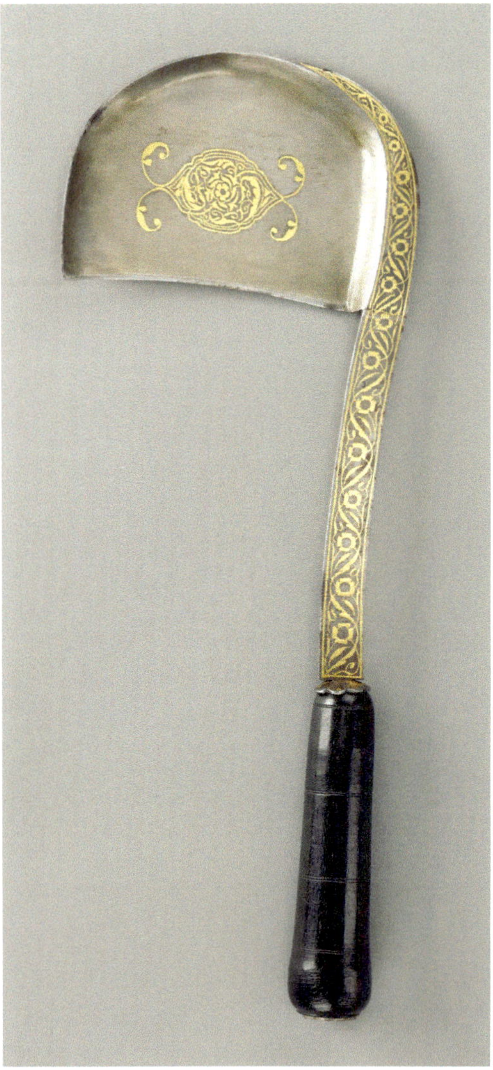

Figure 9 One of the set of tools (Figure 8) which might be a knife for trimming hooves, in the Wallace Collection. © Trustees of the Wallace Collection.

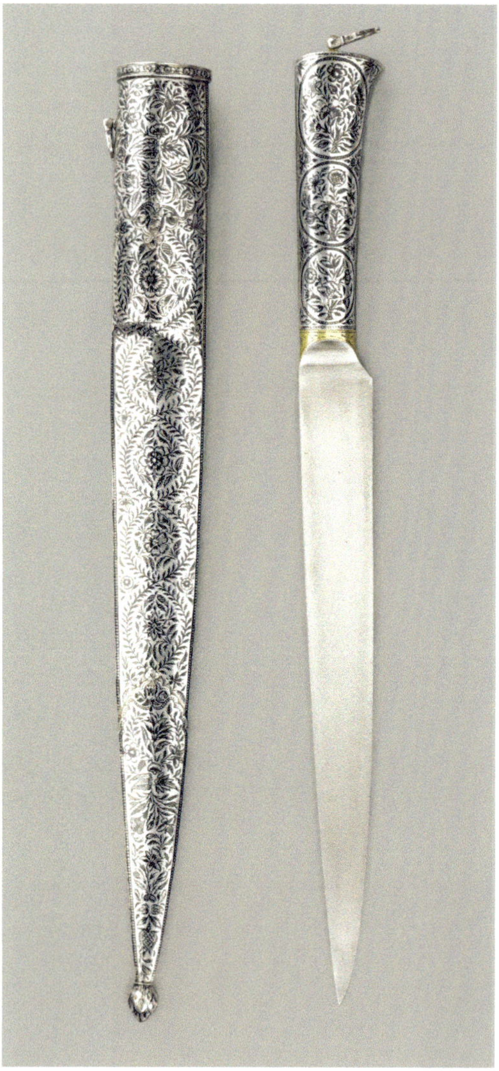

Figure 10 A knife, probably from Lucknow, late 18th or early 19th century, in the Wallace Collection (OA1986). © Trustees of the Wallace Collection.

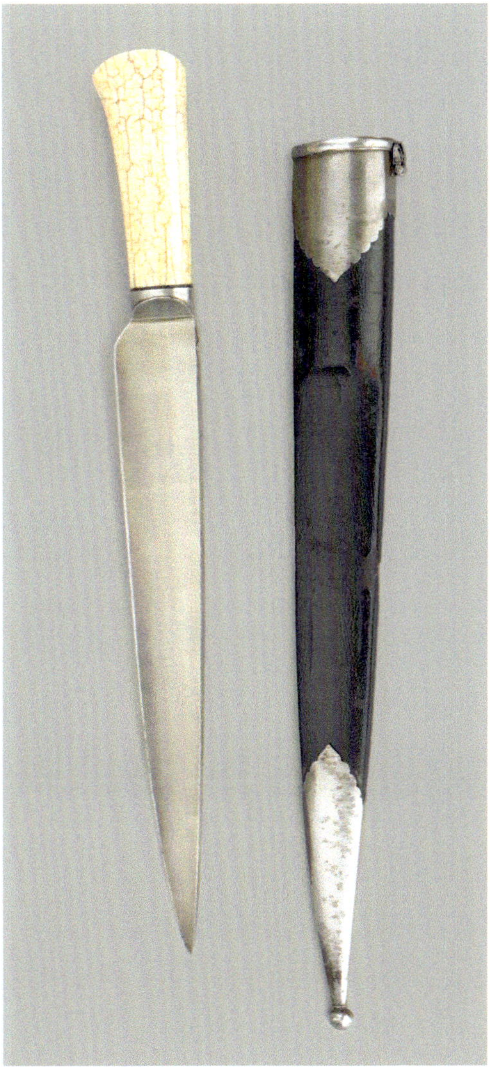

Figure 11 A knife from Iran or India, early 19th century, in the Wallace Collection (OA2260). © Trustees of the Wallace Collection.

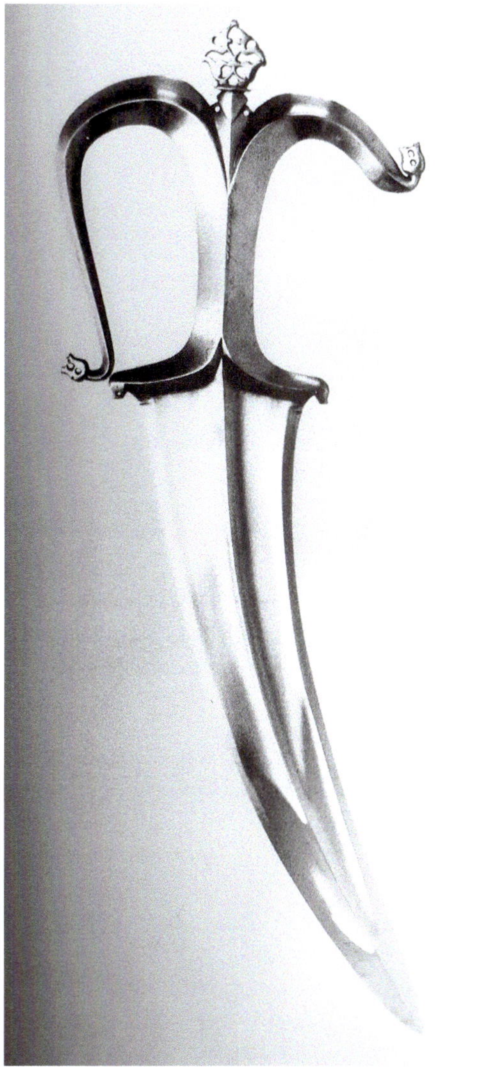

Figure 12 A blade from a private collection also examined.

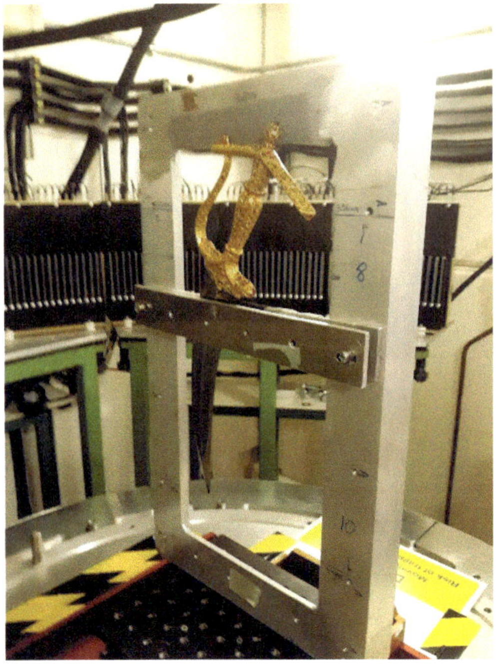

Figure 13 The dagger of Shah Jahan being analysed by neutron diffraction in the neutron source ISIS at the Rutherford Appleton Laboratory, Harwell.

- **Tool, perhaps a knife for trimming a hoof, Iran, 19th century, OA1737** (Figures 8 and 9). The sidebar was examined in three places: the carbon content varied between 0.01% C and 0.46% C. However, the carbon content of the blade, which was examined in five places, varied between 1.29% C and 1.38% C. This is also a high-carbon crucible steel.
- **Knife, probably Lucknow, late 18th or early 19th century, OA1986** (Figure 10). The blade was examined in six places: the carbon content varied between 1.15% C and 1.61% C. The tip was 1.48% C. This is also a high-carbon crucible steel.
- **Knife, Iran or India, early 19th century, OA2260** (Figure 11). The blade was examined in six places: the carbon content varied between 1.15% C and 1.69% C. The tip was 1.47% C. This is also a high-carbon crucible steel.
- **Steel dagger with a steel hilt, private collection** (Figure 12). The blade was examined in six places: the carbon content varied between 1.54% C and 1.61% C. The tip was 1.54% C. This is also a high-carbon crucible steel.

Acknowledgements

We are very grateful to the Science and Technologies Facilities Council (STFC) for its support in carrying out this experiment (RB 1920366).

Notes and references

1. Panseri, C. (1965) 'Damascus steel in legend and reality', *Gladius* IV: 5–66. Available at: https://core.ac.uk/download/pdf/268620601.pdf (accessed 14 June 2021).
2. Lang, J., Craddock. P. and Simpson, S.J. (1998) 'New evidence for early crucible steel', *Journal of the Historical Metallurgy Society* 32: 7–14.
3. Verhoeven, J.D., Pendray, A.H. and Gibson, E.D. (1996) 'Wootz Damascus steel blades', *Materials Characterization* 37: 9–22.
4. London, Victoria and Albert Museum, IM.14-1925 and London, Royal Collection Trust, RCIN 1005025.j.
5. Grazzi, F., Cantini, F., Salvemini, F., Scherillo, A., Schillinger, B., Kaestner,A., Edge, D. and Williams, A. (2018) 'The investigation of Indian and central Asian swords through neutron methods', *Journal of Archaeological Science: Reports* 20: 834–42.
6. Grazzi, F., De Francesco, A., Barzagli, E., Scherillo, A., Williams, A., Edge, D. and Zoppi, M. (2016) 'Determination of the manufacturing methods of Indian swords through neutron diffraction', *Microchemical Journal* 125: 273–8.
7. Published in Norman, A.B.V., Ricketts, H. and Andersen, V.J. (1982) *Islamic Arms and Armour from Private Danish Collections.* Copenhagen: Davids Samling, p. 139.
8. This analysis was carried out on the instrument INES in the neutron source ISIS at the Rutherford Appleton Laboratory, Harwell.

12

Indo-Persian axes and the faking of *wootz*

David Edge, Richard Furrer, Francesco Grazzi, Antonella Scherillo and Alan Williams

Introduction

A unique method to prepare steel was developed in the ancient Middle East: crucible steel. It was made by placing lumps of iron with carbon-rich plant material (or even cast iron) in a covered crucible and heating for many hours or even days. At some point, in the insulated system, enough carbon was absorbed in the metallic matrix to lower the melting range to 1400–1300 °C, melting the new iron-carbon alloy. Eventually the crucible was broken open and an ingot of cast steel obtained from which a sword blade might be forged. Since the liquid metal had separated from the liquid slag, it was a homogeneous and virtually slag-free steel with a high carbon content (usually 1.2%–1.6% C).

Its manufacture is described in a treatise once ascribed to Zosimos, the 3rd-century alchemist, by his editor Berthelot in the 19th century,[1] but this recipe is now regarded (in a more recent critical edition by Mertens[2]) as the work of Zosimos by association only, since it is part of a larger codex (Marciana 299), dating from the 9th–10th century. This seems to have come from a Byzantine compilation which mentions the Emperor Heraclius (610–641), who made war on the Sassanians and besieged the Persian capital Ctesiphon in 627, therefore it might well be a 7th-century manuscript. Some of the other manuscripts in this codex are also ascribed to Heraclius. While this Byzantine emperor did invade Persia, it seems unlikely that he personally penned these recipes, but he was said to have been a patron of alchemists, suggesting a link to Sassanian Persia.

The Persians traded in crucible steel and a Persian blade of the 7th century from Dailaman in the British Museum has recently been analysed and found to be made of crucible steel.[3] France-Lanord also published an analysis of a Luristan blade of uncertain date, but perhaps somewhat earlier, which was made of a very high-carbon steel.[4] In the 9th century, the Arabic writer Ya'cub Ishaq al-Kindī (c.800–870), in his letter to the caliph of Baghdad,[5] provides a detailed description of different types of iron and steel, including crucible steel, and their use in making swords.[6] Two centuries later, Abu Rayhan al-Biruni (973–1048) quoting earlier writers, describes how crucible steel was made around Herat and exported via North India to Persia and other Muslim lands:[7]

> Herat is especially noted for steel, and it is there called eggs on account of its shape. These eggs are long and round-bottomed according to the form of the crucibles in which they are made ... the *narmahan* (soft iron) and *dus* (water of iron) are melted equally in the crucible, and they blend so that one cannot distinguish the one from the other ... (if) the melting is unequal so

> that the two do not mix completely ... this is called watering (*firind*).[8]

This patterned steel might have been *wootz*, sometimes referred to in the Western world as 'Damascus steel'.[9] He then goes on to describe a method which might have been employed to obtain a surface resembling crucible steel (or even *wootz*) on an iron base:

> A person who was in Sind told me that he sat and watched a blacksmith making swords. He looked at the swords closely and their iron was *narmahan*, and the blacksmith used to sprinkle on the iron a highly refined medicament of reddish colour, then threw the sword (into water) and consolidated it by immersion. Then he used to take it out and lengthen it with a hammer and repeat the sprinkling and working many times ... I discovered that it was *dus* which he mixed with the *narmahan* for edging and ornamenting (the sword).[10]

Murda ibn Ali al-Tarsusi dedicated his treatise on arms and warfare, written around 1190, to Saladin. He describes both the methods of al-Kindī and al-Biruni as well as a recipe for steel said to have been taken from Jabir's *Book of the Seventy*.[11]

The 14th-century writer, Izz al-Din Aydamir al-Jildaki, quotes from a book *Kitab al-Hadid* ('On iron') – also ascribed to Jabir and included in the English edition of the latter's works printed in 1678 – which describes the making of both white cast iron and crucible steel.[12] The historical crucible steel industry has been discussed in recent years by various archaeometallurgists, including Bennet Bronson,[13] Paul Craddock,[14] and Anna Feuerbach.[15]

Investigations into the microstructure of 'Damascus steel' have been undertaken largely by metallography. Zschokke carried out detailed chemical and microscopical analyses as well as (destructive) mechanical tests on some 'Damascus' blades,[16] working in collaboration with the traveller Henri Moser who sacrificed six Indo-Persian blades (from his collection of 2,000) in the interests of science. All had a pattern of some sort due to rows of cementite in the hypereutectoid steels (with carbon contents varying from 1.3% to 1.9% C) except for one blade, which had only 0.6% C. In this blade, the pattern was due to rows of ferrite grains in a pearlitic groundmass.

Other works by Nicholas Belaiew,[17] Carlo Panseri,[18] and Jerzy Piaskowski[19] illustrate the microstructures of more 'Damascus' steel blades.[20] The precise mechanism of the pattern formation seems to depend upon the thermal history of the steel and has been the subject of considerable debate.[21] However, it remains clear that *wootz* was only a small part of crucible steel production.[22]

Steel is mainly constituted of two phases: ferrite, a body-centred cubic structure, which is almost exclusively composed of iron atoms, and cementite, an orthorhombic structure containing three iron atoms for every one carbon atom. The cementite is responsible for the hardening of steel when compared with iron, as well as for its appearance. In a patterned crucible steel, the cementite phase exhibits an anisotropic distribution of the domains, which may be revealed by performing neutron diffraction measurements at different sample orientations, and this was the approach applied to the analysis of the set of axes from the Wallace Collection.

Indian axes and their analysis by neutron diffraction

These Indo-Persian saddle-axes (*tabarzin*) were carried by horsemen,[23] and frequently exhibit a surface pattern which appears very similar to that formed by segregated lines of the cementite particles ('damask') in *wootz* or patterned crucible steel.[24] These patterns were originally highly prized as indicators of their metallurgy, but have often been obscured by corrosion and ill-advised cleaning.

In order to study such patterns, 16 Indo-Persian axes (from the 17th and 18th centuries) were taken from the Wallace Collection in 2016 to the ISIS neutron source at the Rutherford Appleton Laboratory at Harwell[25] and studied by time-of-flight neutron diffraction (ToF-ND) using the INES beamline.[26] This is a completely

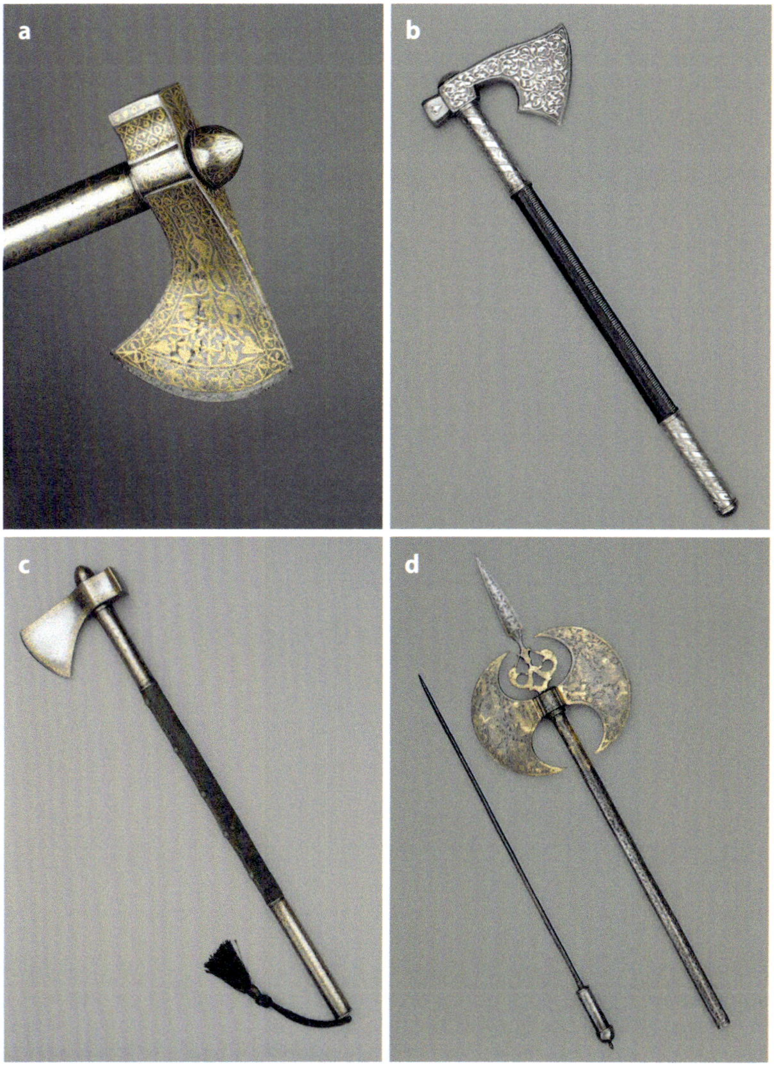

Figure 1 Four axes all made simply of wrought iron: (a) OA1566; (b) OA1568; (c) OA1592; (d) OA1604. © Trustees of the Wallace Collection.

non-invasive technique able to provide quantitative phase analysis and to determine any anisotropic distribution in the cementite.[27]

Measuring the diffraction signals allows the relative concentration of ferrite and cementite to be determined and thereby obtain the carbon content of the iron or steel. But with crucible steels, there is another factor to consider: if it is possible to determine any anisotropy in the distribution of cementite, then the presence of a pattern, even if hidden, is indicated. This information allows museums to identify which blades are made of *wootz* and facilitate decisions on ethical conservation. The several diffraction peaks of cementite vary in intensity with the orientation angle of the blade only when the cementite is anisotropic. If they do not vary, then the steel would be entirely homogeneous (isotropic), and there would be no pattern.[28]

Each axe was measured in several areas to examine the possibility that a different type of iron or steel was used in specific parts. Every

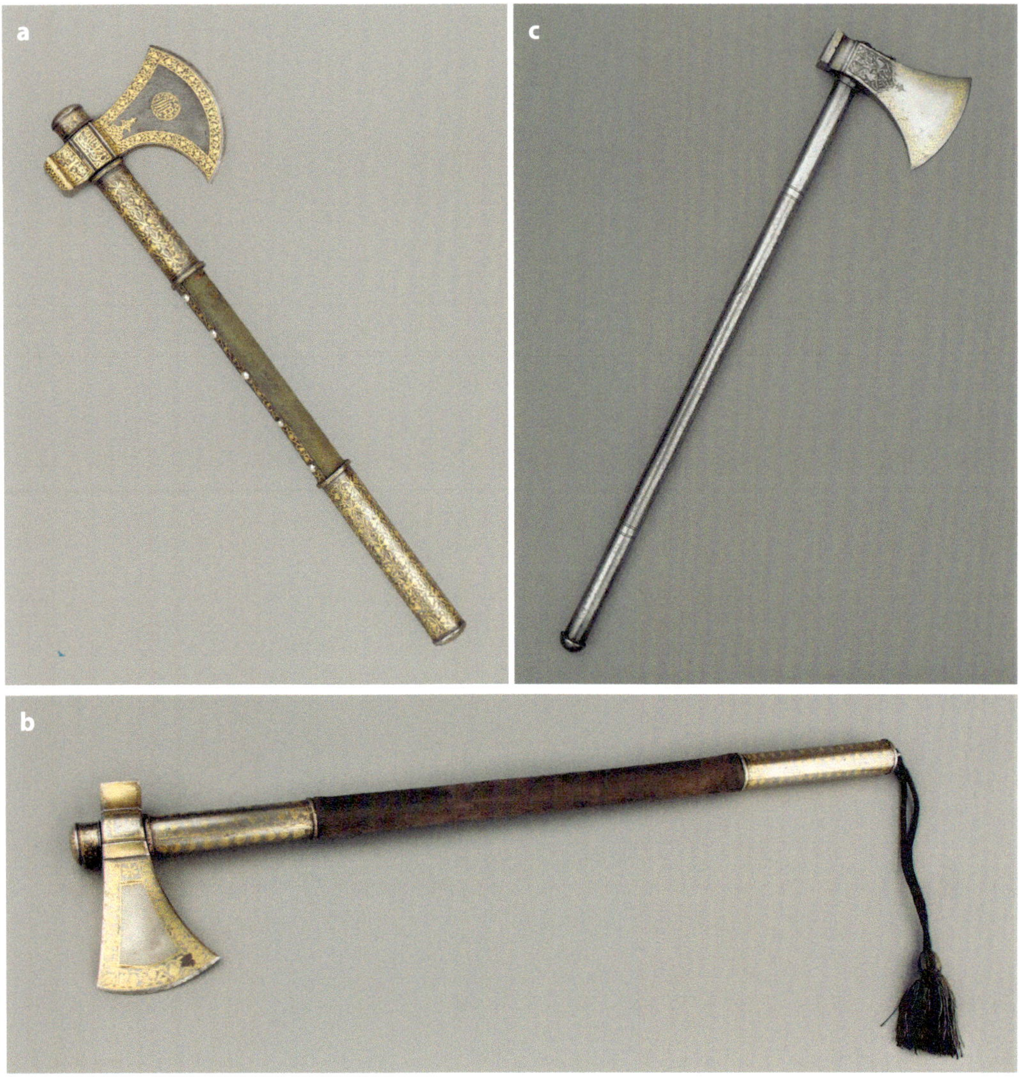

Figure 2 Three axes made of iron with some low-carbon steel in the blade: (a) OA1564; (b) OA1552; (c) OA1596. © Trustees of the Wallace Collection.

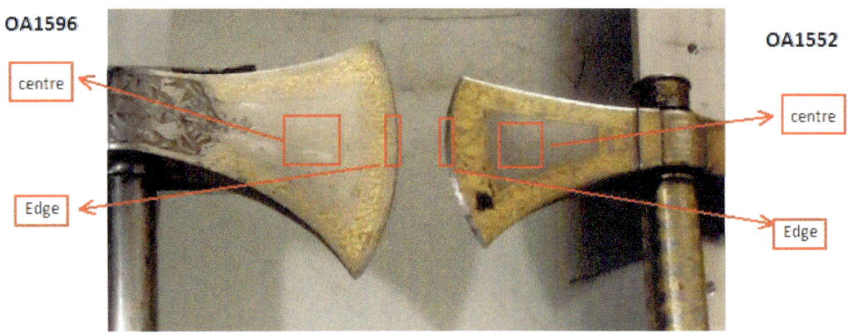

Figure 3 OA1552 and OA1596 being analysed.

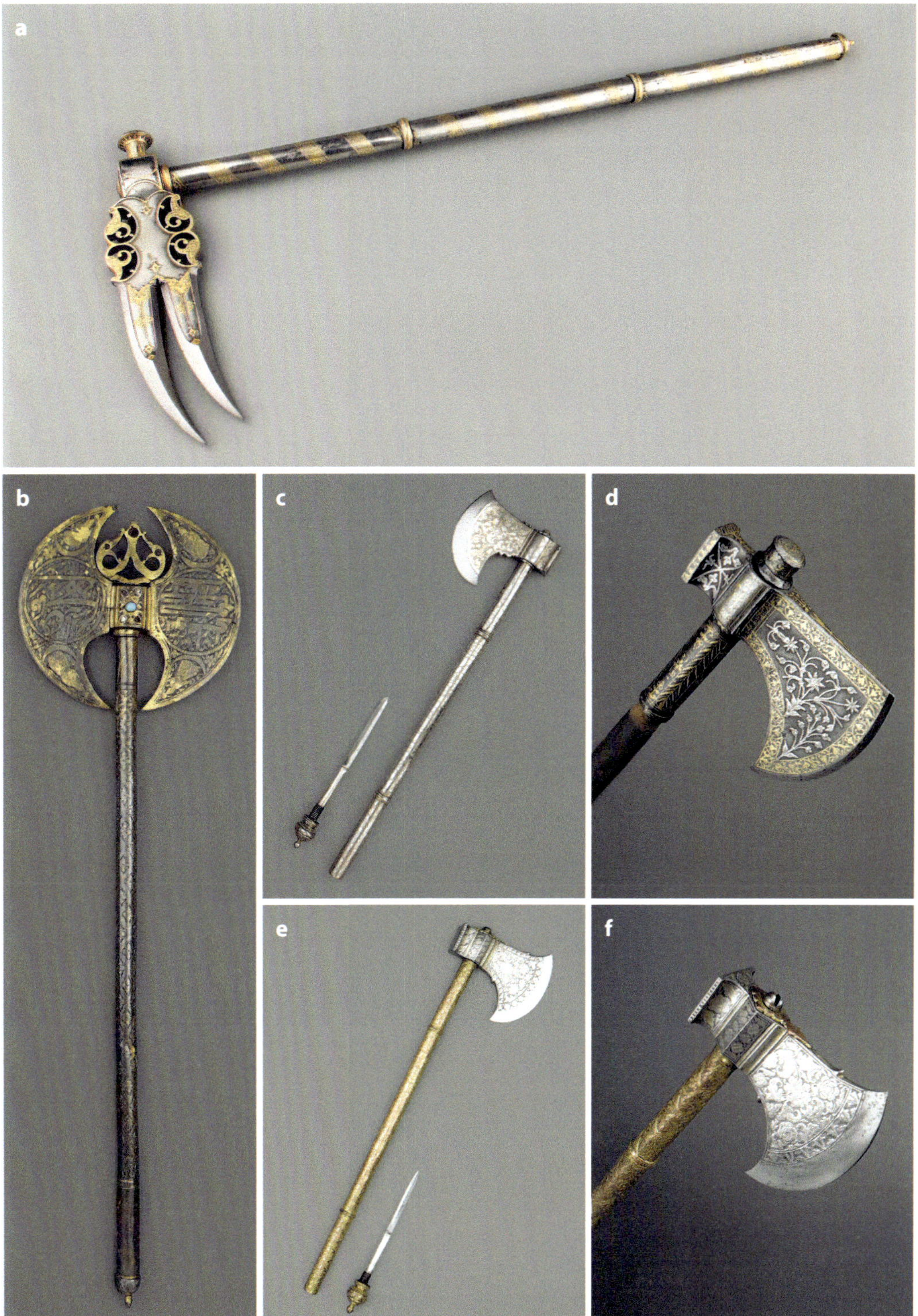

Figure 4 Six axes simply made of low-carbon steel: (a) OA1554; (b) OA1556; (c) OA1562 showing the stiletto which fits inside the hollow shaft; (d) OA1594; (e) OA1598; (f) OA1610. © Trustees of the Wallace Collection.

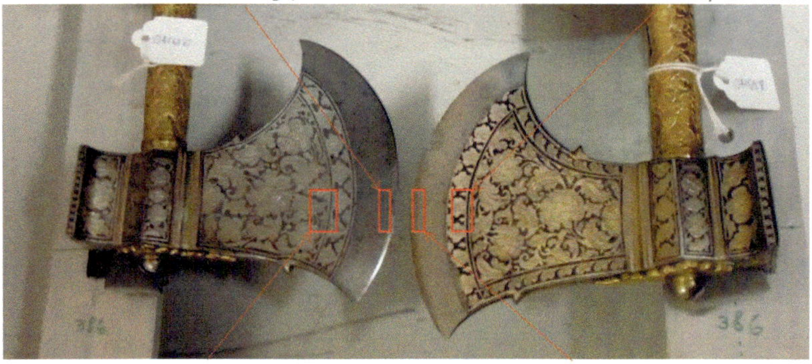

Figure 5 Analysis of OA1598 and OA1610.

measurement needed from 1 to 4 hours to be completed according to the size of the area examined and the thickness of the sample. The results were obtained for phase analysis, carbon content and identification of *wootz* steel areas.

Summary of results

The results show that there were four axes entirely made of wrought iron: OA1566, OA1568, OA1592, OA1604 (Figure 1). Three axes were made with wrought iron in the edge and low-carbon steel in the centre of the blade: OA1552, OA1564 (very low in this case) and OA1596 (Figures 2 and 3). Six axes (OA1544, OA1556, OA1562, OA1594, OA1598, OA1610) were made of low-carbon steel (Figures 4 and 5) and two (OA1614 and OA1616) of low- and medium-carbon steel (Figures 6 and 7).

Notwithstanding their surface appearance, only one (OA1586) was made entirely of a hypereutectoid steel although since it exhibits isotropic cementite it may be described as a crucible steel rather than *wootz*.

It is interesting to note that the cementite in OA1544 shows some anisotropy in both the centre and in the edge, while the cementite in OA1552 indicates some anisotropy in the centre and the cementite in OA 1616 shows very weak anisotropy, also in the centre. The presence of some anisotropic cementite in these three axes when contrasted with their low average carbon content suggests that the cores are mostly ferrite and that the cementite signal was only coming from the outer layers. These axes were probably produced by wrapping a thin layer of genuine *wootz* around a core of wrought iron or low-carbon steel to make them appear to have a desirable surface pattern.

Finally OA 1598 had been quenched on its edge without any subsequent tempering, since its diffraction pattern showed the presence of martensite and also retained austenite. The lattice parameters of these two phases allowed us to evaluate the carbon content in this area as being around 0.5% ±0.1 wt%.

Discussion

To our great surprise, very few of these 16 axes consisted of steel of any significant carbon content: many were simply iron and only one of crucible steel (OA 1586). It appears that at least three (OA1544, OA1552 and OA1616) were made as contemporary fakes with a very thin surface of high-carbon steel on a body of iron in order to give the appearance of *wootz* but intended simply for ceremonial use. How this was achieved is a very interesting question. The high cost of making arms and armour from *wootz* probably encouraged craftsmen from earliest times to produce articles which looked like *wootz* using various other methods. Pattern welding was one possible method (one example being the sword

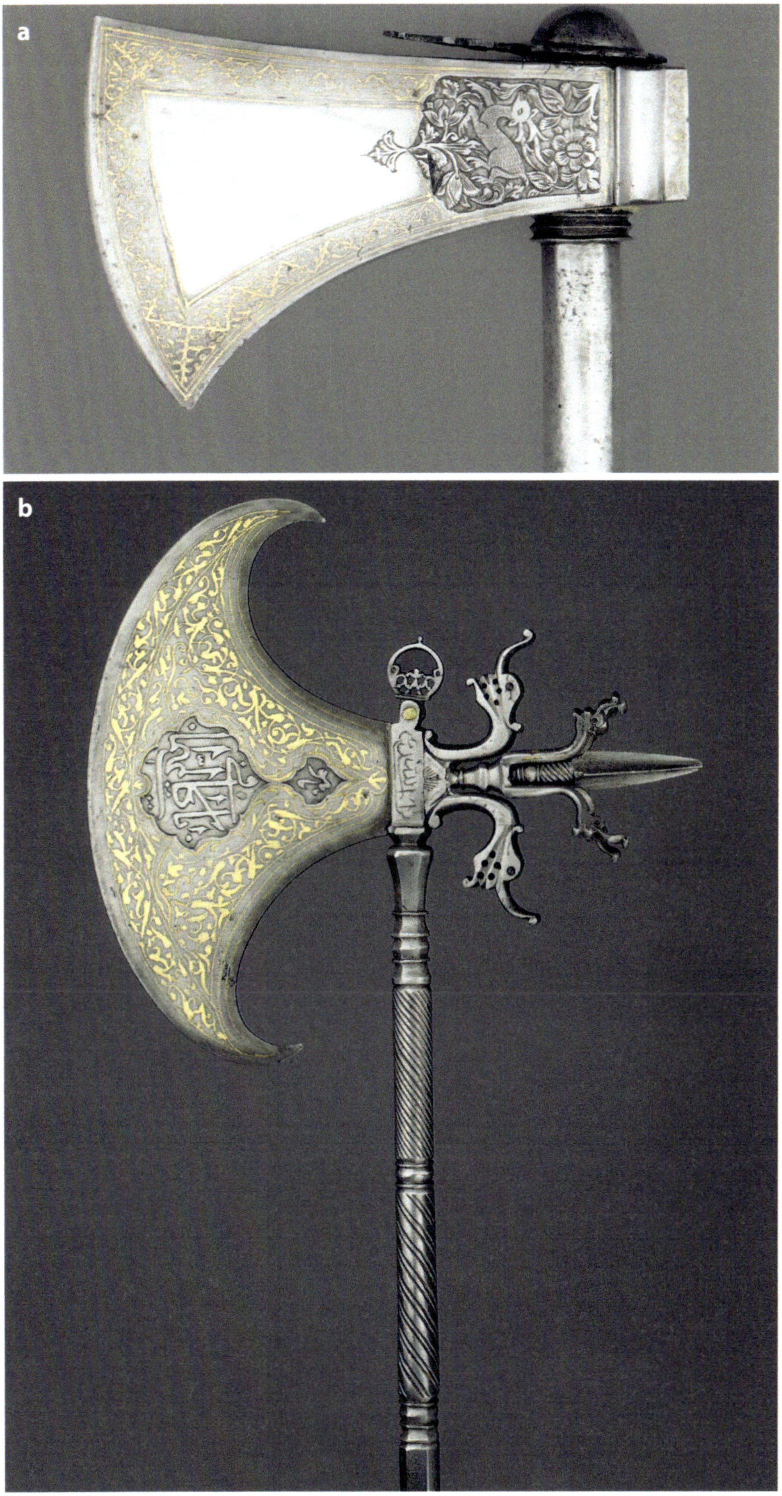

Figure 6 Two axes made of medium-carbon steel: (a) OA1614; (b) OA1616. © Trustees of the Wallace Collection.

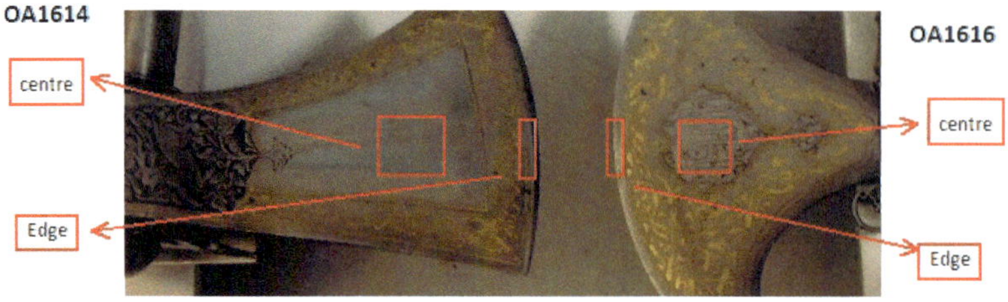

Figure 7 Analysis of OA1614 and OA1616.

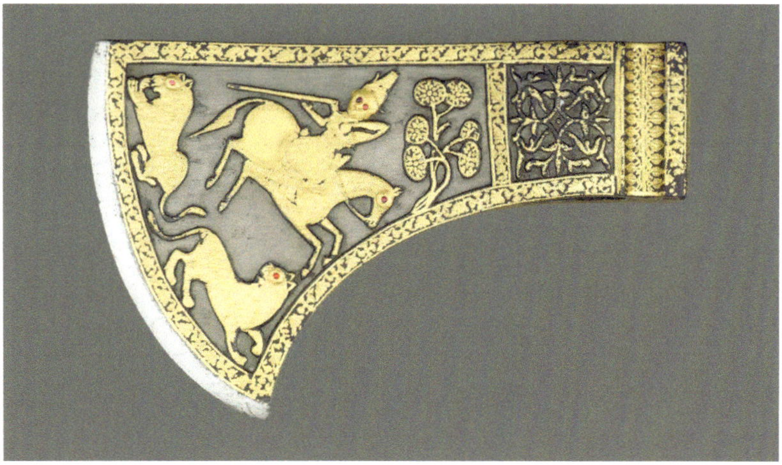

Figure 8 OA1586 is the only axe made of a high-carbon (crucible) steel. © Trustees of the Wallace Collection

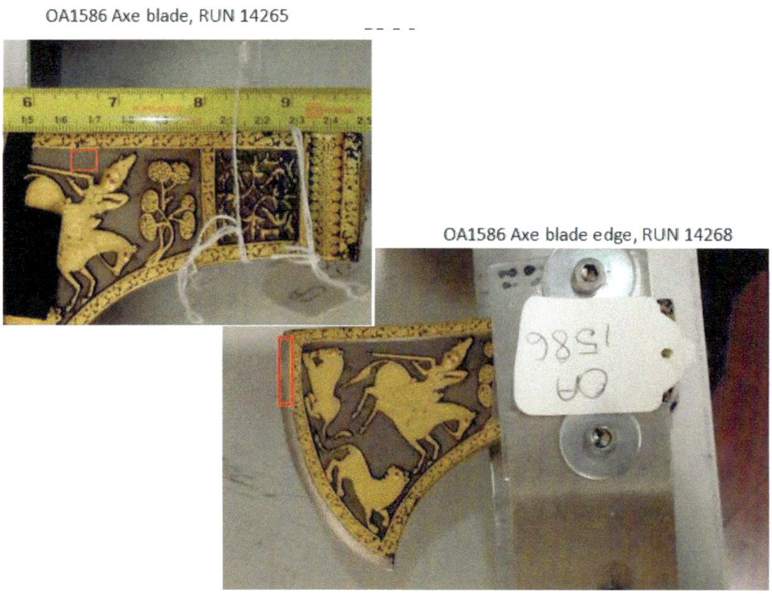

Figure 9 Analysis of OA1586.

Table 1 The elements involved in gilding identified by nuclear resonance analysis in the measured areas of the axes and a list of the areas exhibiting the presence of arsenic which might have been used for decorative purposes.

Sample and position	Gilding				Decoration
	Au	Ag	Cu	Sb	As
OA1544 axe blade edge					
OA1544 axe blade centre	X	X	X		X
OA1544 stiletto					
OA1552 axe blade edge	X				
OA1552 axe blade centre	X				
OA1556 axe blade edge	X				
OA1562 axe blade edge		X			
OA1562 axe blade centre		X	X		
OA1564 axe blade edge	X				
OA1564 axe blade centre	X		X		
OA1566 axe blade edge	X				
OA1566 axe blade centre	X				
OA1566 stiletto					
OA1568 axe blade centre		X			
OA1586 axe blade edge	X		X		
OA1586 axe blade centre	X			X	
OA1592 axe blade edge	X		X		X
OA1592 axe blade centre	X		X		X
OA1594 axe blade edge	X	X			
OA1594 axe blade centre		X			
OA1596 axe blade edge	X				
OA1596 axe blade centre	X				
OA1598 axe blade edge		X	X		
OA1598 axe blade centre		X	X		
OA1604 axe blade centre	X				
OA1604 axe top spike			X		
OA1610 axe blade edge					
OA1610 axe blade centre		X			
OA1610 stiletto		X			
OA1614 axe blade edge	X				
OA1614 axe blade centre	X		X		
OA1614 axe hollow shaft			X		
OA1614 stiletto					
OA1616 axe blade edge	X				
OA1616 axe blade centre	X		X		
OA1616 axe back spike					
OA1616 axe hollow shaft			X		

OA1752 in the Wallace Collection) or even differential etching.[29]

Another possible method is that described by al-Biruni above. Sprinkling particles of cast iron onto a billet of iron and then forging it all together may well have been a successful method of applying a thin high-carbon layer on top of an iron weapon. Forging a veneer of high-carbon steel onto an iron body may have been an alternative method, albeit a difficult one, requiring a high degree of forging skill. Such a surface layer could, after suitable polishing and etching, deceive the unwary, especially if the purchaser was unlikely to use the axe in battle. Attempts have been made to duplicate both of these techniques (see the appendix for the results).

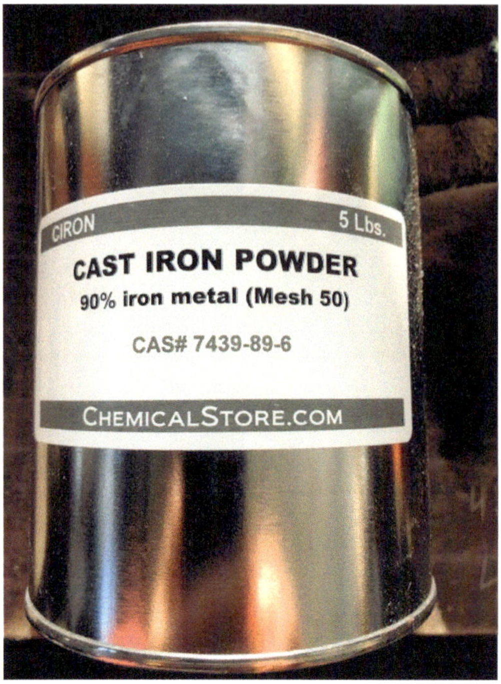

Figure 10 The powdered cast iron employed.

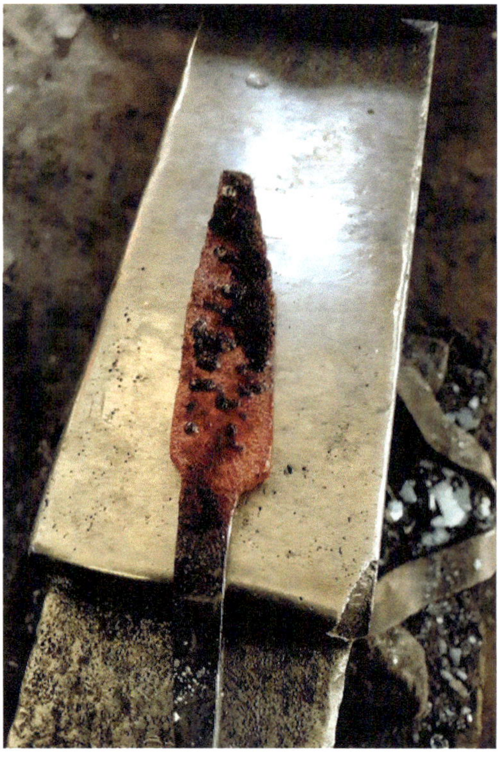

Figure 11 Cast iron sprinkled onto the billet of iron.

Elemental analysis

The INES beamline is also equipped with a set of detectors able to measure nuclear resonance induced by high energy neutrons.[30] This technique can give semiquantitative information on the presence and relative abundance of several minor or trace elements. This measurement is performed at the same time as the diffraction measurement so we also carried out resonance data analysis to identify the minor elements present in the iron-carbon alloy as well as the elements involved in the surface decoration (etching and gilding). Concerning the Fe-C core, it turned out that there is a steady presence, with the same concentration ratio (determined semiquantitatively), of manganese, cobalt and chromium. The results concerning the other elements which can be ascribed either to surface etching or gilding are reported in Table 1.

According to these results, the composition of gilding varies among the samples and also within each one. In the axe heads, when present, copper may be found in an alloy with gold or silver or applied as a preparatory layer on the steel surface before gilding. Copper found in the shaft or in a spike is ascribed to rings or other decorations. Arsenic is present in two axe heads and might have been used as an alloying element to enhance a surface pattern, since iron-arsenic alloys were also employed for this purpose in medieval Europe.[31]

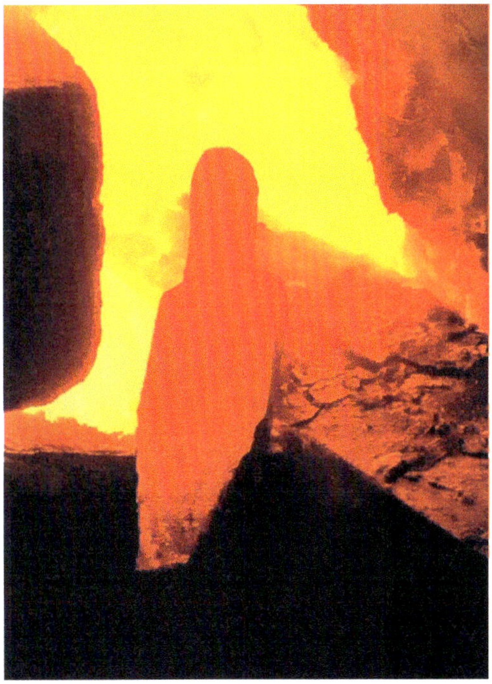

Figure 12 The composite billet heated.

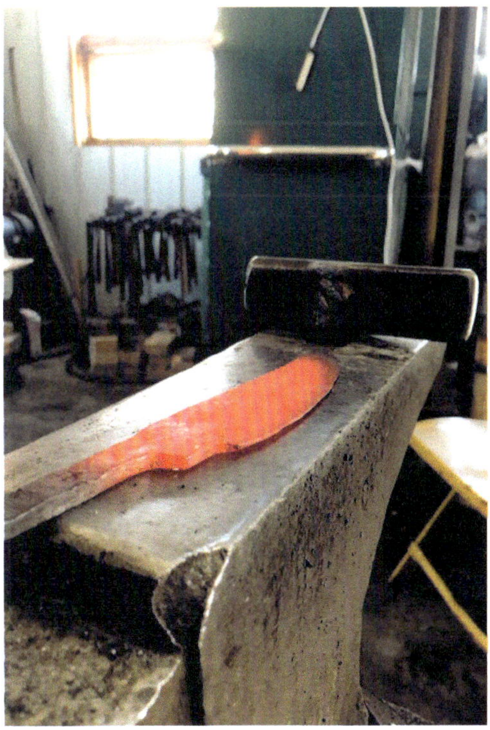

Figure 13 The billet forged into a knife.

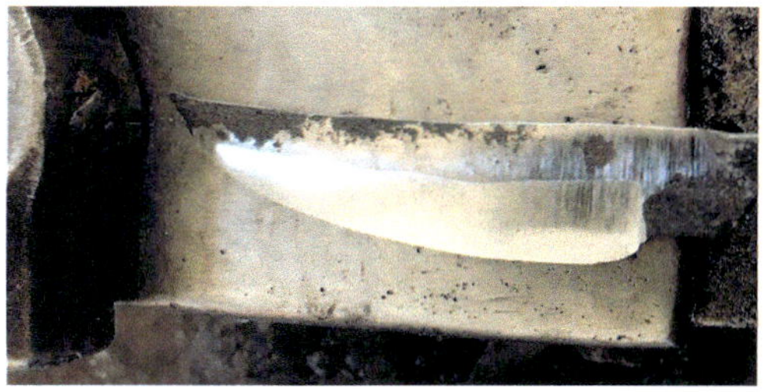

Figure 14 The knife after being ground.

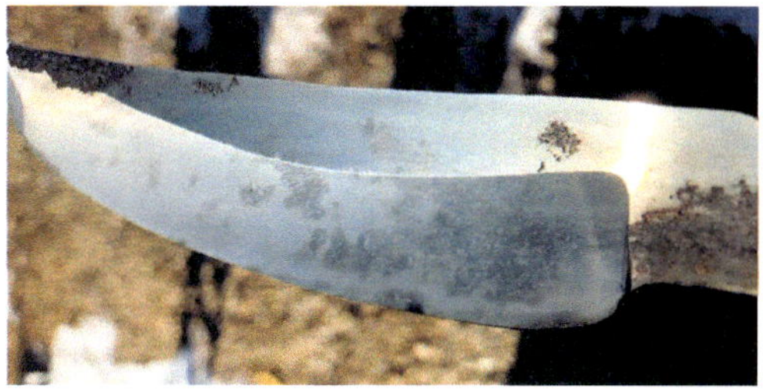

Figure 15 The surface of the knife polished and etched.

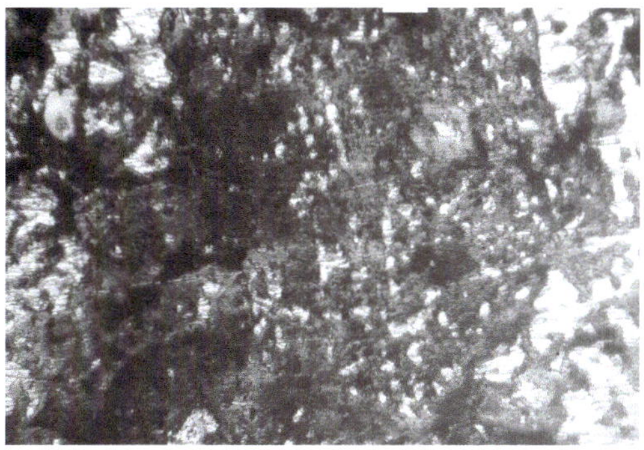

Figure 16 Photomacrograph showing the surface of the knife after etching (scale bar = 200 μm).

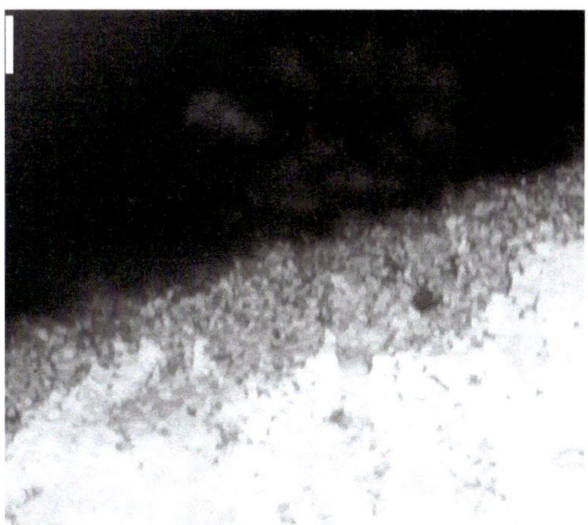

Figure 17 The microstructure of the knife in section (scale bar = 100 μm).

Appendix

An attempt to reconstruct al-Biruni's method for faking wootz steel was carried out by the blade-smith Richard Furrer. Many craftsmen may have attempted to make a small amount of *wootz* (the more expensive material) cover the whole surface of an iron weapon, and mislead the customer into thinking that the whole artefact was made of *wootz*. One possible method of doing this might have been to cover an iron body with a thin veneer of *wootz* welded on, as suggested by some of the results from neutron diffraction analysis. The obstacle would have been the difficulty of welding a thin sheet onto a larger body of iron without excessive decarburisation of the *wootz*.

An alternative method suggested by al-Biruni is another possibility. He had described the making of crucible steel by mixing together *narmahan* (pure iron) and *dus* (white cast iron with 2%–3% C) and then melting them together in a crucible to form an ingot of perhaps 1.5% carbon steel.[32] Adding a layer of *dus* to a surface of *narmahan* and then mixing them to give an outer layer appearing to be *wootz* could have been a logical application of his recipe for making an ingot of *wootz*. Of course, like many craftsmen, al-Biruni's interlocutor left the time and quantity unspecified in the recipe! Both of these methods were attempted in this trial.

Method 1: al-Biruni

A roughly knife-shaped billet (about 8 × 42 × 150 mm) of pure iron (ArmCo) was coated with cast iron powder (3.6% C) and then heated. This was repeated over six welding cycles of heating and hammering which took about 1 hour 45 minutes. The thickness was about 10 mm at the tang and 2.5 mm the blade centre. The forging of the blade to shape took six heats. This forging was done at around 1200 °C with a flux of borax, heated and cooled several times to refine the grain and relieve stress, and finally water quenched then flame tempered to about 250 °C. The edge was found to be hardened to about Rc 45. One side was polished to 1200 SiC then 3 µm diamond paste, and given a nital etch for photomacrography. The surface appearance now suggests a patterned steel (Figures 10–17).

The knife was sectioned and a section (taken 30 mm from the tip) was mounted in epoxy and polished and etched for metallography. The microstructure shows that a carburised layer (of martensite and ferrite) had formed along one surface. Its thickness was estimated with a measuring graticule eyepiece and found to vary between 36 and 57 µm. By contrast, case carburising at 900 °C might yield a case depth of 0.76 mm after 2 hours.[33]

Repetition of this procedure would make the outer layer thicker, although since al-Biruni's method is a welding rather than a case-hardening technique, the thickness of the outer layer is immaterial because its presence is cosmetic rather than functional. A skilled craftsman would have aimed to give a layer of high-carbon steel on the outside of the iron object, enough to avoid decarburising it by excessive heating.

Method 2: a 'veneer'

A piece of modern '*wootz*' (a replica cast steel with a carbon content of 1.6% C) was welded to a larger piece of bloomery iron to form a composite billet which was then sawn in two. One half had a cross-section of some 5 mm thickness polished and etched as before (the other half was made into a knife). A weld seam was visible to the naked eye on the section (Figures 18–25).

Metallography revealed a microstructure consisting of ferrite and slag inclusions in the centre with a layer of fine pearlite at the interface, becoming mixed with more carbides near the outer surface. Metallography also suggested that there was considerable decarburisation near to the weld, but if decarburisation had taken place on the outside as well, the decarburised layer could have been removed by a final 25 mm grinding. The outer surface of the blade would still have had a high-carbon layer of around 2 mm deep, thereby preserving the appearance of *wootz* – or at least of a hyper-eutectoid steel.

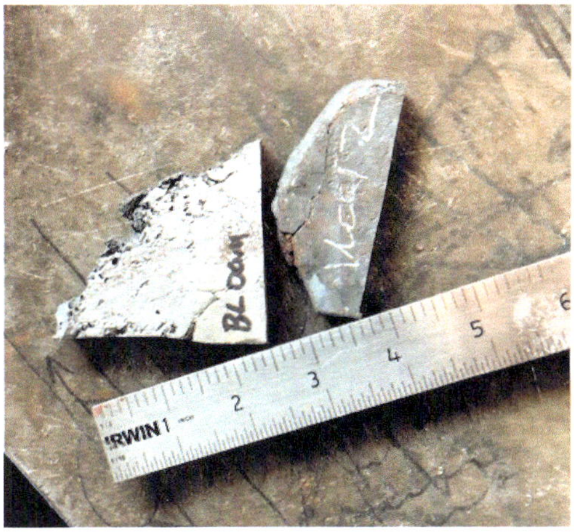

Figure 18 The pieces of steel and iron employed to make the 'veneer'.

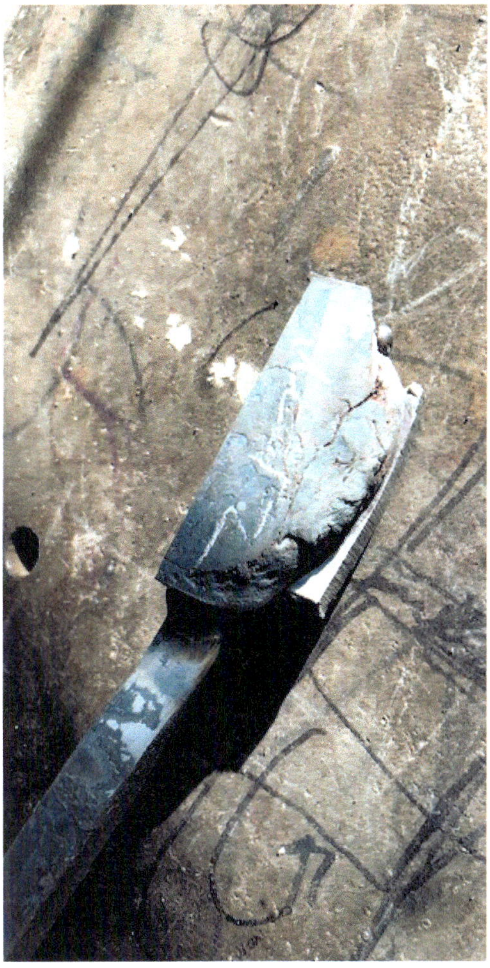

Figure 19 The forming of the composite billet.

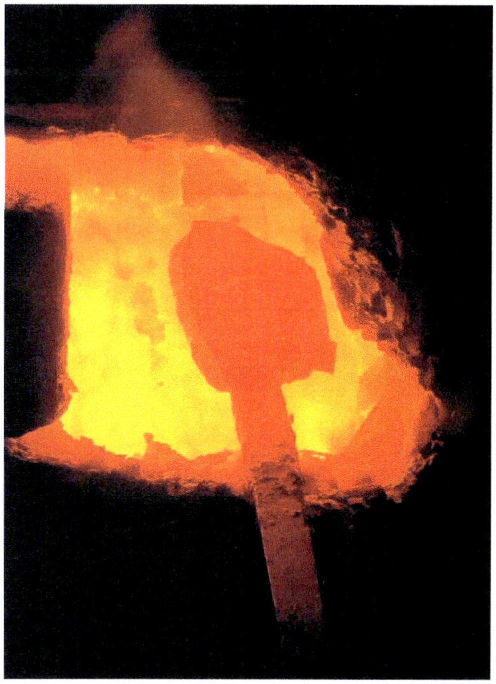

Figure 20 The heating of the composite billet.

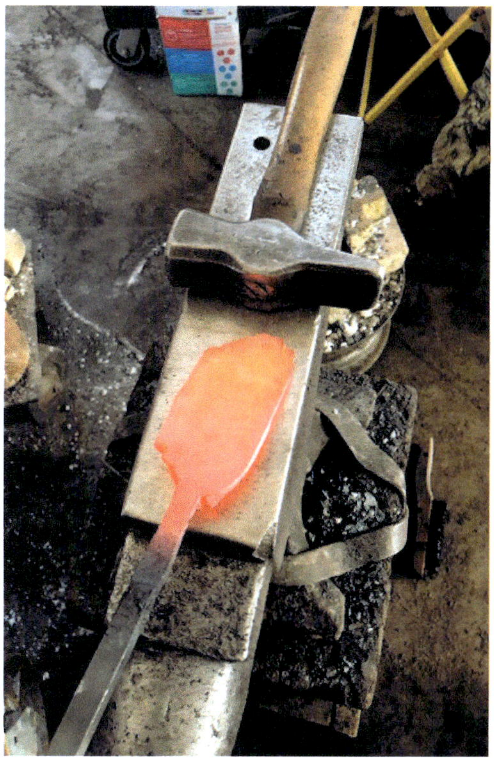

Figure 21 The forging of the composite billet.

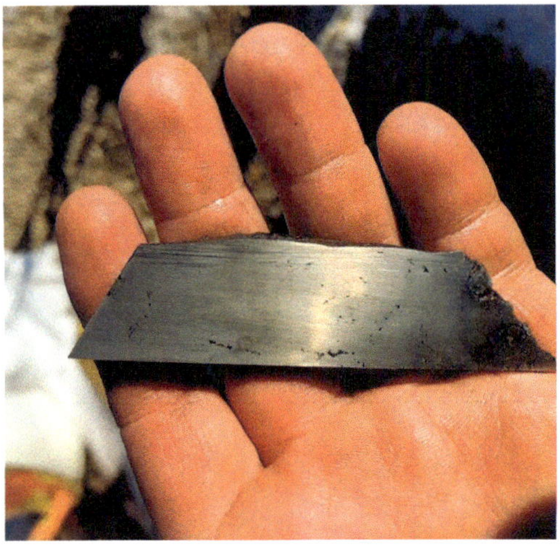

Figure 22 The side view of the composite billet.

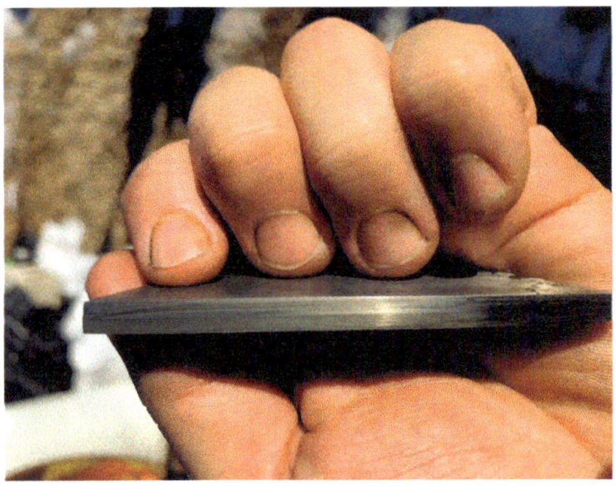

Figure 23 The section of the composite billet.

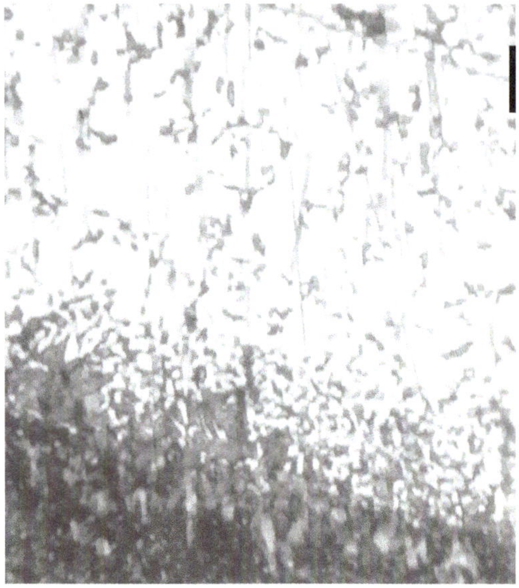

Figure 24 The microstructure of the composite billet (scale bar = 50 μm).

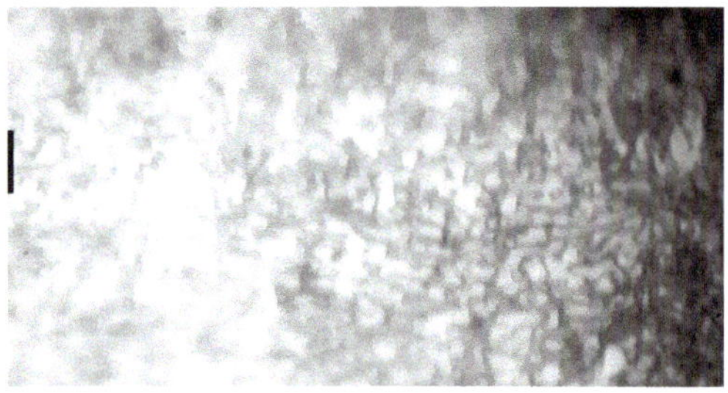

Figure 25 The microstructure of the composite billet (scale bar = 10 μm).

Conclusions

It is evident that both of these methods of simulating (or 'faking') *wootz* would have been successful in at least adding a layer of hypereutectoid steel onto an iron base. Leaving a pattern in any way resembling 'watered silk' on the surface of the steel would have required a high level of skill to avoid homogenising the *wootz*. But such was evidently the case, and the appearance of the resulting surface is frequently convincing.

Depictions of swordsmiths from this period (16th-century Iran) show them forging blades on what is a small scale by modern standards, and apparently at low temperatures, which would have entailed a much slower method of working (Figure 26). It is possible that such a different approach was the key to the successful faking of *wootz*. Their products may still be gracing many museum collections, and can only be detected by the employment of an analytical technique such as neutron diffraction.

Acknowledgements

These experiments RB 1661004 –1661006 were carried out at the Italian Neutron Experimental Station. We are very grateful to the Science and Technologies Facilities Council (STFC) for its support.

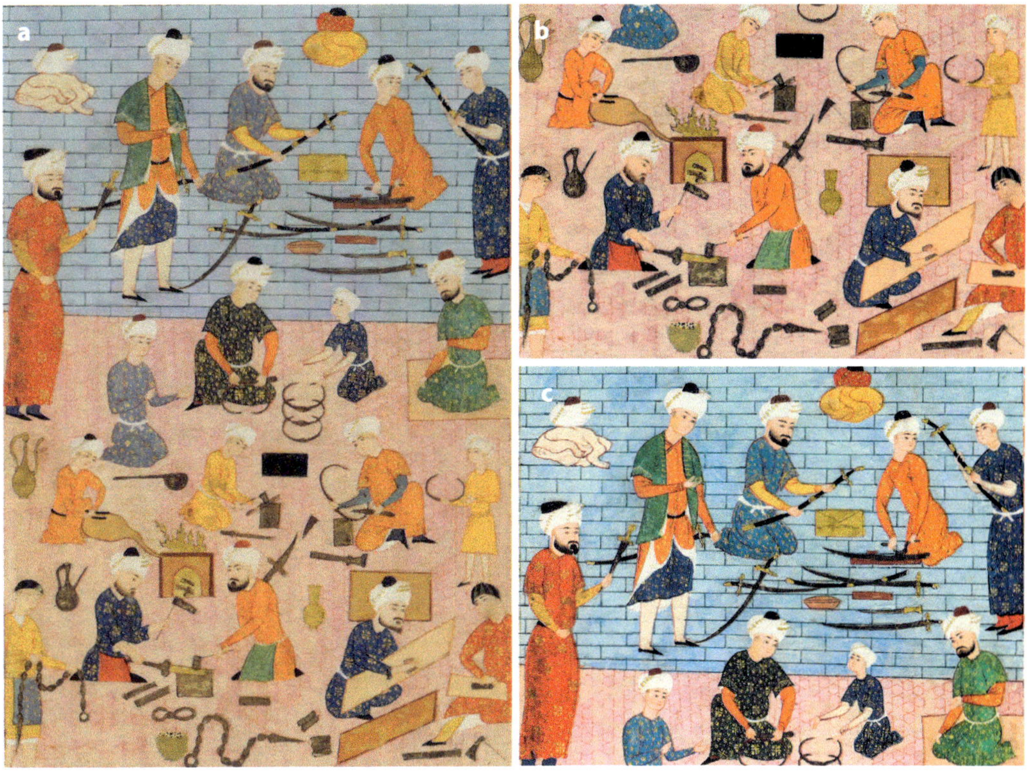

Figure 26 (a) This reproduction of a miniature painting of Persian sword makers from a 16th-century Persian manuscript in the Metropolitan Museum of Art, New York, which measured only 7½ × 5 ¼ inches, offers a rare depiction of an arms-making workshop. (b) In the foreground, two smiths stand in pits that allow them to work upright while hammering a blade; behind them, an assistant pumps a bellows to keep the forge hot. At the lower right, workmen appear to be burnishing sheets of leather, perhaps for scabbards or quivers and bow cases. (c) In the middle ground, craftsmen are engaged in bow making and, at the top, they examine completed swords and daggers that are probably intended for sale. ('An Arms Workshop', ink, colors and gold on paper, Iranian, Safavid period, mid-16th century, Metropolitan Museum of Art, New York. Gift of Edith Macy Schoenborn-Buchheim, in memory of Valentine E. Macy, Jr., 1981, 1981.473.)

Notes and references

1. Berthelot, M. (1883) *Collection des anciens alchimistes grecs*, vol. 3. Paris, p. 328.
2. Mertens, M. (1995) *Les alchimistes grecs, vol. IV, Pt 1 Zosime de Panopolis*. Paris: Les Belles Lettres.
3. Lang, J., Craddock. P. and Simpson, S.J. (1998) 'New evidence for early crucible steel', *Journal of the Historical Metallurgy Society* 32: 7–14.
4. France-Lanord, A. (1969) 'Le fer en Iran au premier millénaire avant J-C', *Revue d'histoire des mines et la métallurgie* 1: 75–126.
5. Wiedemann, E. 'Über Stahl und Eisen bei den muslimischen Völkern', *Beiträge zur Geschichte der Naturwissenschaften* 15. Published as *Sitzungsberichte der Physikalisch-medizinischen Sozietät in Erlangen*, vol. 43 (1911), pp. 114–31. See also vols 34–37 (1902–1906).
6. For a new edition of al-Kindī's treatise see Hoyland, R.G and Gilmour, B. (2012) *Medieval Islamic Swords and Swordmaking*. Cambridge: Gibb Memorial Trust.
7. Allan, J.W. and Gilmour, B. (2000) *Persian Steel: The Tanavoli Collection*, Oxford Studies in Islamic Art 15. Oxford: Oxford University Press, p. 62.
8. Ibid., p. 151.
9. For a detailed description of *wootz* see Chapter 11 in this volume.
10. al-Biruni, chapter on iron in his *Kitab al-Jamahir*, F. Krenkow (ed.), Hyderabad, (1936), pp. 247–57. Quoted in Allan and Gilmour 2000 (cited in note 7), p. 79.
11. Cahen, C. (1947/48) 'Un traité d'armurerie composé pour Saladin', *Bulletin d'études orientales* XII: 103–63.

12. *The Works of Geber, Englysh'd* by Richard Russell (1678, reprinted 1928), p. 231. See also Hoyland and Gilmour 2012 (cited in note 6), p. 144.
13. Bronson, B. (1986) 'The making and selling of wootz: a crucible steel of India', *Archeomaterials* 1: 13–51.
14. Craddock, P. (1995) *Early Metal Mining and Production*. Edinburgh: Edinburgh University Press, ch. 7 and on crucible steel, pp. 275–83.
15. Feuerbach, A.M. (2003) 'Early Islamic crucible steel production at Merv, Turkmenistan', in P. Craddock and J. Lang (eds), *Mining and Metal Production through the Ages*, London: British Museum Press, pp. 258–66.
16. Zschokke, B. (1924) 'Du damassé et des lames de Damas', *Revue de métallurgie* 21: 635–69.
17. Belaiew, N.T. (1921) 'Damascene steel', *Journal of the Iron & Steel Institute* 104: 181–4.
18. Panseri, C. (1965) 'Damascus steel in legend and reality', *Gladius* IV: 5–66. Available at: https://core.ac.uk/download/pdf/268620601.pdf (accessed 14 June 2021).
19. Piaskowski, J. (1978) 'Metallographic examination of two Damascene steel blades', *Journal for the History of Arabic Science* 2: 3–30.
20. For attempts to recreate this process see also Verhoeven, J.D., Pendray, A.H. and Berge, P.M. (1992) 'Studies of Damascus steel blades', *Materials Characterization* 29: 335–41 and 30 (1993): 175–87.
21. See, for example, Wadsworth, J. and Sherby, O.D. (1983) 'Damascus steel-making', *Science* 216: 328–30; and taking a different view, Verhoeven, J.D., Pendray, A.H. and Gibson, E.D. (1996) 'Wootz Damascus steel blades', *Materials Characterization* 37: 9–22.
22. Craddock, P. (2003) 'Cast iron, fined iron, crucible steel: liquid iron in the ancient world', in P. Craddock and J. Lang (eds), *Mining and Metal Production through the Ages*. London: British Museum Press, pp. 231–57.
23. Melikian-Chirvani, A.S. (1979) 'The tabarzins of Lotfali', in R. Elgood (ed.), *Islamic Arms and Armour*. London: Scolar Press, pp. 116–35.
24. Hales, R. (2013). *Islamic and Oriental Arms and Armour* Farnham: Robert Hales C.I. Ltd., pp. 302–3.
25. See http://www.isis.stfc.ac.uk.
26. Grazzi, F., Celli, M., Siano, S. and Zoppi, M. (2007) 'Preliminary results of the Italian neutron experimental station INES at ISIS: archaeometric applications', *Il Nuovo Cimento* C 30: 59–65.
27. Kardjilov, N. and Festa, G. (eds) (2017) *Neutron Methods for Archaeology and Cultural Heritage*. Springer Verlag.
28. Grazzi, F., Cantini, F., Salvemini, F., Scherillo, A., Schillinger, B., Kaestner,A., Edge, D. and Williams, A. (2018) 'The investigation of Indian and central Asian swords through neutron methods', *Journal of Archaeological Science: Reports* 20: 834–42.
29. Allan, J.W. (1979) *Persian Metal Technology 700–1300 AD*. Reading: Ithaca Press, p. 81.
30. Pietropaolo, A., Festa, G., Grazzi, F., Barzagli, E., Scherillo, A., Schooneveld, E.M. and Civita, F. (2011) 'A multitask neutron beam line for spallation neutron sources', *EPL* 95(4): 48007.
31. Williams, A. (2009) 'A note on liquid iron in medieval Europe', *Ambix* 56: 68–75.
32. Hoyland and Gilmour 2012 (cited in note 6), p. 148.
33. *ASM Handbook*, 1981, IV, p. 142.

13

Fakes and their detection by analysis

Brian Gilmour

Introduction

Fakes and forgeries have been a problem for the arms and armour world for a long time. Perhaps the most common motive is financial gain: when a particular artefact is perceived to have a market value, as an antique, then unscrupulous buyers may be quick to offer whatever is valued irrespective of its provenance. However, this is not the only motive and other reasons why false claims are made may include a grudge on the part of the perpetrator – perhaps to establish a rival theory where there is no firm evidence – or occasionally a practical joke that gets out of hand.

Definition of a fake

Plain copies, reproductions or stylistically similar versions of an object may be made for good reasons, and there may also be genuine cases of mistaken identity as opposed to intentional fraud. Copies, reproductions and stylistically similar versions are perfectly legitimate creations until one is sold as claiming to be something other than a copy, at which point it becomes a 'fake'. This essentially involves deception of some kind and broadly covers two categories: counterfeits and forgeries with dishonesty as the motive. A 'counterfeit', implying something made as an exact copy of a specific artefact, is much rarer in the arms and armour field. A 'forgery' is the creation of an original object which purports to have a different identity. 'Restorations', in which an incomplete or broken object is made 'whole' again, range from anything between the cleaning of a lightly corroded object or repairs to an object slightly damaged to the creation of a whole new object around an original surviving fragment.

The problem of fakes and forgeries (but rarely counterfeits), especially in the field of arms and armour, can be traced back at least 1200 years to the early 9th century when the caliph in Baghdad ordered his chief scientist, Ya'cub Ishaq al-Kindī, to investigate how one might tell the difference between swords of greater and lesser quality as well as genuine, as opposed to fake, blades. Thus, unwittingly, the caliph's enquiry gave rise to one of the most valuable early metallurgical treatises on iron and steel: al-Kindī's treatise, 'Of Swords and their Kinds'.[1]

The treatise of Ya'cub Ishaq al-Kindī

Ya'cub Ishaq al-Kindī (c.800–870) was one of the foremost scholars of his day and might be considered the chief scientific advisor to the caliphs of Baghdad during the middle of the 9th century. It was the eighth Abbasid caliph, al-Mu'tasim (833–842) who instructed al-Kindī to investigate and report on an already widespread problem of fraud, in this case the misrepresentation of swords as being of better quality than they actually were.

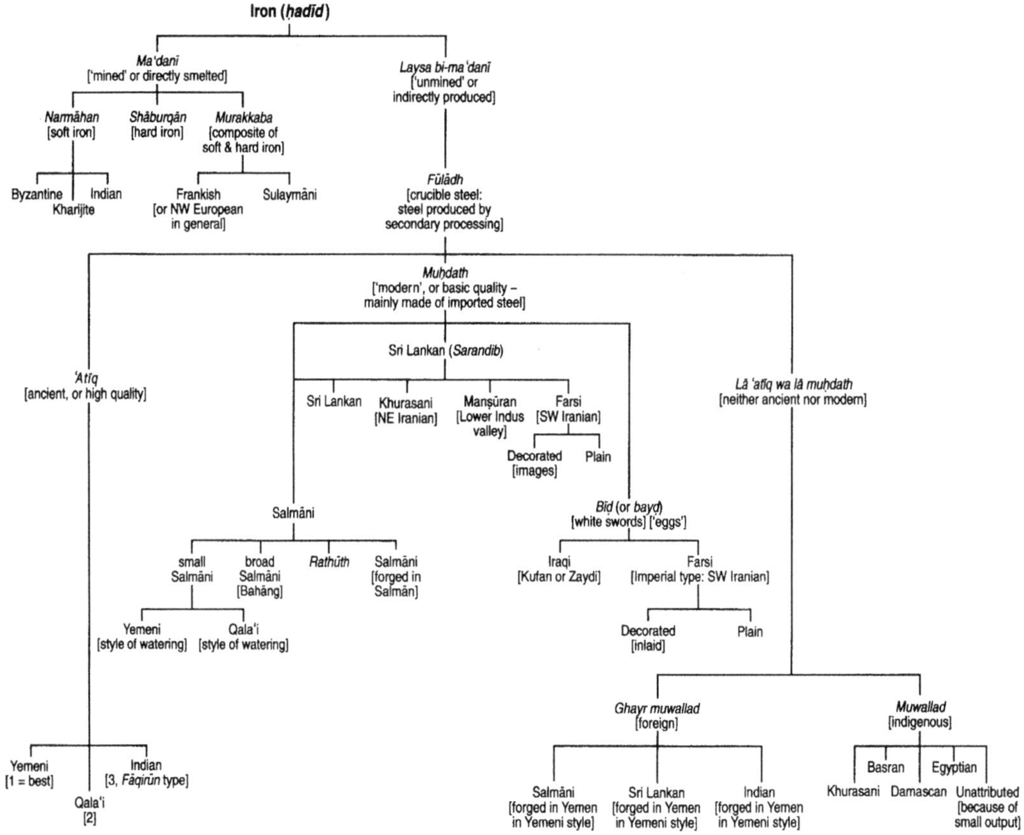

Figure 1 Diagrammatic representation of Ya'cub al-Kindī's sword treatise.

This sword treatise is essentially a set of instructions on how to discern the difference between high quality swords and lesser quality blades, with particular emphasis on imitations of one kind or another, as well providing background information on the different types of iron and steel that were used, where these different types of swords were made and how this industry operated.

As can be seen from the 'family-tree' representation (Figure 1) of the different types of swords described in al-Kindī's treatise, the common ancestor is simply iron from which all types of iron and steel derive. The European blades all come under the minor or subsidiary branch of the diagram, *ma'dani*, or 'directly produced' iron or steel (i.e. bloomery smelted metal). For al-Kindī, these were less important foreign swords and much less relevant than the swords which were made using crucible steel that comprise the much larger subgroup entitled *laysa bi-ma'dani*, meaning 'indirectly produced' iron or steel. In practice, this refers to steel that has been produced indirectly, in this case crucible steel made by heating in small (lidded) clay crucibles mixtures of plain (probably bloomery) iron, and carbon-rich material such as wood, leaves or even cast iron.

Al-Kindī's sword treatise gives the reader a set of guidelines on how to discriminate between different types of sword and to judge their quality and relative merits in order to avoid being duped. It seems that the chief problem in the caliphate lay with the sword sellers rather than the makers. Swords in one region were sometimes made in the style of those from elsewhere, but not necessarily with deception as the prime aim. This problem is likely to have persisted for many centuries as in the case of the famous Persian sword-maker Asadallah Isfahani (active

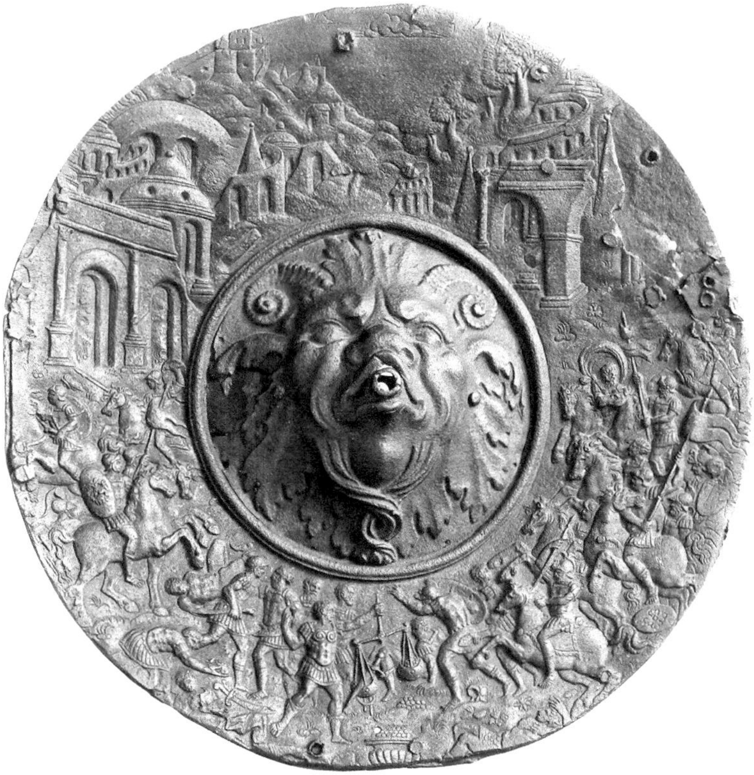

Figure 2 Front view of the decorated iron buckler known as Dr Woodward's shield. © Royal Armouries.

c.1600), although many of the swords that bear his name also carry anachronistic dates varying between 1408 and 1808. A few of the earlier swords may well have been made by this famous swordsmith (or his son) but not the later ones:[2] these may well belong to the date inscribed but with the name 'Asadallah' added to boost their value. Even if they were of perfectly good quality, and perhaps earlier, they became fakes as soon as a fraudulent signature was added. The value attached to patterned crucible steel (or *wootz*) meant that misleading materials which appeared to be entirely *wootz* were employed to make weapons in order to sell them at a higher price.[3]

A genuine item can become a fake depending on circumstances, as in the celebrated instance of Dr Woodward's shield, now in the British Museum, which became well known in London towards the end of the 17th century (Figure 2).

Dr Woodward's shield

In 1693, Dr John Woodward bought a small iron shield – perhaps more accurately called a buckler – from the daughter of an acquaintance, John Conyers, an apothecary of Fleet Street who had recently died.[4] It appears that John Conyers had himself purchased it from an ironmonger in the City of London sometime before 1665. Dr Woodward soon became convinced that his shield was a genuine Roman relic with depictions of the sack of Rome by the Gauls led by Brennus following the Battle of the Allia in 390 BCE. One obvious anachronism was the depiction of a monument that bore a clear likeness to the Colosseum (the building of which only began under Emperor Vespasian during the 1st century CE) as a ruin – as indeed it was by the 16th century. It is unclear whether this was pointed out to Woodward at the time but in any case it was overlooked. Other

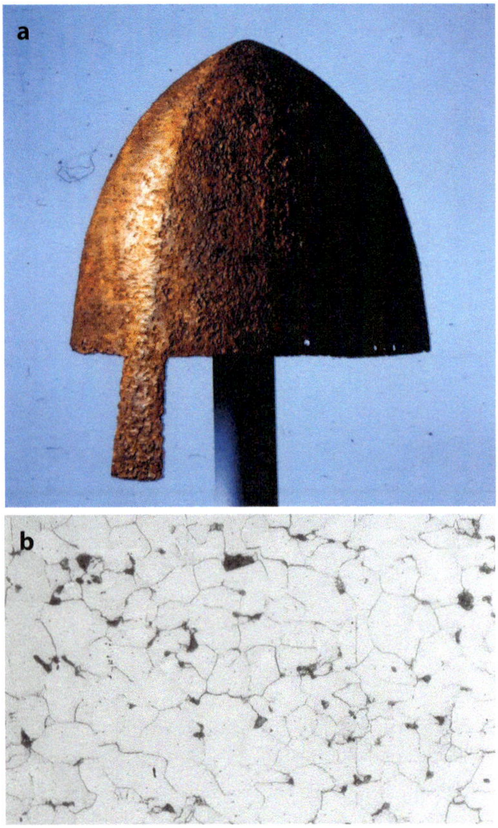

Figure 3 (a) A one-piece nasal helmet that came onto the UK antique arms and armour market in the early 1990s. © Royal Armouries. (b) Photomicrograph showing a homogeneous, fine grain, slag-free low-carbon alloy typical of modern ingot steel (original magnification ×320; etched with 2% nital).

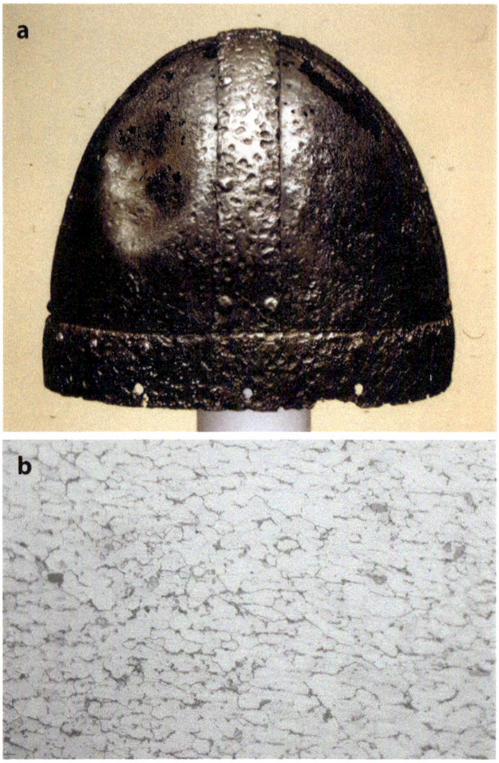

Figure 4 (a) Front view of the *spangenhelm* with clear indications of a deep and rapid short-term, corrosion-pitted surface but no sign of any long-term corrosion effects consistent with burial in the ground or in a waterlogged environment. © Royal Armouries. (b) Photomicrograph showing a homogeneous, fine grain, slag-free low-carbon alloy typical of modern ingot steel (original magnification ×320; etched with 2% nital).

anachronisms included weapon forms that were unknown until much later.

The more doubt that was cast on this claim, the more entrenched Woodward became in his identification. He should perhaps have known better given that he was an accomplished scientist who had been elected to the Royal Society and appointed Professor of Physick at Gresham College (City of London) in the same year that he acquired his shield.[5] But in this, as in many other cases of mistaken belief, Woodward seems to have suffered from a common psychological malaise: the following of an idea that was attractive to him and that he wanted to believe. In other words, he was obsessed by what seems to have been his own insight that the scene represented by the rich repoussé decoration on his shield might be a near contemporary record of the events depicted.

Perhaps Woodward thought he was making an important discovery but a little research would have revealed that armour decorated by repoussé comparable to this was not unusual, at least in 16th-century Italy. His shield might have been inspired by the 'Medusa' shield now on display in the Imperial Armoury of the Kunsthistorische Museum, Vienna, and possibly made in the Negroli family workshop in the period following 1545.[6] Woodward's shield was thus a case of an artefact which appears to be a not untypical piece of decorated iron armour from the mid-16th century but effectively became a fake because of Woodward's claim that it was a Roman relic.

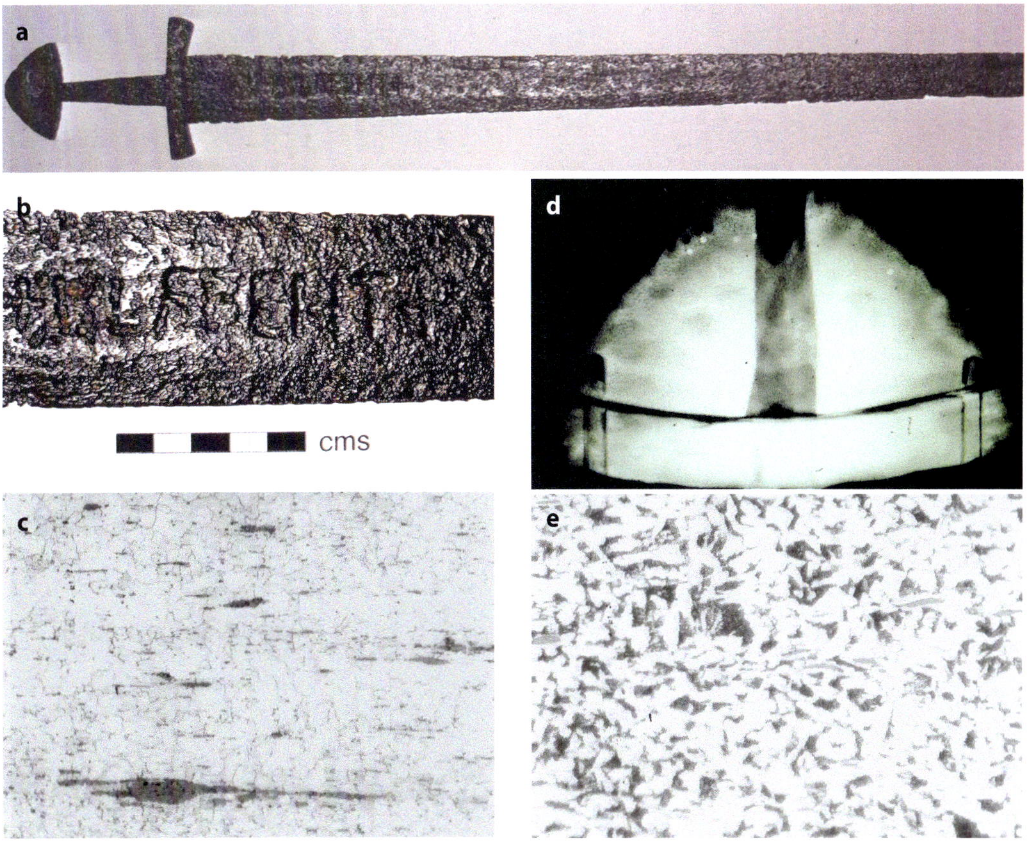

Figure 5 (a) Part of a sword of Viking Age style with (b) an inscription +VLFBEHT+ and (c) photomicrograph of the sword blade showing ferrite, some large slag inclusions and numerous smaller inclusions of circular cross-section. (d) An X-radiograph of the pommel and upper guard. (e) Photomicrograph of the crossguard showing ferrite and pearlite corresponding to low-carbon 'mild steel' of around 0.2% C and some very small, elongated, grey, non-metallic, manganese sulphide inclusions, typical of 'mild steel' (original magnification ×320).

Examination of 'medieval' items

A number of items purporting to be medieval have been examined by the author and colleagues at the Royal Armouries and elsewhere in recent years:

- a nasal helm
- a *spangenhelm*
- a sword with an 'Ulfberht' inscription
- a great helm
- a Negroli helmet

In the arms and armour world, the most popular items – and not surprisingly those that fetch the most money – are helms and swords, particularly those that were easier to make and hark back to a heroic view of the late Middle Ages. There are simpler but earlier forms of helmet, such as the iconic form of nasal helm worn by both sides in the Battle of Hastings and well known from the Bayeux Tapestry. These are sometimes referred to as 'Norman' nasal helmets but their actual survival is extremely rare although interest in owning them is much less rare (see Figure 6). The classic examples were apparently made from a single piece of metal because this was preferable both from a production and protection point of view. The nasal helmet illustrated in Figure 3 was made in the traditional way – hand forged from a single piece of iron – but the iron that had been used was a homogeneous, very fine grain (and modern) low-carbon iron (see Figure 3b).

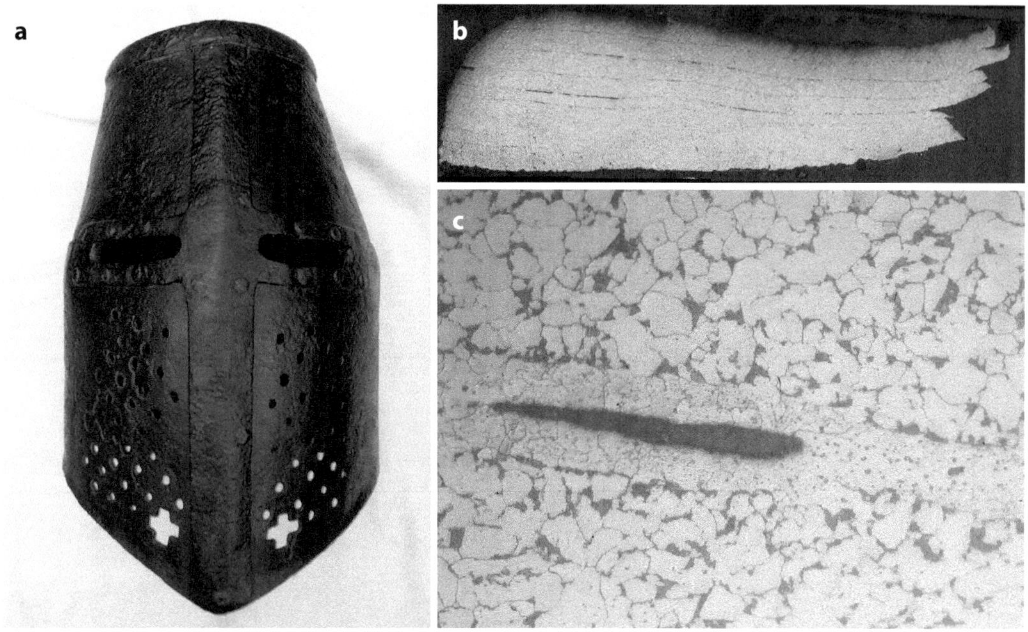

Figure 6 (a) A great helm of 13th/14th-century form. (b) Photomicrograph of the helm showing a uniform microstructure of ferrite and pearlite, corresponding to a modern 'mild steel' of around 0.1-0.2% C without visible inclusions, but arranged in bands separated by rows of large slag inclusions (original magnification ×40). (c) A large slag inclusion between two layers of steel showing a layer of other micro-constituents associated with the slag. This may have been the result of forge welding with a flux (original magnification ×400).

Unlike medieval bloomery iron that invariably contained non-metallic slag (left over from the smelting process which reduced the ore but did not melt the iron), this iron showed no visible traces of slag but did contain a dispersion of tiny manganese sulphide inclusions, typical of this form of modern iron. This is the same material that is used for modern structural steel, either in bar or sheet form, and is now readily available in very large quantities. The *spangenhelm* illustrated in Figure 4 has a similar microstructure.

Both the helms featured here were reported to have been 'probable river finds from eastern Europe', a similar claim having been made for a sword that was examined at much the same time (Figure 5). The blade was found to be made of post-medieval, charcoal-smelted 'fined' iron, typically low in sulphur but rich in slag inclusions, these having a high potassium content. This metal also appeared to have been rolled, all of which suggested a 17th- or early 18th-century date. The metal of the lower crossguard was later and consisted of a typical fine grain, low-carbon 'mild steel' with 0.5% manganese and contained many small manganese sulphide inclusions but no slag. This is a modern type of iron not invented until c.1860.[7]

A great helm supposedly of 13th- or 14th-century origin is illustrated in Figure 6. Renewed interest in jousting during the earlier 19th century[8] meant that helms like this were becoming more popular (and still are), suggesting a manufacturing date of mid-19th century or later. But it appears uncertain as to whether or not this helm was simply made as a replica which was then suggested to be a late medieval original. Many armours were supplied to collectors, jousters and even the Tower Armouries by the dealer Samuel Pratt, who also supplied armours for the Eglinton tournament.[9] This helm was made of a modern low-carbon steel which has been rolled and forged into a layered material, probably to simulate a medieval iron.

Perhaps the most noticeable feature in the appearance of all these items is their similar and very evenly corroded condition. In section,

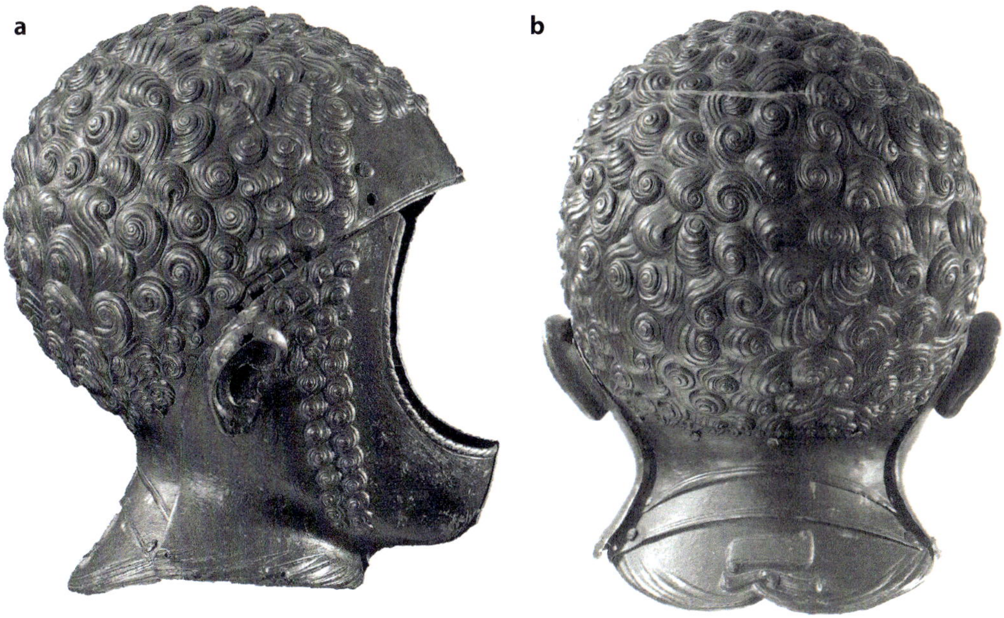

Figure 7 (a) Right side view of the Negroli-style burgonet comprising a main skull piece, a lower jaw, cheek and front neck guard (two parts), and two rear lower neck guard plates, and (b) a rear view.

all showed no sign of the uneven or deep corrosion pitting that would be inevitable with any long-term burial, even underwater or in river mud. The only way that they could have been preserved in anything like as complete a state as this is if they were buried in a relatively oxygen-free, waterlogged environment, but the corrosion is too even and too light overall, suggesting that the metal is homogeneous which indeed it is. The homogeneous, fine grain, slag-free iron or low-carbon 'mild steel' of the helmets is clearly modern as evidenced by the presence of very small manganese sulphide inclusions typical of the modern industrial process that gave rise to the metal. The detection of high manganese levels in such an object would probably indicate a fake, although the absence of manganese does not guarantee authenticity.

A technique useful for identifying the metallic elements present (such as manganese) is energy-dispersive X-ray fluorescence (EDXRF). In the case of the nasal helmet, a surface survey using EDXRF showed the surface corrosion to contain brass, suggesting that this helmet was packed, perhaps in a cloth bag of some kind, with brass filings, and buried so as to induce accelerated corrosion. An obvious aspect to both the nasal helmet and the sword is that for objects that have been supposedly buried, even in river mud, they are in a suspiciously evenly corroded condition. To use an old adage, they seem 'too good to be true' and therefore are unlikely to be genuine.

Another interesting object examined was the 'Negroli' burgonet (Figure 7) bought by a major national museum in the belief that it represented a previously unknown example of the work of the famous Milanese armour-maker Filippo Negroli (c.1510–1577). There is a similar burgonet made by Filippo Negroli in 1533 for King Charles V, now in the Real Armería, Madrid (Figure 8).[10] The main skull piece and the two neck guard plates – the lower neck plate also showing the same style of monogram as used by Negroli – which now form part of this burgonet do appear to be quite a close copy of those parts of the burgonet made by Filippo Negroli for Francesco Maria I, Della Rovere, Duke of Urbino in 1532 (now in Vienna)[11] but who made it and when is quite a different matter (Figure 9).

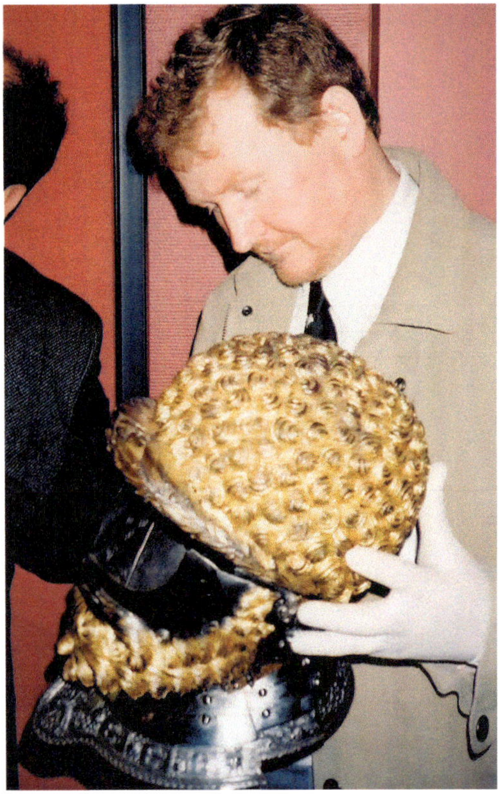

Figure 8 David Edge examining the Madrid helmet made for Charles V by Filippo Negroli, signed and dated 1533.

It is perhaps the quality – particularly of the main part of the skull – that is the most important factor by which this burgonet can best be judged, and it seems to be most likely at least partly on the basis that it was rejected as the work of Filippo Negroli by a panel of scholars in New York. Although it looks quite competent it was clearly not of the same high standard expected of Negroli or other master armour-makers. One obvious oddity of this burgonet – with all five pieces in place – is the peculiar angle of the right ear (the left being much the same), plus the slightly awkward angle between the lower edge of the cheek pieces and the lower edge of the back lower neck plate (Figure 9c). A possible reason for this could be that the two cheek pieces, the rear neck plates and the main skull part come from similar helmets, but not the same helmet overall. This is also suggested by the microstructure of the iron used for these different parts.

A noticeable feature of the main decorated skull piece was that surface blemishes (small splits) showed up in places which made it possible to detach metallic flakes and identify the metal used for this main part of the helmet. In section this was found to consist of largely slag-rich, variable grain, plain (carbon-free) iron

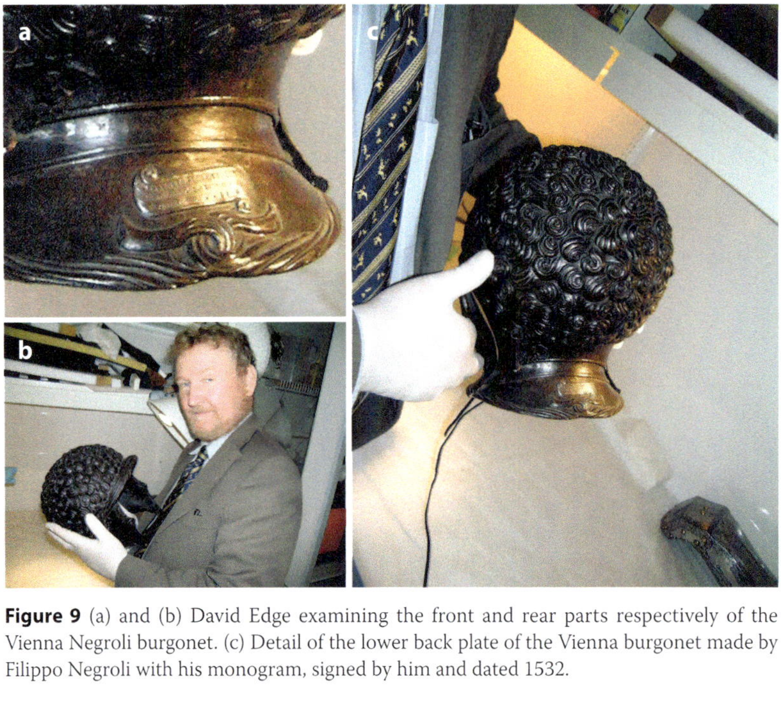

Figure 9 (a) and (b) David Edge examining the front and rear parts respectively of the Vienna Negroli burgonet. (c) Detail of the lower back plate of the Vienna burgonet made by Filippo Negroli with his monogram, signed by him and dated 1532.

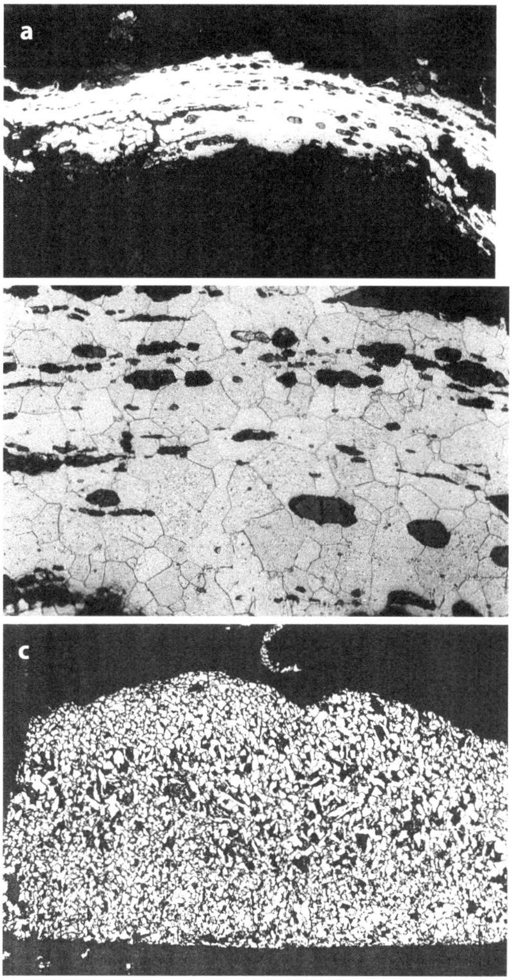

Figure 10 Photomicrographs showing the appearance in section of a flake from (a) the main skull part of the burgonet showing both its overall corroded nature and a microstructure of ferrite and slag, present in numerous inclusions (original magnification ×100); (b) the ferrite and slag microstructure of the same flake at higher magnification (original ×400); (c) the pearlite and ferrite microstructure of the lower neck plate with many fewer inclusions (original magnification ×400; etched with 2% nital).

(Figure 10a and b), the slag being the reason for the blemishes or splits visible in places. Vickers microhardness testing results varied between 90 and 104 HV (0.1), showing the iron to be relatively soft, consistent with its structure as seen in section. Thus although the main skull piece has been skilfully made – a relatively easy process with a soft iron – because of its high slag content it could never have looked as good as it might have done had the better type of metal been used. One other notable feature of the structure of this slag-rich iron is that in places it is comparatively deeply penetrated by corrosion, both intergranular and along the elongated slag inclusions, indicating that the manufacture of this helmet is not recent, and the forging to shape is likely to have taken place some centuries ago.

By contrast the pair of cheek/ear pieces and the two rear lower neck guard plates were made of very different metals, more consistent with what one might expect a master armourer to have used. In the case of the two neck plates, they consist of fairly homogeneous, fine grain, low-to-medium carbon steel with only a few, relatively small, elongated slag inclusions visible in places (Figure 10c). This material would have been preferred by master armourers because of the finer surface detail that could be achieved with less risk of surface defects. The cheek pieces were found to consist of low-carbon iron of variable grain size with only a few, smaller elongated slag inclusions visible in section.

It is difficult to put more than an approximate date on the iron used for the main skull piece of this burgonet. Apart from its very high slag content, the iron is reasonably carbon free and consists of relatively homogeneous ferrite (pure iron). The variability of the slag form and distribution of this microstructure is most typical of fined iron – earlier decarburised cast iron that was made (in Britain) between the early 16th and the late 18th century when puddled wrought iron was introduced and subsequently replaced fined iron. In this case, electron probe microanalysis (EPMA),[12] carried out at Oxford, also showed the iron to be low in sulphur, indicating that the iron was made before coke-fired smelting was introduced by Abraham Darby in 1709. We can therefore state that the slag-rich iron used here appears most likely to have been made sometime during the 16th and 17th centuries.

Perhaps all that we can conclude from this is that the main skull part of this burgonet is an inferior copy of that form of helmet made famous by Filippo Negroli in c.1538, produced probably sometime between the mid-16th and late 17th century, unless it was made later using older, charcoal-fined wrought iron. It seems clear from the published photomicrographs of the Negroli armours that despite the cost, the Negroli workshops favoured steel of bloomery origin rather than the inferior slag-rich 'fined iron' which was developed at about this time.[13]

This contrast in the form of metal used for the main skull, and that of the other plates now attached to it, may help to explain two less obvious oddities in the way the whole assemblage fits together. As assembled, the angle of the ear is tipped forward from the vertical, and the lower edge of the cheek/ear pieces and that of the lower neck guard plate do not line up as might be expected. One obvious explanation for these discrepancies is that the five plates of the burgonet were not originally intended to belong together and thus may be a 'marriage' of parts from different original armours, and possibly of different dates. In common with many apparently complete armours in museums, this does appear to be a composite. For example, metallography of many of its components has recently shown that the famous, indeed iconic, armour for a mounted knight and his horse in the Wallace Collection (A21) is a composite comprising elements of at least four different armours.[14] Another armour (A20) in the Wallace Collection is the product of no less than five workshops.[15]

Conclusions

This article has shown that fakes and forgeries have been a problem for the arms and armour world for a long time and are likely to continue to be so. Resolving which objects are authentic will require both curatorial expertise and scientific analysis.

Acknowledgements

I would like to thank the editors of this volume and current members of the Royal Armouries curatorial staff for their careful review of the manuscript.

Notes and references

1. Hoyland, R. and Gilmour, B. (2006) *Medieval Islamic Swords and Swordmaking: Kindi's Treatise 'On Swords and their Kinds'*. Oxford: Gibb Memorial Trust, passim.
2. Allan, J.W. and Gilmour, B.J.J. (2000) *Persian Steel: The Tanavoli Collection*, Oxford Studies in Islamic Art 15. Oxford: Oxford University Press, p. 103.
3. See chapter 13 in this volume describing the analysis of such a group of Indo-Persian axes from the Wallace Collection.
4. Levine, J. (1991) *Dr Woodward's Shield: History, Science and Satire in Augustan England*. Ithaca, NY and London: Cornell University Press, pp. 150–51.
5. Levine 1991 (cited in note 3), pp. 19 and 30.
6. Pyhrr, S. and Godoy, J. (1999) *Heroic Armour of the Italian Renaissance: Filippo Negroli and his Contemporaries*. New York: Metropolitan Museum of Art, p. 216.
7. It may be interesting to compare the radiograph of the pommel with those illustrated in Chapter 9 of this book.
8. Watts, K. (1992) 'Samuel Pratt and armour faking', in M. Jones (ed.), *Why Fakes Matter: Essays on Problems of* Authenticity. London: British Museum Press, pp. 100–107, esp. p. 100.
9. Reid, W. (2014) 'A ceremonial soaking: the Eglinton Tournament, 1839', in D.J. La Rocca, *The Armorer's Art: Essays in Honor of Stuart Pyhrr*. Woonsocket, RI: Mobray Publishing, pp. 25–32.
10. Pyhrr and Godoy 1999 (cited in note 6), p. 116.
11. Ibid., p. 126.
12. Electron probe microanalysis of the various metal parts, and that of the non-metallic slag within it, was carried out by C.J. Salter (then) of the University of Oxford, Department of Materials and for this work the author is very grateful.
13. Williams, A. (1999) 'The steel of the Negroli', *Metropolitan Museum Journal* 34: 101–24, esp. pp. 112–13.
14. Williams, A. and Edge, D. (2001) 'A study of the German "Gothic" 15th century equestrian armour (A21) in the Wallace Collection, London', *Gladius* 21: 233–56.
15. Williams, A. and Edge, D. (2013) 'A suit of armour produced by five workshops: Wallace Collection A20', in R. Cordoba (ed.), *Craft Treatises and Handbooks: The Dissemination of Technical Knowledge in the Middle Ages*. Turnhout: Brepols, pp. 197–206.

14

Metal woodscrews during the flintlock years: West European metal woodscrews from 1650 to 1830: an archival and metallurgical study

Yannick Chastang

Introduction

The study of decorative arts and, specifically, of furniture, has provided considerable understanding of materials and manufacturing techniques used during the age of enlightenment. However, a vitally important component, particularly of firearms and furniture, is generally overlooked: these decorative arts could not exist without the humble metal woodscrew. Prior to the development of woodscrews, cabinetmakers and other craftsmen relied on wood pegs or iron nails and rivets. A metal woodscrew presents many advantages over these other fixings: a screw can hold together pieces of metal and wood without adhesive, a screwed piece can easily be assembled or dismantled for maintenance and last, but not least, screws do not loosen with the small vibrations from daily use of firearms or furniture hinges, drawers and folding parts.

There have been many studies on screws but the majority have failed to distinguish the different types of woodscrews from bolts.[1] Early woodscrews (Figure 1) were generally tapered or parallel shank finished with a blunt or a sharp threaded tip called a 'gimlet', and previous studies have not differentiated the manufacture of screws with parallel shanks from that of tapered screws or gimlet screws. Although the difference might appear subtle, today screws are generally parallel with a threaded gimlet and that shape, as its longevity attests, has always been regarded as superior.[2] However, mechanising the manufacture of the tapered screw or gimlet was complex: the first mechanical screw manufacture developments were restricted to parallel shank screws and the production of handmade tapered screws continued apace.

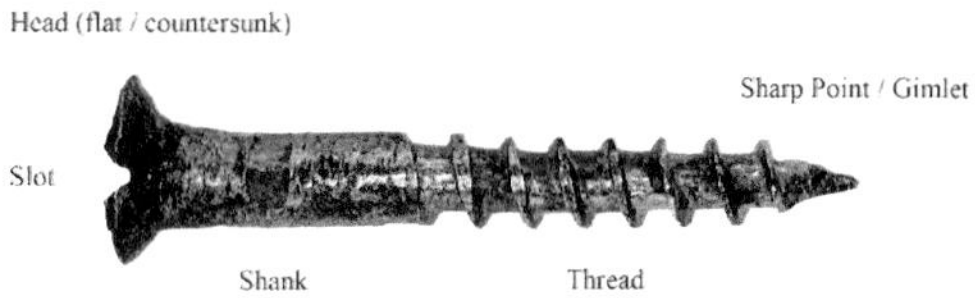

Figure 1 French woodscrew, c.1770.

It is still uncertain when the woodscrew was first invented but, by the second half of the 17th century, metal woodscrews were common. In 1831, woodscrews were still considered 'of much more recent invention' than nails.[3] The woodscrew's design, origin and manufacture are so closely related to those of nails, however, that woodscrews were (and are) sometimes called 'screw nails'.

Previous studies of screw manufacture have concentrated on the developments in

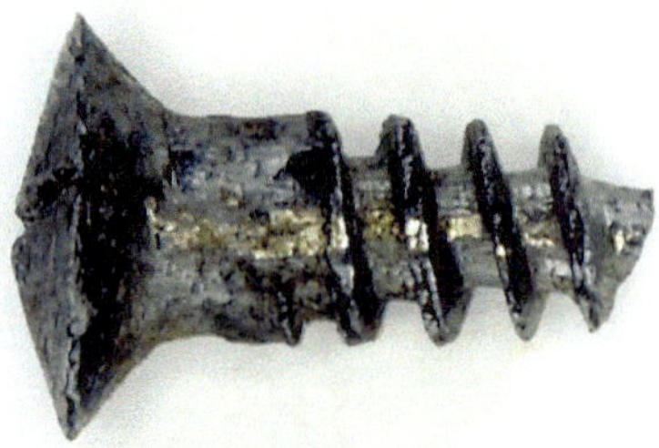

Figure 2 Screw from a drop-front secretaire stamped by Jean-Henri Riesener, c.1770–80, private collection. The screw, entirely handmade, is only 8 mm long, the shank and head are forged, and the thread has been cut by hand using a file.

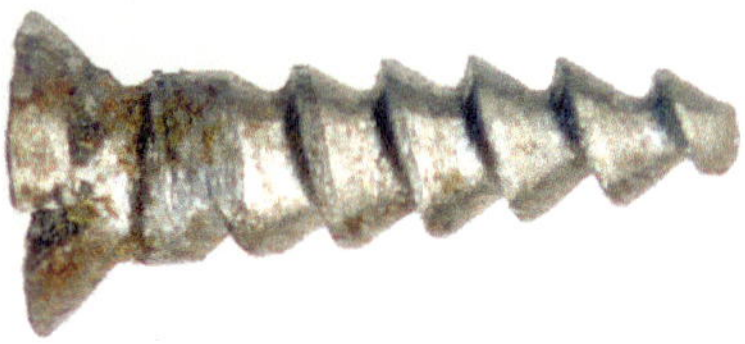

Figure 3 Screw from a coffer on a stand attributed to A.C. Boulle, The Lewis Walpole Library, c.1690. The screw, handmade, is only 9 mm long, the shank and head are forged, and the thread has been cut by hand using a file.

manufacturing techniques with particular interest in the numerous patents from 1760 onwards. This interest presupposes that mechanisation must have been desirable and would have been achieved as soon as increased demands from the industrial revolution manifested, but studying original metal woodscrews from dated furniture demonstrates surprising resistance to industrialisation and mass production. Virtually all the screws that I and my conservation colleagues have studied over the past 30 years demonstrate that, from the 17th to at least the first quarter of the 19th century, woodscrews were handmade in a labour-intensive process instead of by lathe or machine. The main characteristics of a handmade woodscrew are an irregular

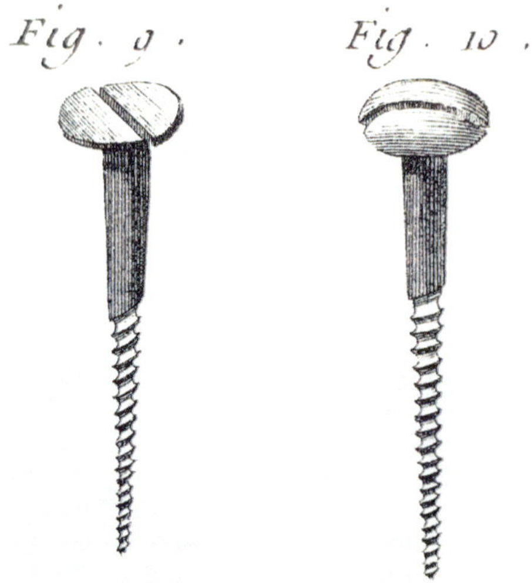

Figure 4 'These screws are supplied to the carpenter by the locksmith who sells them by the dozen' (André Jacob Roubo, *L'art du menuisier*, seconde partie, 1770, pl. 98).

hand-forged head, a tapered shank, an irregular off-centre head slot and, above all, an irregular thread clearly made using a file, with different depths of the thread cut and a faceted thread (Figures 2–4). It is not until the second quarter of the 19th century that mechanically produced, consistently shaped screws become common.

Metal woodscrews or nuts and bolts?

Metal woodscrews should not be confused with bolts. The distinction between a bolt and a screw is commonly misunderstood, resulting in some difficulties in sourcing the origins of the woodscrew in treatises and literature. There are several practical differences, but most have some degree of overlap. The defining distinction is in their intended purpose: bolts are used for the assembly of two unthreaded components, with the aid of a nut. Bolts have parallel threaded shanks whereas screws have a tapered threaded shank or a parallel shank with a gimlet. The woodscrew takes no nut because it threads directly into the substrate with the thread cutting grooves in the fastened material (generally wood), pulling fastened materials together and preventing pull-out (Figure 5).

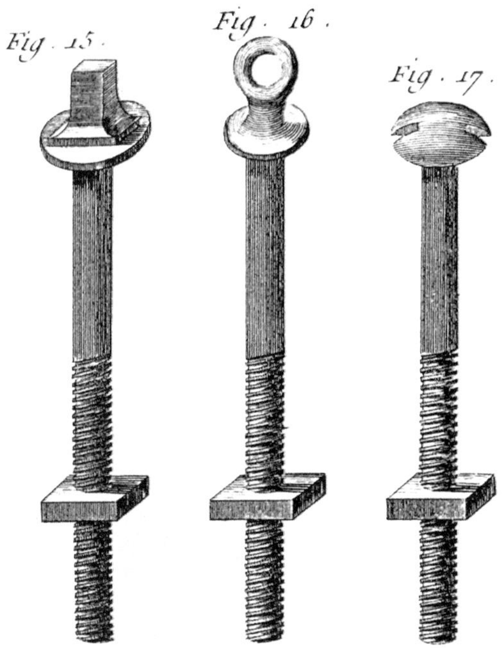

Figure 5 Typical nut and bolt (André Jacob Roubo, *L'art du menuisier*, seconde partie, 1770, pl. 98).

Manufacturing processes: handmade or machine-made?

The misunderstanding of the differences between woodscrews and bolts is not limited to their shape and function but also to their manufacture. Bolts were made successfully by machine, lathe, tap and die, screw plate or screw stock considerably earlier than woodscrews. Resistance against developing a mechanised process can be the result of both human reluctance and technological difficulties. The human factor should never be underestimated: until the beginning of the 19th century, guild regulations and workers' communities were eager to protect their trades, secrets and jobs. Change was usually very slow unless expedited by war or revolution. In France, the powerful guilds that had governed the French trades, controlling all stages of many manufacturing processes, were finally dissolved by the Le Chapelier Law in 1791. However, in the case of woodscrews, it seems that technological barriers were the most difficult to overcome. The manufacture of a woodscrew can be divided into three main stages:

- Forming or forging the shank and head
- Cutting of the thread
- Finishing the head by cutting the slot

Forming or forging the shank and head

The round and tapered screw blanks, of the desired finished length and finished diameter, with a countersunk flat head on one end of the blank, were first formed from square iron bars and later from round iron bars (Figure 6). By 1831, the woodscrew blanks, commonly known as 'screw-forgings'[4] were still said to be 'generally forged by the nailors, much in the same manner as the nails themselves'.[5] The art of the nailor, also known as nail-maker or *cloutier* in French, is well documented and has been widely discussed and illustrated in Diderot's *Encyclopédie*[6] to name just one historical source, as well as in more recent publications.[7] Although there were numerous developments in the USA and in England that mechanised the manufacture of nails (mostly by stamping iron sheets or drawing an iron wire to make wire nails), during most of the 19th century, many of the nails manufactured in Europe were produced by hand in small workshops. A nailor's workshop was generally a 'cottage industry', employing whole families, including women and children. It is estimated that in Birmingham (England) over 50,000 workers were involved in the hand manufacture of iron nails during the 1830s, using simple, reliable and well-proven techniques that required minimal equipment and rough materials.[8] The number of nails they could produce was staggering. In 1828, James Leighton won a bet to forge 17,000 double flooring nails by hand in less than two consecutive weeks. With such efficiency, it is not surprising that, by 1831, screw blanks were still being forged in the same traditional way as nails (Figure 7).[9]

Rounding off the screw blank when forging it from a square bar was made easier with the use of a 'swage block'.[10] By the beginning of the 19th century, the process was further improved by using a special vice developed in 1817.[11] Additionally,

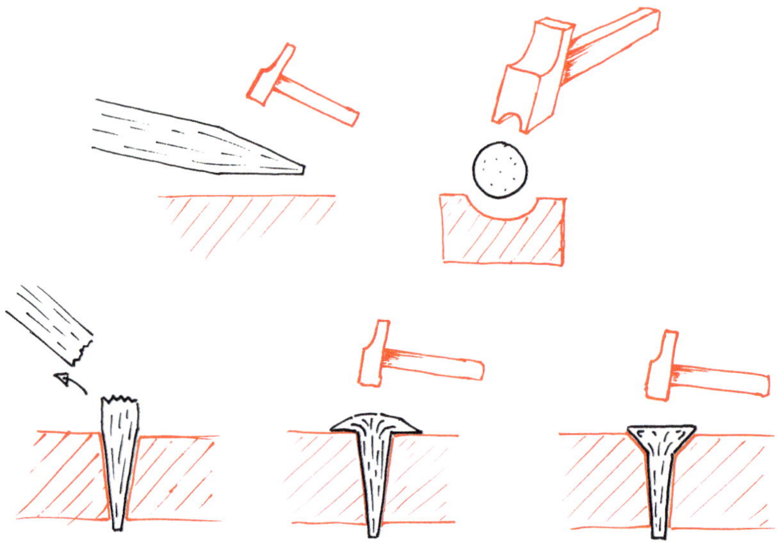

Figure 6 Schematic drawings showing the forging of a nail or screw blank from an iron bar (in black).

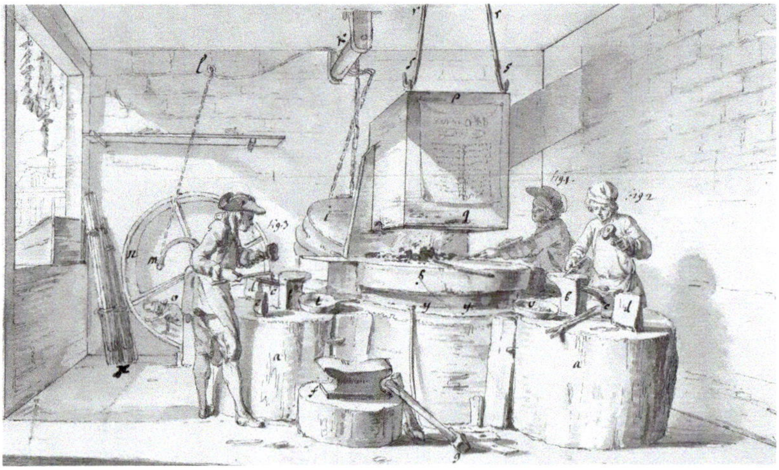

Figure 7 Jean-Michel Moreau le Jeune, *Le cloutier grossier* (The Nail Makers), c.1763. The J. Paul Getty Museum, object number 2004.87.

the quality of the forged blanks might have been slightly improved by using round iron bars instead of square bars.[12] If we believe Edward Wolley of Bilston's patent of 1818, 'screw-forgings' from round iron may not have occurred before 1818.[13] Early threading lathes required the blanks to be 'cleaned-up' with a file while holding the blank 'by a pair of clams (sic) fixed at the end of a spindle'. These screw-forgings also needed annealing 'as they come from the hammer possessing a degree of hardness which would render the process of cutting harsh and difficult'.[14]

Metallurgical studies of antique 18th-century screws and nails, as decribed below, confirm that screw blanks were indeed forged from hot iron bars and not stamped cold or machined, as is common today. While the small technical improvements listed above may have helped to produce better quality forged blanks, the nail/screw-making process was still mainly a manual one, with a heavy reliance on talented nailors.

Cutting of the thread

Cutting threads on parallel-shanked bolts was well established by the 15th/16th century if not before, using a die plate, a threading die or on a lathe (Figure 8).[15] However, the die plate commonly used for manufacturing bolts is impossible to apply to the tapered shank of a woodscrew (Figure 8). Common die plates are designed to cut a thread on the cylinder of a parallel bolt, not on the conical shank of a woodscrew. Additionally, the thread of a woodscrew is much coarser than on a bolt. As opposed to the sharp modern die, early die plates did not form the thread by cutting and removing material but mostly by forcing the thread into the iron bar.

Forming or cutting the deep woodscrew thread in early iron was particularly difficult as noted in the 1839 Roret *Encyclopaedia*:

> Two kinds of processes are used to make wood screws: the die and the lathe. The die has the disadvantage of destroying the ductility of the iron: the threads of the screw are of uneven strength. Certain parts of the nail presented to the die being more or less hard, more or less nervous, more or less resistant, the threads are all the more affected by this inequality, as the strong compression of the die, and the heat which it develops, still tend to increase these defects and make the iron more brittle: also chapping, straws and other accidents, make reject these kinds of screws by many workers.[16]

Alternative options to the threading die are described in the 1831 *Treatise on the Progressive Improvement and Present State of the Manufacture in Metal*:

> The methods adopted for the production of the worm, or thread, of the screw, are various; the first and most simple course, and that which would most obviously suggest itself in the infancy of screw-cutting, was to bring out the threads with a three-square file. Such a plan was too tedious, expensive, and imperfect long to continue; and accordingly it has for many years been entirely superseded by cutting and tapping.[17]

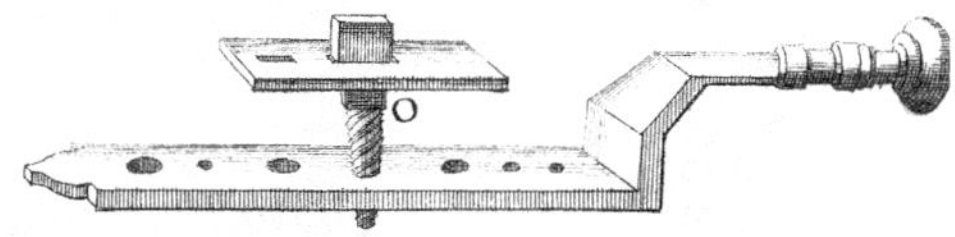

Figure 8 A typical antique die plate used to thread bolts (A. Félibien, *Les principes de l'architecture, de la sculpture, de la peinture, et des autres arts qui en dependent*, 1673, J-B. Coignard, Paris, p. 241).

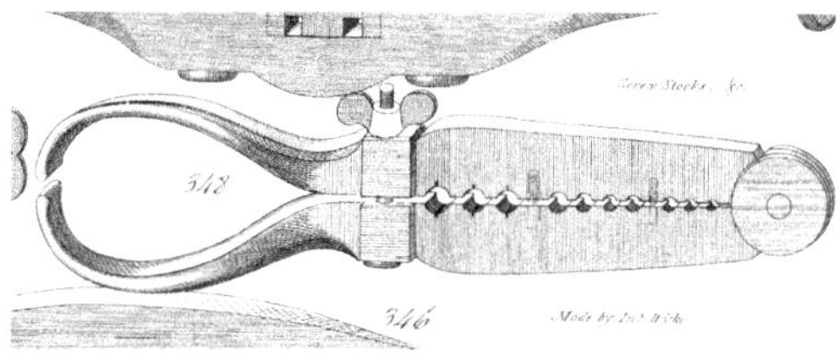

Figure 9 A 'screw stock' (*A Catalogue of Tools for Watch and Clock Makers, by John Wyke of Liverpool*, c.1770).

Based on the 1831 and 1839 treatises and on the study of historical screws, three threading techniques were possible:[18]

- Threading die
- Hand file
- Lathe

One historical variant of the die plate called a 'screw stock', is illustrated in John Wyke's tool catalogue c.1770 (Figure 9).[19] This small, simple device enables the die opening to be adjusted by applying hand pressure via two handles so that the die continuously applies pressure to form a thread on a bolt. The ease and speed of adjustment of the screw stock also make it suitable to form the thread into the tapered body of a woodscrew blank. In 2011, the gunsmith James Wilson Everett successfully produced a decent woodscrew with a tapered shank using an antique version of John Wyke's screw stock.[20]

Everett's distinctive thread is the same as that on historical screws produced by early

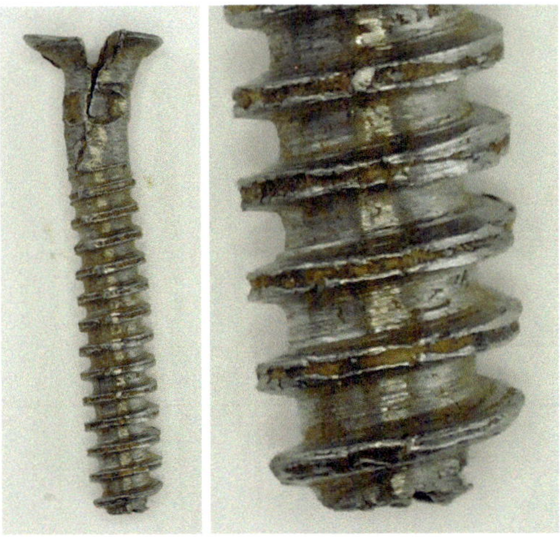

Figure 10 Woodscrew, probably produced on a machine, using the early Job and William Wyatt 1760 patent (length 35 mm).

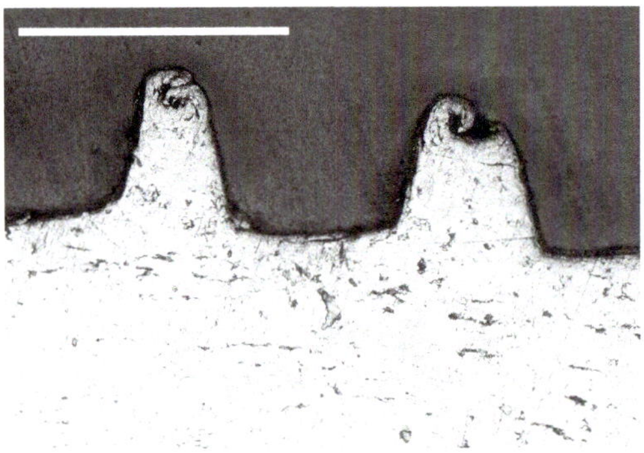

Figure 11 Screw thread from a late 18th-century English wood screw: cross-section microscopic view.

screw-cutting lathes and the distinguishing of one process from the other is currently a difficult one, and based on the resultant look of the threads. However, the supposed lathe-made screws studied have a distinctive 'cleaning up' of the shank and they all have parallel shanks, as described in early patents. The 'cleaning up' described in early patents may be identifiable by filling marks and rounding off the shank just below the head. Although Everett produced a tapered screw in 2011, to date no 18th-century examples of tapered screws with a thread produced mechanically have been found: it is fair to assume that John Wyke's screw stock was not a great commercial success or widely applied to the manufacture of woodscrews with a coarse thread.

The screw has a parallel body, typical of early patents. The shank, just below the head was reworked and 'cleaned up', as well as the area below the head, as described in the Wyatt patent (Figure 10). It should be noted that the damage below the head is not original and is the result of dismantling. The thread is regular. The cutting of the thread is the result of cold working and hence distorting the metal, resulting in a burr on the upper sharp edge of the thread. As a result, the diameter of the unthreaded shank is slightly smaller than the diameter of the threaded part of the screw (the root diameter). The increase in diameter is due to metal being displaced when threading. The distinctive thread is the same as that found on historical screws produced by early screw-cutting lathes or screw stocks, and distinguishing one process from the other is currently still difficult if based only on the resultant appearance of the threads.

The screw was probably produced on a lathe using the early Job and William Wyatt 1760 patent (Figure 10).[21] The cutting of the thread has distorted the metal and created a burr on the upper sharp edge of the thread. A small, distorted burr arising from the compression of the thread can also appear on modern 20th-century screws. This distortion is less pronounced than earlier screws, probably the result of better raw materials and improved thread-cutting tools.[22]

The threads produced using a screw stock or an early lathe are usually regular, smooth and of equal depth, making them easy to identify (Figure 11). In contrast to a high-speed, modern steel-cutting lathe, the thread on historical screws was 'cut' by scraping metal away using great force and at low speed. The cutting was clearly not efficient: it resulted in the metal being pushed aside rather than removed. Machine-made woodscrews, of all centuries, have a distinctive metal burr on the sharp edges of the thread as a result of the considerable force required to produce the thread. Earlier screws usually have a very pronounced burr although, over time, this becomes minimal and, by the 20th century, is scarcely visible to the naked eye. Screws produced using a hand file do not possess this burr as a hand file puts less stress

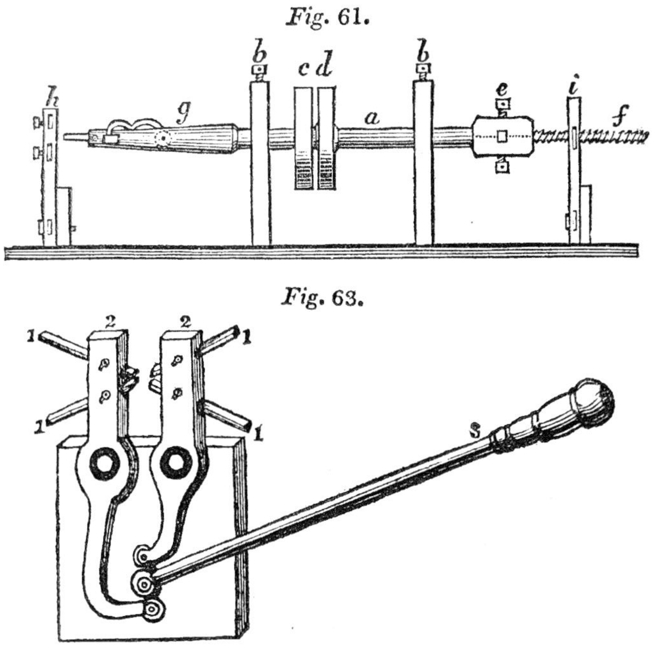

Figure 12 Drawings of a device used to cut the thread of wood screws with detail drawings of the 'cutter and follower' tool (J. Holland, *A Treatise on the Progressive Improvement and Present State of the Manufacture in Metal*, 1831, p. 202).

on the metal and does not distort it to the same extent. An artisan can easily remove the burr while cutting the thread by hand.

There were numerous other patents and attempts to mechanise the making of woodscrews using complex tools and lathes.[23] The first important development is the above-mentioned 1760 English patent by the brothers Job and William Wyatt. Unfortunately, there were no drawings submitted with the Wyatt specification, but the methods described appear to have formed the basis of production methods for several decades afterwards and later drawings (Figure 12) from other similar patents may give us the best representation of the Wyatt patent.

By the 1790s, the Wyatt brothers' factory employed 50–60 workers and produced approximately 25,000 screws per day. The screws produced had a parallel shank and, while it appears that the Wyatt factory was relatively successful,[24] English furniture makers were slow to adopt the new machine-made screw: English furniture conservators have found only a few examples of the Wyatt brothers' parallel screws.[25] In addition, no Wyatt-style screws have been found on European furniture but this might be due to war, trade blockage or guild protection. Eventually factories similar to that of the Wyatt brothers were established in Europe, including the factory established in 1806 by the French Japy brothers. By the 1830s, the Japy factory was producing about 2.5 million screws a month.[26] These screws, like their English counterparts, were still made with a parallel shank.

The first fully mechanised making of a screw with a tapered shank and gimlet was made possible by Thomas J. Sloan's invention of 1846, a United States patent owned by the American Eagle Screw Company. Sloan's screws with a gimlet, combined with major improvements in the quality of iron production, brought to an end two centuries of handmade screw production. Although in 1831 'to bring out the threads with a three-square file' had been considered 'too tedious, expensive, and imperfect', without satisfactory mechanised alternatives, it proved to be the most successful and common technique used until the mid-19th century.

Finishing the head by cutting the slot

Compared to the challenge of threading the woodscrew, cutting a slot in the head using a hand saw and finishing it presented only a minor challenge. The process of cutting the slot was mechanised and improved at the same rate as the cutting of the thread. According to the Wyatt patent, after cleaning up the shank and head, the screw was placed into a special chuck with a rotating cutting tool. To date, all the historical screws studied have a sawn instead of a stamped slot. The major difficulty for early hand screw-makers was to centre the slot correctly. As a result, most handmade screws made before the 1760 Wyatt patent display a slightly off-centre slot.

While the manufacturing processes involved in the making of woodscrews appear to have remained stagnant over many years, despite the numerous inventions described above, the shape and design of the thread did change over time (Figure 13). Three different types of threads were progressively favoured during the 17th and 18th centuries in order to ease either the screws' penetration into the wood or to improve its holding power.[27] The Nettlefolds Screw Company briefly reintroduced the early 18th-century thread form in a slightly more elongated and flatter shape, patenting it in 1826 as 'buttress' thread. The buttress had the self-proclaimed advantage of making the screw easier to screw into the timber and increased the resistance to withdrawal by force.[28] The machine-made buttress thread did not survive long and these 19th-century versions are easily distinguishable from 18th-century handmade screws being machine-made and regular.

The change of thread patterns is clearly related to the file used to cut the thread. The earliest thread shape suggests the use of a triangular file (an equilateral triangle with 60-degree angles), the second type indicates the use of a fine file with a rounded edge, while the third shape suggests the use of a square file with sharp edges and is similar to virtually all machine-made screws made subsequently. Distinguishing 19th-century screws of the third type from modern screws can be challenging.

The chronology below should be treated with a degree of caution as when screws were both developed and made redundant overlap: there are examples of old-style screws being used long after the screws had supposedly ceased being made. For example, the c.1755 double desk at the J. Paul Getty Museum (inv. no. 70.DA.87), marked by the Paris ébéniste BVRB II (Bernard II van Risenburgh, born after 1696, died about 1766) is made using all three different types of screws described above.[29] Screws of the earliest design are generally known to have disappeared between 1720 and 1730, the second type of screw is common during the middle of the 18th century, while the third type, having been invented later, might have been applied during restorations. The Van Risenburgh Paris workshop was started by Bernard I (died before 1738) and continued by his son Bernard II. It is likely that the latter inherited an old stock of screws that could have been used as late as 1755. It is also probable that some craftsmen, in particular gunsmiths, may have produced their own screws and their own designs. The chronology below should therefore be regarded as a work in progress which future studies of antiques may help to refine.

A fascinating aspect of the study of screw design, and historical technologies in general, is the speed at which knowledge and innovation transferred across regions, countries and workshops in an era with very limited numbers of trade publications.[30] The designs of the screws illustrated below, while sourced primarily from French furniture, appear to have been used throughout Europe at more or less the same time. Political instability, religious intolerance and the forces of change and opportunism during the Age of Enlightenment caused the displacement and emigration of many people. This population movement resulted in the quick exchange of ideas and technologies, disseminating knowledge of how to improve screw manufacture speedily throughout Europe.

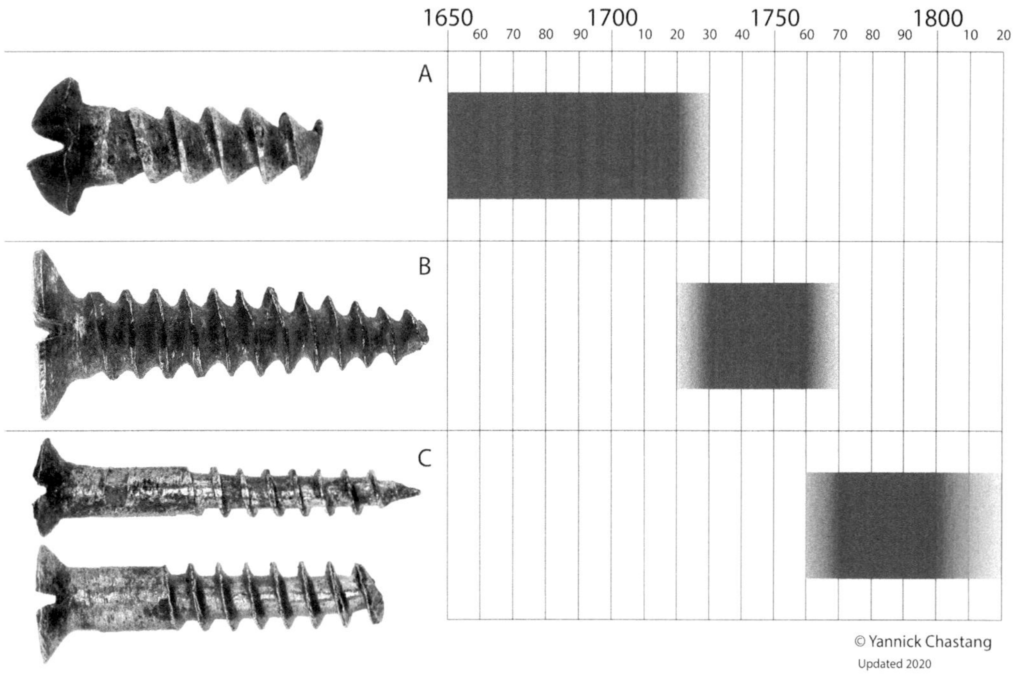

Figure 13 Chart showing the chronology of the 18th-century woodscrew.

Figure 13 shows three generations of the most common screws (note the difference in the thread caused by the use of the differently shaped hand files):

- A: Tapered with buttress thread
- B: Tapered with rounded thread
- C: Parallel shank with square thread with a hand-made gimlet or a blunt tip.

Hand-threaded gimlets first seem to appear during the second half of the 18th century, replacing the tapered shank. After the 1846 Thomas J. Sloan invention, most screws have a parallel shank with a gimlet similar to model C above. The last two screws illustrated in this chronology originated from the same object, demonstrating that the two models were used at the same time.

Metallurgical analyses of historical woodscrews

Most archival research has concentrated on the potential mechanisation of screw manufacture and the evidence of numerous patents and attempted technical developments. Almost all 20th- and 21st-century publications have failed to observe that the majority of the screws made before the beginning of the 19th century were hand-forged and hand-finished. Also overlooked is the poor quality of the iron material that was available. Strength is an essential component of a screw and the inherent strength of the material is important for a successful mechanisation of the manufacture process. Marrying the results of metallurgical analysis with archival resources may offer an explanation as to why screws were made by hand for much longer than might be expected. Metallurgical examination of the iron (metallography) generally reveals the base metal to be of poor quality and this flawed raw material may have made mechanisation impractical. Additionally, it is sometimes difficult to simply believe that certain screws could even be handmade, in particular some of the smallest screws (see Figures 1 and 2). A late 18th-century screw's thread is identical to that of a modern-day thread, and the quality of execution can be so good that only by using metallography at high magnifications can clues as to its origins be revealed.

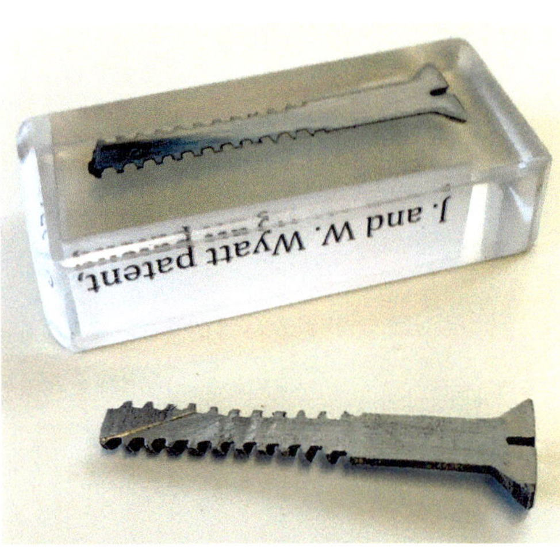

Figure 14 A wood screw cut in half and mounted in hard epoxy resin, polished and etched, ready for microscopic analysis, next to the offcut.

The screws in this study originated mainly from my collection or were donated by independent conservators. Until the end of the 1990s, it was not uncommon to replace screws on an antique object or to buy severely damaged antiques (called a 'breaker') from which to salvage material to repair other antiques. Many of the screws used in this study had little or no provenance. However, comparing them with dated examples still *in situ* on antique furniture at the Wallace Collection and other collections, it was possible to attribute a date to the screws prior to analysing them.

Specimens for metallography

The screws selected for analysis included 11, probably from 18th-century objects, 7 from the 19th century and 2 from 20th-century objects. In addition, 2 nails from the 14th century and 5 from the 18th century were analysed. As none of the screws offered for analysis belonged in an artefact, it was decided that cross-sectioning them (a destructive technique) was an acceptable sacrifice. Cutting the screw right through the middle would help us not only to identify the metal used, but also to see the eventual metal distortion resulting from the forging and manufacture processes. Cross-sectioning is not recommended for historically important screws. It is hoped that this article will enable future historians to identify screws without the need to sacrifice them. Polishing a small edge of the metal screw, an acceptable alternative to cross-section, could also aid identification.

Under the guidance of Alan Williams, Wallace Collection archaeometallurgist, the screw and nail samples selected for analysis were first cut in half using a jeweller's handsaw fitted with the finest blade possible. The better half was mounted in a two-part epoxy resin. The mounted half-screw was then ground to optical flatness with silicon carbide lubricated with paraffin oil (200, 400, then 800) and then taken to a fine polish, with 6 μm (and finally 1 μm) diamond paste (Figure 14). As the polished surface appears bright without detail, the samples were etched using a solution of 2% nitric acid in alcohol. The etched surface shows the grain boundaries of the metal. The samples were studied with a Vickers metallurgical microscope using 160× and 800× magnifications and an Olympus BH2-UMA using 50×, 100× and 200× magnifications.

Metal identified[31]

The microstructure of wrought iron (Figure 15) is easily identifiable by the presence of slag inclusions and the extremely low carbon content (usually below 0.1% C) compared to modern steel. The microstructure of wrought iron comprises large polyhedral crystalline grains of iron (ferrite) with irregular slag inclusions. These slag inclusions are formed from the non-metallic elements found within most iron ores and are created during the smelting process, but not separated, since the iron was not melted in primitive (bloomery) furnaces. Slag is a glass-like material that softens with heat so the inclusions would be extended into stringers by forging operations.

The presence of slag is beneficial for some blacksmithing operations as it gives the material its unique fibrous structure. Bending a bar of wrought iron will result in the iron cracking and splitting open along the grain (the rows of the slag inclusions) as opposed to modern types of irons that have a tendency to break cleanly (being slag free). This simple technique is still used to

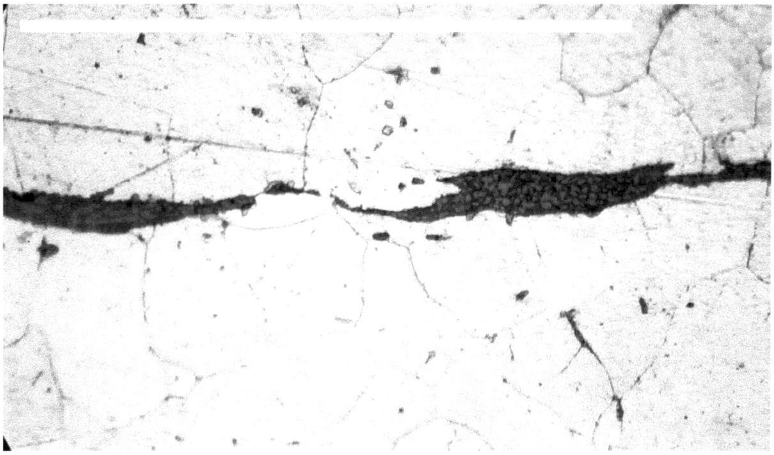

Figure 15 Screw made of wrought iron with large slag inclusions (the irregular and large black areas), c.1750. The large polyhedral grains are ferrite. Wrought iron does not contain any carbon. The metal chosen is of poor quality but it was widely available and cheaper than steel.

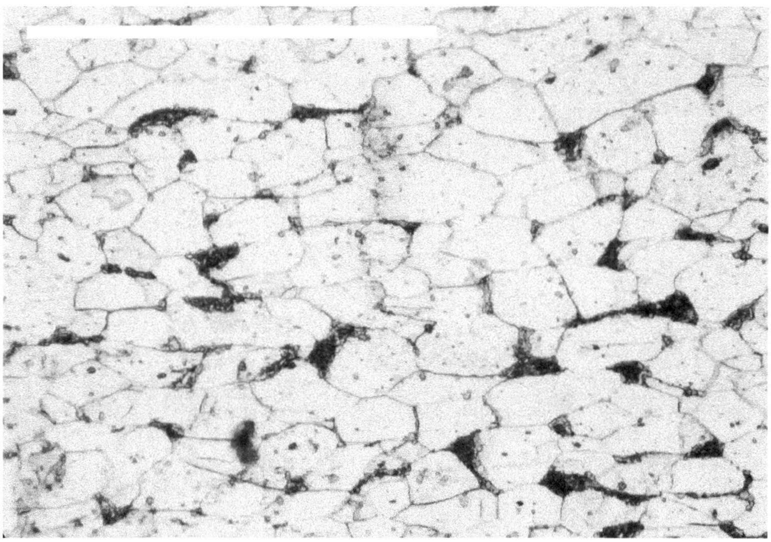

Figure 16 Japy frères screw, c.1940. This metal is mild steel with a low carbon content. The metal is extensively refined and has no slag inclusion. The microstructure consists chiefly of ferrite with some dark areas of pearlite (a mixture of iron and iron carbide).

identify wrought iron by modern blacksmiths. The silicate stringers of the slag are also believed to protect the iron from corrosion. However, based on my experience, wrought iron is not a metal suitable for modern machining techniques on a high-speed lathe – the uneven grain with large slag inclusions makes it very difficult to machine the metal without cracks or, indeed, complete breakage of the metal.

Compared to wrought iron, the microstructure of modern mild steel (Figure 16) consists chiefly of ferrite without slag. At the borders of the ferrite grains are small areas of a dark etching material known as pearlite, a mixture of iron and iron carbide: the higher the carbon content, the greater the proportion of pearlite in the steel. This improved metal would be more durable under stress and is better suited to machining.

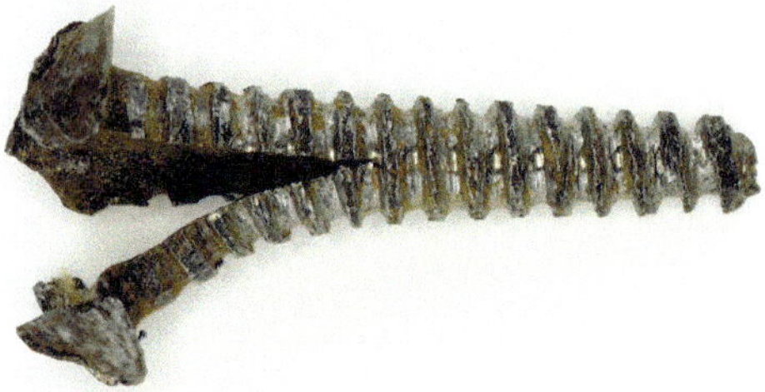

Figure 17 A typical crack following the direction of the slag inclusion. This is the result of poor quality metal combined with a major design flaw: the slot for the screwdriver is placed to start a crack that, when put under stress, can result in this disastrous malfunction (length 32 mm).

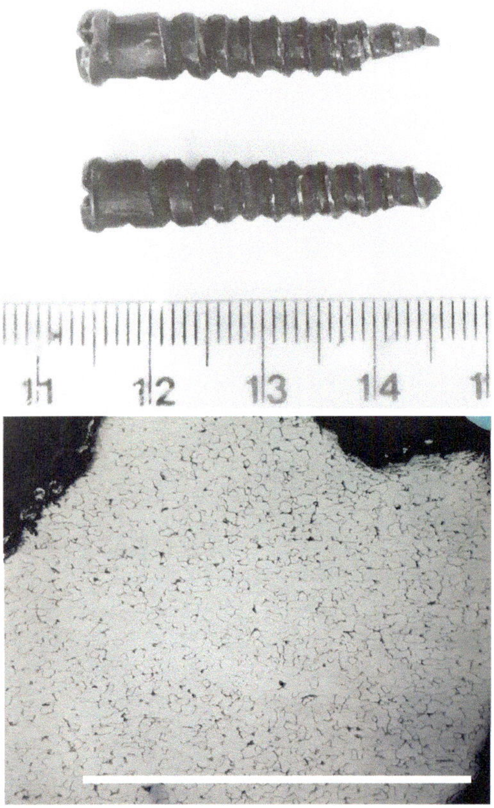

Figure 18 Screw (c.1990–95) made of mild steel without any impurity, typical of the 20th century. The screw thread is (unconvincingly) handmade with a file in the style of the 17th century, reusing a modern screw blank. This screw originated from a pair of cabinets on stand, supposedly from the collection of the Ducs de Feltre and now belonging to the AXA group in Paris. These cabinets are generally believed to be fake and to have been made in the 1990s (scale bar = 1 mm).

Thanks to the higher carbon content, steel can be hardened by quenching (plunging when red-hot into cold water) when it forms a different, very hard, crystalline material known as martensite. Before the development of effective methods of steelmaking, wrought iron was the most common form of malleable iron. It was given the name 'wrought' because it was hammered, rolled or otherwise worked while hot enough to expel molten slag. The modern equivalent of wrought iron is mild steel or *acier doux*, also called low-carbon steel (perhaps 0.2% C). Neither wrought iron nor mild steel contains enough carbon to be hardenable by quenching. All the nails and screws produced from 1650 to the second half of the 19th century were identified as being made of wrought iron.

Following the introduction of commercial puddled iron at the end of the 18th century, it slowly replaced wrought iron during the 19th century. Puddled iron was made by decarburising cast iron rather than by reducing iron ore in a bloomery. Since it was made in large quantities, it was cheaper rather than of higher quality. It also contained slag inclusions, so distinguishing wrought from puddled iron can be extremely difficult, indeed speculative. Puddled iron sometimes has fewer slag inclusions but this small difference in quality cannot be used as a basis of dating as, by the end of the 18th century, round wrought iron bars were used in the manufacture of screw blanks (see above). These round bars, of various diameters are also known in France as *fil d'archal*.[32] In all likelihood, the smithing process involved in producing *fil d'archal* would have resulted in the disposition and size of the slag inclusions in wrought iron being similar to puddled iron.

As steel production improved during the 19th century, so too did the quality of the base metal available. There is a noticeable fall in slag inclusion content in all screws from the second half of the 19th century, when the quality of the metal improved and, concurrently, screws started to be made exclusively by machine. Although the present study would need to be extended to examine more screws dating from 1850 to 1950 to establish a more precise date when the metal improved, early results do follow a known trend in iron production.

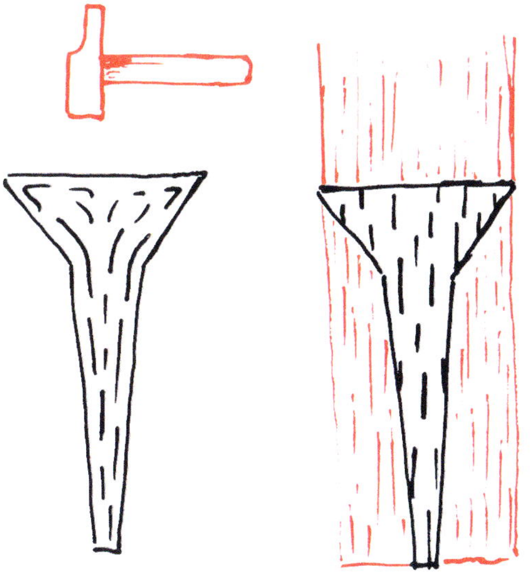

Figure 19 Schematic drawing of a forged screw blank (*left*) and a machined screw blank (*right*). The black lines of slag inclusions follow the shape of the head when forged and are interrupted or cut off if machined.

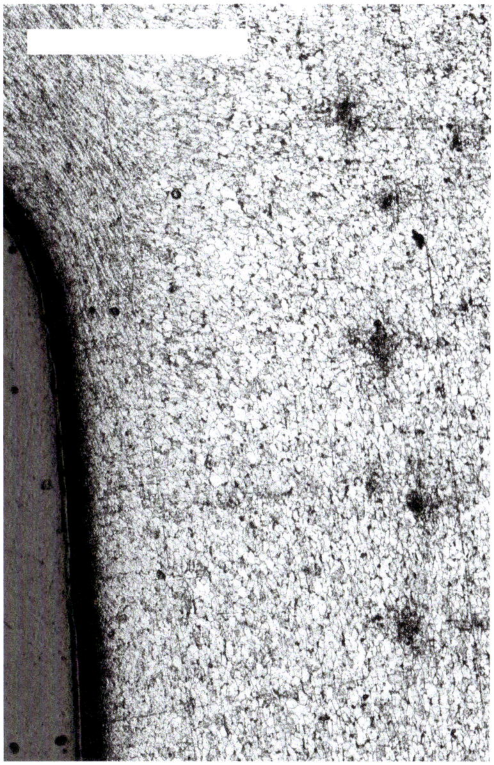

Figure 20 Japy Frères screw made of mild steel, c.1940. The deformed ferrite grains of the head (*top left corner*) are indicative of cold working, i.e. cold stamping. The screw blank was not annealed after stamping the head as the grains retained their distortion (*right side*). The ferrite grains of the shank are equiaxed and indicate that the metal was not reworked cold.

The present study established that all the screws produced during the 18th century and the beginning of the 19th century were made from an iron rich in slag inclusions and without carbon. As noted above, in 1839 Roret regarded the quality of the metal as being absolutely critical for the success of machining screws. The poor quality of the wrought iron blank may also have had further ramifications for the reputation of any screw business. It is not unusual to find historical screws that have cracked, with a crack starting at the screwdriver slot in the head (Figures 17 and 18). The screwdriver applies pressure in the direction of the grain (the rows of slag inlusions), meaning that the head slot is located in a position where it is particularly likely to start a crack in the body of a screw that may well render the screw unusable and impossible to retrieve. Screws made of inherently low-grade wrought iron are therefore not only difficult to machine but their durability is also greatly influenced by the necessity to include a slot in the head. By comparison, usable nails can be made of low-grade iron as hammering the head puts little torsional stress on the wrought iron.

Manufacturing techniques: forging or cold stamping?

Archival resources reveal that historical screw shanks were forged (i.e. hot worked) whereas modern screws are stamped cold (a process sometimes called 'cold forging'). The microstructure of ferrite grains worked cold will appear distorted. In contrast, hot-worked ferrite grains would be able to recrystallise into equiaxed grains. All the historical screws analysed displayed equiaxed grains of ferrite, implying that the shank had been forged hot. A few screws, however, exhibited small areas of distorted ferrite grains, suggesting that the metal had started to cool down before the head or the shank had been fully forged. Although annealing iron at 450–700 °C would have enabled grains

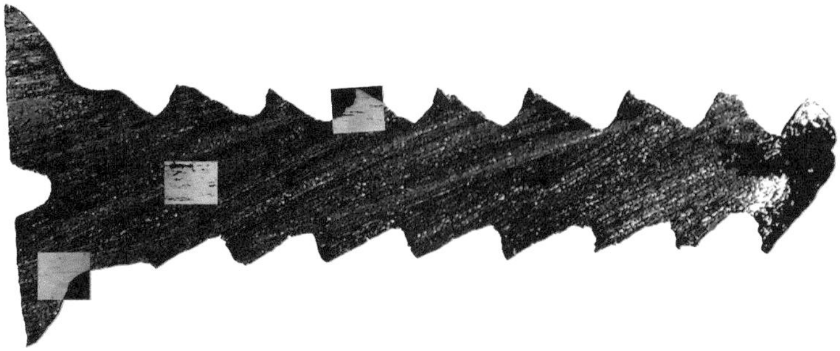

Figure 21 Cross-section of a screw, c.1700 (for detailed microstructure photos of the screw head, the thread and the centre of the screw above see Figures 22 and 23 below).

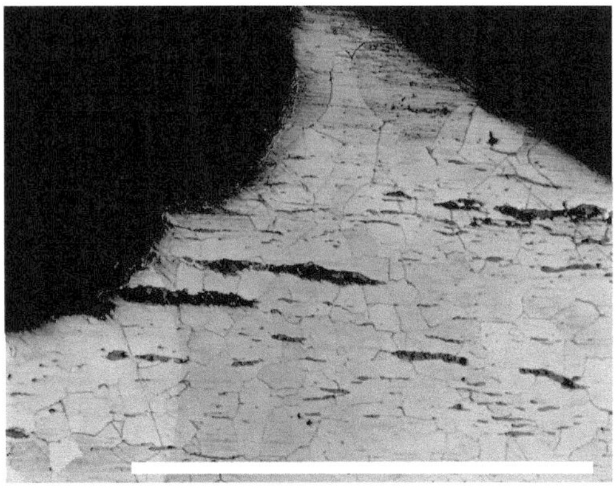

Figure 22 Microscopic view of the thread. This photo demonstrates that the thread was cut and not stamped or forged. The slag inclusions and ferrite grains are interrupted abruptly by the cutting of the thread.

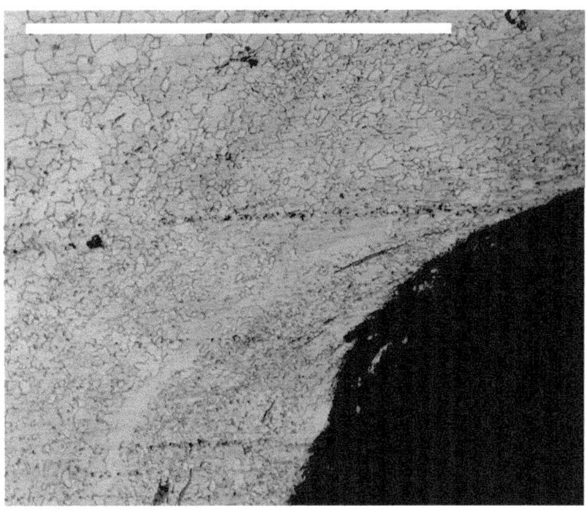

Figure 23 Microscopic view of the corner between the screw head and the shank. The head was formed by forging. The slag inclusions and the ferrite grains follow the shape of the head, implying that the head was forged or bent instead of being machined. The equiaxed shape of the ferrite grains also demonstrates that the metal was forged when hot, enabling the grains to recrystallise.

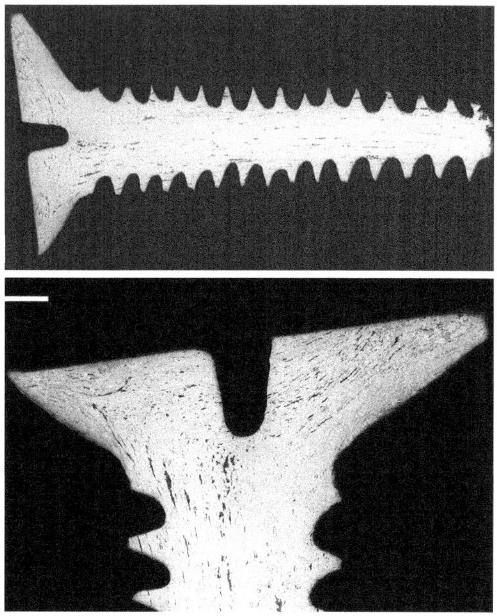

Figure 24 Screw made of wrought iron, c.1750. The slag inclusions clearly follow the shape of the head. The head was forged instead of being machined.

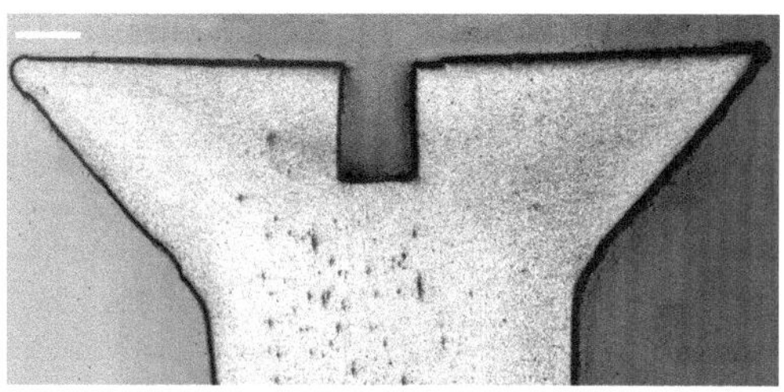

Figure 25 Microstructural details of a late 20th-century screw head and thread. The distorted ferrite grains are indicative of cold working, i.e. cold forged.

to recrystallise, it is unlikely that the screw blanks were annealed after forging as some distorted grains are still visible (Figures 19 and 20).

The metallography also confirmed that all the screw heads studied had been forged rather than machined on a lathe. The 2011 replica screw made by the gunsmith James Everett (discussed above) was machined instead of being forged: machining was a logical 21st-century choice as it enabled long bars to be securely maintained in the jaw of the vice during the cutting of the thread, but machining today is easier and faster than before. Metallography of historical screws demonstrates that the slag inclusions follow the shape of the screw head: had the head been machined on a lathe from a solid bar, the slag inclusions would simply run parallel to each other before being interrupted by the machining. However, such interruptions of the slag inclusions are visible on the screw's thread and at the slot on the head, demonstrating that both the head slot and the thread were cut instead of being stamped. To date, despite some patents mentioning stamping the head's slot, all the 18th-century screws analysed have a sawn rather than a stamped slot (Figures 21–25).

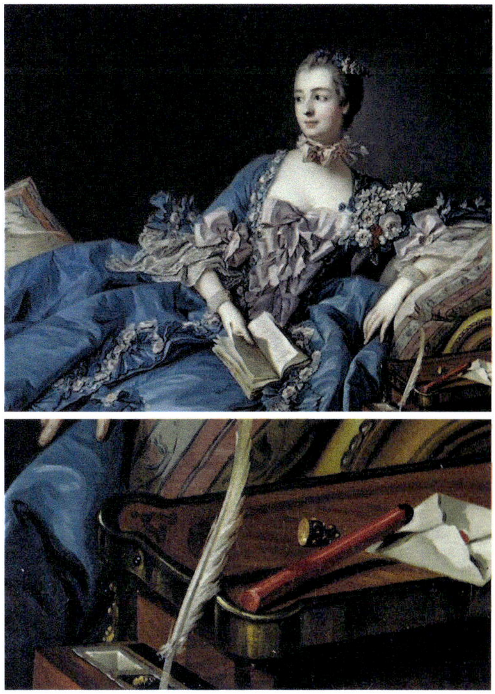

Figure 26 Portrait of Madame de Pompadour attributed to François Boucher, 1756–58, depicting blue-headed screws. National Galleries Scotland, acc. no. NG429.

The iron screw: grey iron, blue iron or gilded head?

For most situations the natural grey colour of the iron screw was perfectly suitable. Francis Watson, a former director of the Wallace Collection, noted 'the crude way in which most mounts are attached to the wooden body of the furniture with highly visible steel-headed screws contrasting strangely with the gilding of the mounts themselves.'[33] However he also observed that a very small number of screws were coloured blue or gilded as in the 1756–58 portrait of Madame de Pompadour at the National Galleries Scotland (Figure 26). A second portrait at the Alte Pinacotek in Munich is more detailed. These two portraits show a small worktable attributable to Bernard II van Risenburgh, c.1750 and on both portraits, the screw heads are dark blue.

To date, the only example of blue screw heads used to attach gilt bronze I have encountered is on the pair of marriage coffers attributed to A.C. Boulle, c.1690 at Boughton House, in the collection of the Duke of Buccleuch and Queensberry (Figure 27). The screws on the cabinet were painted with gold paint but there was no evidence of previous gilding. The rarity of this discovery, however, does not mean that blue screws were not more common during the 18th century than they are today. Francis Watson, who would have been well aware of, at least, the Wallace Collection workshop's practices, also stated 'Today it is all too common for restorers to replace steel-headed screws by gilt ones in order to obtain a supposedly original effect of unity.' Although not well documented, replacing old iron screws was common practice for 20th-century conservators: the Wallace Collection Conservation Department has several small boxes of unidentified handmade screws that were removed by previous generations of restorers.

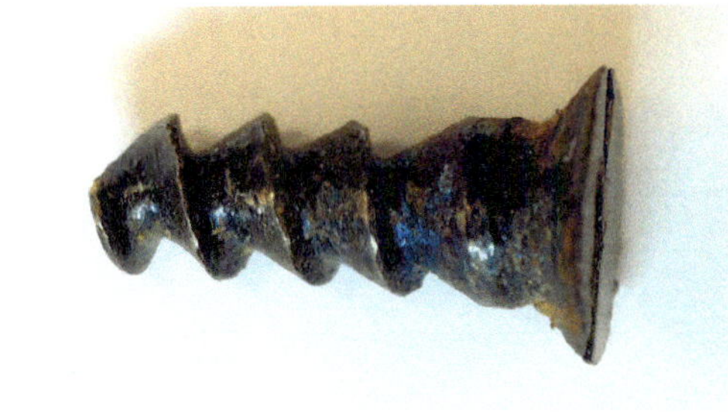

Figure 27 A blue-coloured screw from a marriage coffer attributed to A.C. Boulle, c.1690, at Boughton House (Collection of the Duke of Buccleuch and Queensberry). It was used to fix gilt bronzes and showed no sign of previous gilding.

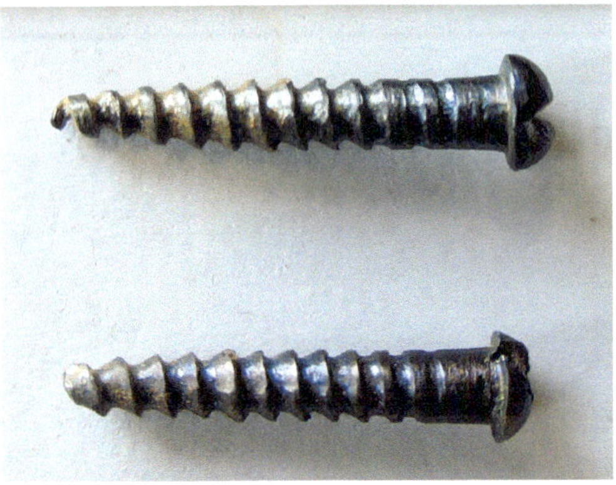

Figure 28 Two screws with blued iron on the upper part. The blue colour is the result of a heat treatment. Screws from the lock on the double secretaire marked by the Paris ébéniste BVRB II (Bernard II van Risenburgh), c.1755. The J. Paul Getty Museum, inv. no. 70.DA.87.

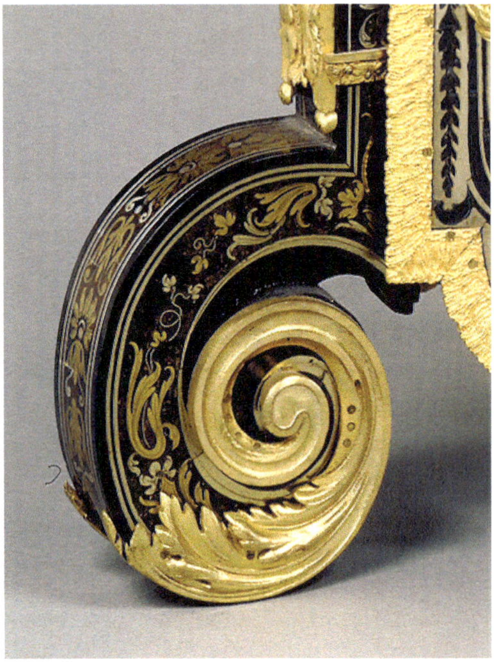

Figure 29 Detail of a gilt bronze scroll on a pair of torcheres attributed to A.C. Boulle, c.1715. Three nails are used next to each other for decorative reasons. The J. Paul Getty Museum, inv. no. 87.DA.5.

The blue colour of the screw is the result of heating the iron. Iron or steel will form a surface layer of oxides that appear blue if heated to around 250–300 °C. Such heat treatment of arms and armour is well documented and was common for gun barrels and other parts. Blueing the steel increases its resistance to corrosion. The wrought iron used to manufacture screws is itself naturally more corrosion-resistant. It is notable that 300-year-old woodscrews used in the acidic oak of a piece of furniture are seldom corroded, possibly the result of using wrought iron, of bluing the iron or simply the consequence of the furniture being stored in a dry environment. On some historical gilded-head screws, only the head and the upper part of the shank are blue, suggesting that the heat treatment must have been localised at the head (Figure 28).

Blue screws were clearly not to everyone's taste – even in the early 18th century, A.C. Boulle tried to resolve the issue of the screw heads disfiguring his gilt bronze decorations by brazing threaded lugs or screwing short threaded rods at the back of the decorative bronzes. These rods passed through holes in the woodwork and were secured by nuts inside the carcase. This technique was clearly limited to high-end furniture and was only suitable for bronzes that were large enough to have perpendicular bolts attached to them.[34] When fixing the bronzes from the back was not an option, Boulle would use the unavoidable screw heads as a decorative element. On many of his acanthus scrolls, three screws or nails are used next to each other: as only one would have been sufficient to secure the bronze in place, clearly three in a line is a purely decorative feature (Figure 29). Interestingly, makers of copies of these candle-stands did not understand

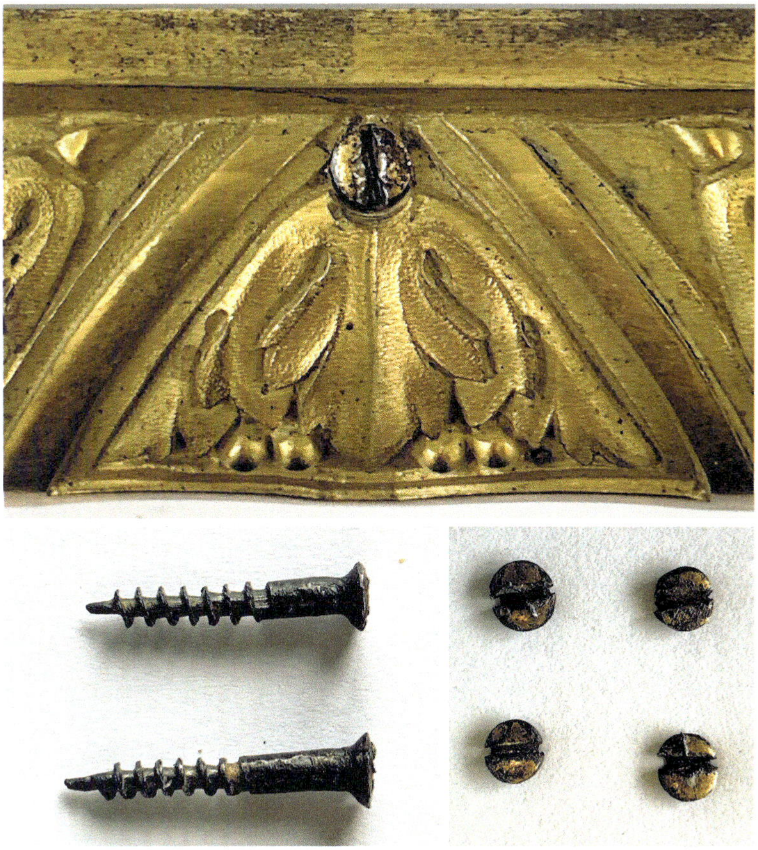

Figure 30 Gilded head screw used to attach the gilt bronzes on a lacquer commode attributed to Adam Weisweiler, c.1780, private collection.

the decorative origins of these three screws and often only fitted one.

Invisibly fixing bronze mounts from the back would become increasingly popular, reaching its peak by the end of the third quarter of the 18th century. However, for most furniture-makers, fixing bronze mounts via the front was the most practical option and, at some time during the second half of the 18th century, screws with gilded heads generally replaced blue-headed screws. Examples of original gilt head screws are more common as they fit modern restorers' expectations and therefore tend to have been preserved. Gilded head iron screws share with the early blue-headed screw a blue-coloured body (Figures 30 and 31).

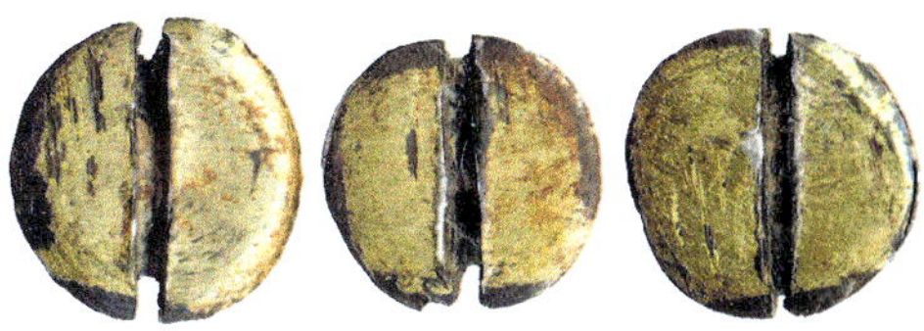

Figure 31 Gilded heads of screws from the double secretaire marked by the Paris ébéniste BVRB II (Bernard II van Risenburgh), c.1755. On careful observation, it appears that the heads were gilded using small squares of gold leaf. Possibly for economical reasons, the head is not entirely covered with the gold. The J. Paul Getty Museum, inv. no. 70.DA.87.

The process of gilding iron screws differs from the techniques employed on wood or copper alloys (gilt bronzes). Oil gilding would easily become damaged when screwing the screw with a screwdriver and it cannot be burnished to a high level. Mercury gilding ('fire gilding') only works on copper alloys or on iron previously

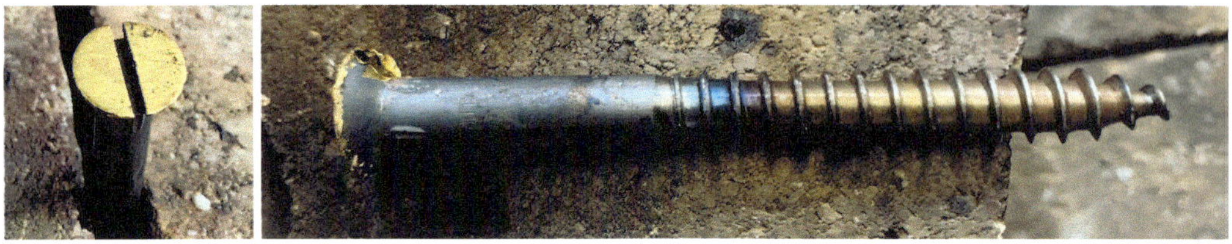

Figure 32 Fire gilding tests carried out in 2020 without mercury. The upper part of the large screw turned blue after heating the screw at about 310–330 °C. The gilding is strongly bonded to the steel.

coated with a layer of copper. X-ray fluorescence (XRF) analyses carried out on these screw heads did not reveal any mercury or significant copper either on the head or the shank, suggesting that the screws were not mercury gilded. The screw heads must have been gilded using a process called *dorure au feu* (fire gilding) that involved heating the screws to 310–330 °C. *Dorure au feu* should not be confused with the usual 'fire-gilding' technique that employed a mercury/gold amalgam. This technique of fire gilding without mercury is described in early treatises:[35]

> Fire gilding with gold leaf. To prepare the iron or copper to receive this gilding, it is necessary to scrape them well with the scraper, & polish them with the iron burnisher, then put them in the fire to turn them blue, that is to say to heat them, until they take a blue colour. When the metal is blued, we apply the first layer of gold, which we lightly burnish with a polisher, and which we then put on a low heat. ... Usually only three or four layers are given, each layer being one gold leaf in common works, & two in fine works; & with each layer that we give, we put them back into the fire. After the last layer, the gold is in a state of being burnished.[36]

The process described by Diderot and others relies on the adhesion of the gold to iron using 'solid state diffusion', an ancient and simple technique. Tests carried out in 2020 using the technique described produced a blue-coloured iron screw with a decent gold layer on its head that could be burnished (Figure 32). While the majority of iron screws retained their natural grey colour, blue-headed screws or gilded screws were commercially available. Blue-headed screws can therefore be original to an 18th-century piece of furniture and it is to be hoped that restorers will not now be tempted to replace them.

Conclusions

For those of us in the 21st century accustomed to buying cheap mass-produced screws, it must seem surprising to read that screws were made by hand for such a long time. Handwork, namely the expertise of generations of nailors and blacksmiths, had been able to meet extensive demand using simple, yet efficient, processes that were perfectly adapted to the materials available. Study of archival resources, combined with microscopic analysis of the metal, demonstrates that until quality metal became available, cutting the thread using a lathe or employing a machine to make screws simply did not prove as effective as doing it by hand. Furthermore, developing a lathe to cut a thread on a tapered blank was clearly more challenging than one might imagine. Progress requires both the materials and the technology to be available at the same time, and such mechanisation was eventually achieved at more or less the same time as metallurgical progress resulted in the production of suitable quality iron.

This study has unfortunately failed to unearth any names of individuals or shops that specialised in the making of screws. We know that nails and screw blanks were made by families of nailors but it remains unknown whether the cutting of the thread by hand was done in the same workshop or by anonymous specialists elsewhere. Whoever they were, the screw-makers' expertise was key to the success of many industries, particularly furniture and gunmaking. Where the products of the latter are concerned, study of their fastenings may well provide insight into the dates of their construction. If the screws which hold guns together can be removed and analysed non-destructively, then the age of the guns' last assembly can be deduced – which may lead to a few surprises.

Acknowledgements

I am grateful to my former colleagues at the Wallace Collection without whom this study would not have been possible. Thanks to David Edge, long-time Wallace Collection Arms and Armour conservator and curator and latterly Head of Conservation. Also to Dr Alan Williams, Wallace Collection archeometallurgist who instructed me in basic metallurgy and allowed privileged use of his Vickers microscope. Thanks to intern, Vidar Thijssen, for his help with some of the microphotographs published in this article. The metallurgical study requires cross-sectioning antique wood screws and this was only made possible with the gift of unusable antique screws by independent conservators Michel Jamet (formerly Ebénisterie des Cloÿs, Paris), Peter Holmes (formerly Arlington Conservation, London) and William Patrick Edwards (Antique Refinisher, San Diego). Special thanks are extended to the numerous colleagues who have shared information including Nicolas Boucher and Pierre Costerg (formerly Musée des Arts Décoratifs, Paris), Paul Tear (formerly Wallace Collection, London), Carolyn Sargentson (formerly V&A, London), David Williams (independent scholar) and last, but far from least, Arlen Heginbotham (J. Paul Getty Museum, Los Angeles) for his time and expertise, for editing this article and sharing his extensive knowledge on the subject.

Notes and references

1. Jenkinson, G.G. (1999) *Metal Wood Screws: The Evolution and History* (published by the author); C. White, 'Observations on the Development of Wood Screws in North America'. Available at: http://www.wag-aic.org/Am_Wood_Screws.pdf (accessed April 2020); How, C. (2017) *Historic French Nails, Screws and Fixings: Tools and Techniques*. Surry Hills, NSW: Furniture History Society (Australasia); Rybczynski, W. (2001) *One Good Turn: A Natural History of the Screwdriver and the Screw*. London: Scribner; Bennett, M. (1988) 'Development of the machine-made wood screw in England, 1760–1860', *Conservation News* 35. Available at: https://wag-aic.org/Am_Wood_Screws.pdf (accessed 16 June 2021).
2. In some patents for the manufacture of parallel shank screws, the advantages of the tapered woodscrew are often diminished in favour of the greatness of the parallel shank screws made from these patents. However, the claims of patents should be regarded with caution. A patent document is above all a document written with a desire to elevate the new process as superior to all previous processes or inventions. The opinion expressed in a patent document may be biased. A critical mindset was used for this study of patents and published treatises or books.
3. Holland, J. (1831) *A Treatise on the Progressive Improvement and Present State of the Manufacture in Metal. Vol. 1: Iron and Steel*. London, pp. 192–219.
4. Ibid., p. 201.
5. Ibid.
6. Diderot, D. and d'Alembert, J. (eds) (1751–1772) *Encyclopédie, ou dictionnaire raisonné des sciences, des arts et des métiers*. See *Cloutier* and *Epingliers-Aiguilletiers* (needle makers).
7. How 2017 (cited in note 1), passim.
8. Willetts, A. (1987) *The Black Country Nail Trade*. Dudley: Dudley Leisure Services.
9. Although nowadays nails are machine-made they were still hand-forged throughout most of the 19th century and even at the beginning of the 20th. In 2003, I was taught to forge nails by the late Mr Joseph Mevel, artisan *charron* (wheel maker) in Plourivo, Britanny. Mr Mevel had trained as a wheel maker and nail maker just before WWII. Despite his advanced age and his lack of practice, he was still able to forge a perfect nail with a multifaceted head in only a couple of minutes. Even as a 'new apprentice' I was surprised at the ease with which nails can be forged.
10. A swage block is a solid iron block, serving as an anvil, with semicircular grooves on the surface. The iron bar is hammered repeatedly into the groove, resulting in a rounding off of the bar.
11. Patent for a 'Vice in which the screw blanks are formed', patent number 4117, taken in 13 May 1817, by John Gerard Colbert of Winsley Street, St Marylebone.
12. Duhamel du Monceau, H.L. (1768) *L'art de réduire le fer en fil connu sous le nom de fil d'archal*.
13. Holland 1831 (cited in note 3), p. 201.
14. Ibid., p. 202.
15. Besson, J. (1571–72) *Theatrum instrumentorum*; Bossert, H. Th. and Storck, W.F. (eds) (1483) *Das mittelalterliche Hausbuch* (Leipzig 1912).
16. Dessables, M. (1839) *Nouveau Manuel du Tourneur, Manuels Roret* (Imprimerie de Cardon, Troyes), vol. II, p. 288: 'deux sortes de procédés sont en usage pour fabriquer les vis à bois: la filiere

et le tour. La filière a l'inconvénient de détruire la ductilité du fer: les pas de la vis sont d'une force inégale. Certaines parties du clou presenté à la filière étant plus ou moins dures, plus ou moins nerveuse, plus ou moins résistantes, les filets se ressentent d'autant plus de cette inégalité, que la forte compression de la filière, et la chaleur qu'elle developpe, tendent encore à augmenter ces défauts et à render le fer plus cassant : aussi les gerçures, les pailles et autres accidens, font rejeter ces sortes de vis par beaucoup d'ouvriers.'
17. Holland 1831 (cited in note 3), pp. 201–2.
18. A fourth method of rolling the thread between two hardened steel plates grooved to the form of the thread was developed in the 1830s but was not used commercially until the 1860s–80s. 'Cold-forged' wood screws with thread rolling is the most common technique used today in the mass manufacture of woodscrews.
19. *A Catalogue of Tools for Watch and Clock Makers, by John Wyke of Liverpool*, c.1770, reprinted in 1978 by the Smithsonian Libraries with an introduction and technical commentary by Alan Smith.
20. http://americanlongrifles.org/forum/index.php?topic=18831.0 (accessed April 2020).
21. English patent by the brothers Job and William Wyatt of Staffordshire, registered on 14 May 1760 (patent number 751), 'A better method of cutting screws of iron than heretofore practiced'.
22. The retention of the edge on the cutting tool and how firmly the screw is held are key factors in the quality of the thread.
23. Both Jenkinson 1999 (cited in note 1) and How 2017 (cited in note 1) reproduce the drawing of a small machine called the 'domestic screw girdlers machine', supposedly dated c.1730. This little machine could therefore be the first ever machine to produce screws mechanically. Strangely, the drawing resembles the later 1760 patent by Job and William Wyatt. Jenkinson attributes the screw girdler's design to Holland 1831 (cited in note 3), p. 209. But in the 1831 treatise, even though there is a good description of the screw girdler, it is not dated. As I have not found any other evidence of screw girdlers before 1831, in this article I consider the Job and William Wyatt 1760 patent to be the first mechanical development in the making of machine wood screws.
24. The factory eventually closed in the 1790s.
25. Tapered screws with threads cut with a hand file were used on a cloth press by Mayhew and Ince, c.1780 and handmade screws were not uncommon. My thanks to Peter Holmes from Arlington Conservation and Timothy Garland for sharing photos and providing historical screws for cross-section analyses.
26. Smith, M.S. (2006) *The Emergence of Modern Business Enterprise in France, 1800–1930*. Cambridge MA: Harvard University Press, p. 203.
27. Early thread designs have never been tested with a view to fully understanding their pros and cons, in particular the ease with which they penetrate wood and their retention or 'gripping' power. The early 18th-century handmade screw and 19th-century 'buttress' thread design has been clearly angled to allow it to penetrate the wood with greater ease. Further research is needed, but it can be assumed that, if a thread such as the buttress shape was abandoned twice and not revisited by 21st-century manufacturers, the shape of the thread probably did not deliver on its expectations.
28. Jenkinson 1999 (cited in note 1), p. 53.
29. With thanks to Arlen Heginbotham, Getty Museum conservator, for pointing that out.
30. Marquetry techniques are good examples. The bevel cutting technique, also known as conic cutting, was an innovative improvement on a traditional marquetry technique. This 'new technique' appeared almost simultaneously in both Germany and France around 1760–70 without being recorded in any publication. The large number of German immigrants working in the Paris furniture workshops might easily explain this.
31. Sauveur, A. (1912) *The Metallography of Iron and Steel*. New York: McGraw-Hill. See lesson III.
32. Duhamel du Monceau (cited in note 12).
33. Watson, F.J.B. (1972) 'Puzzles and problems in French furniture mounts, *Apollo* XCV(21): 196–200.
34. Large wardrobe attributed to A.C. Boulle, c.1715 (inv. no. F429): Y. Chastang, unpublished conservation report, 2000.
35. Diderot and d'Alembert 1751–1772 (cited in note 6): see 'Dorure au feu'. See also: Savary des Brulonv, J. (1748) *Dictionnaire universel de commerce*.
36. Diderot and d'Alembert 1751–1772 (cited in note 6): 'Dorure au feu avec de l'or en feuille. Pour préparer le fer ou le cuivre à recevoir cette dorure, il faut les bien grater avec le grateau, & les polir avec le polissoir de fer, puis les mettre au feu pour les bleuir, c'est-à-dire pour les échauffer, jusqu'à ce qu'ils prennent une espece de couleur bleue. Lorsque le métal est bleui, on y applique la premiere couche d'or que l'on ravale legerement avec un polissoir, & que l'on met ensuite sur un feu doux. ... On ne donne ordinairement que trois couches ou quatre au plus, chaque couche étant d'une seule feuille d'or dans les ouvrages communs, & de deux dans les beaux ouvrages ; & à chaque couche qu'on donne, on les remet au feu. Après la derniere couche, l'or est en état d'être bruni clair.'